COASTAL GEOTECHNICAL ENGINEERING IN PRACTICE

PROCEEDINGS OF THE INTERNATIONAL SYMPOSIUM / IS-YOKOHAMA / JAPAN / 20-22 SEPTEMBER 2000

Coastal Geotechnical Engineering in Practice

Edited by

Akio Nakase
Nikken Sekkei Co., Ltd, Japan

Takashi Tsuchida
Port and Airport Research Institute, Japan

Volume 2

A.A. BALKEMA PUBLISHERS LISSE / ABINGDON / EXTON (PA) / TOKYO

Published by: A.A. Balkema, a member of Swets & Zeitlinger Publishers
www.balkema.nl and www.szp.swets.nl

ISBN 90 5809 151 1 (set)

Printed in the Netherlands

Coastal Geotechnical Engineering in Practice, Nakase & Tsuchida (eds)
© 2002 Swets & Zeitlinger, Lisse, ISBN 90 5809 151 1

Table of contents

Special discussion session

Late paper for Volume 1

Coastal Geotechnical Engineering in Practice, Nakase & Tsuchida (eds)
© 2002 Swets & Zeitlinger, Lisse, ISBN 90 5809 151 1

Dedicated to Prof. Nakase, Akio (1929-2000)

Professor Akio Nakase, the chairperson of ISSMGE TC3O Coastal Geotechnical Engineering, passed away in the morning of November 18, 2000. He had made many valuable contributions to geotechnical engineering, especially in the area of soft ground engineering and coastal geotechnics. In 1956, he started his career at the Geotechnical Engineering Division of the Port and Harbor Research Institute (PHRI), Ministry of Transport. After serving as the head of the Soil Mechanics Laboratory, he was promoted to the director of Geotechnical Engineering Division in 1972. In 1961, he went to London and received the Diploma of Imperial College for his study on the bearing capacity of cohesive soil. He received a Doctor of Engineering degree from the University of Tokyo in 1967 on "A Study of Stability Analysis by $\phi_u=0$ method". In 1973, he delivered a special lecture on oceanic geotechnical engineering at the 8th International Conference on Soil Mechanics and Foundation Engineering.

He joined Tokyo Institute of Technology in 1974, and he made great contributions to the 9th ICSMFE Conference in Tokyo 1977 as the secretary general. After retiring from the university in 1990, he was appointed Professor Emeritus and later established a new geotechnical laboratory Nikken Sekkei Nakase Geotechnical Institute (NNGI). In 1991, he organized the International Conference on Geotechnical Engineering for Coastal Development (Geo-Coast '91) in Yokohama. After Geo-Coast '91, Technical Committee (TC-30) on Coastal Geotechnical Engineering was established in 1994 by ISSMGE and he served as Chairperson of TC30 from 1994. Although he had been working extensively for IS-Yokohama 2000 as a top of the organizing committee, he could not attend the symposium because of the illness.

The keyword of Prof. Nakase's activities was geotechnical practices on soft ground. While he was working at the Geotechnical Engineering Division of PHRI, he was virtually a leader in the fields of soft ground engineering in Japan. In the days at Tokyo Institute of Technology and NNGI, he had been a core member of the technical committees of major projects, such as Kansai International Airport, the Offshore Development of Tokyo International Airport or numerous seaports projects. As a member of the Japanese Geotechnical Society, he served as the Vice President from 1988 and as the President from 1990 for two years.

He had a number of wonderful hobbies and will be remembered as accomplished cellist. His memory will live forever with his colleagues and friends.

On the special issue of Soils and Foundations, October 2001

Takashi Tsuchida
Secretary, International Symposium on Coastal Geotechnical Engineering in Practice

In conjunction with IS-Yokohama 2000, a special issue of *Soils and Foundations Vol.41, No.4*, was published in October 2001 by Japanese Geotechnical Engineering Society. This issue includes the following ten technical papers that cover the same research themes of the symposium such as the determination of design parameters, use of geo-materials from solid wastes in the coastal areas, consolidation and compaction techniques, and case studies in soft ground construction:

TECHNICAL PAPERS

David Nash : Modeling the Effects of Surcharge to Reduce Long Term Settlement of Reclamations over Soft Clays – A Numerical Case Study

Bo Myint Win, V. Choa and X. Q. Zeng : Laboratory Investigation on Electro-Osmosis Properties of Singapore Marine Clay

Hisao Aboshi, Yoshiji Sutoh, Toshiyuki Inoue and Yutaka Shimizu : Kinking Deformation of PVD under Consolidation Settlement of Surrounding Clay

Norihiko Miura, S. Horpibulsuk and T.S. Nagaraj : Engineering Behavior of Cement Stabilized Clay at High Water Content

Yoichi Watabe and Takashi Tsuchida : Comparative Study on Undrained Shear Strength of Osaka Bay Pleistocene Clay Determined by Several Kinds of Laboratory Test

Toshiyuki Mitachi, Yutaka Kudoh and Masaki Tsushima : Estimation of In-Situ Undrained Strength of Soft Soil Deposits by Use of Unconfined Compression Test with Suction Measurement

Masaaki Katagiri, Masaaki Terashi and Akira Kaneko : Back Analysis of Reclamation by Pump-Dredged Marine Clay – Influence of Ground Water Lowering –

T. N. Lohani, Goro Imai, Kazuo Tani and Satoru Shibuya : G_{MAX} of Fine-Grained Soils at Wide Void Ratio Range, Focusing on Time-Dependent Behavior

T.-S. Tan, P.-L. Leong, K.-Y. Yong, R. Kamata, J. Wei, K.-C. Chua and Y.-H. Loh : A Case Study of the Behaviour of a Vertical Seawall

S. Leroueil, D. Demers and F. Saihi : Considerations on Stability of Embankments on Clay

TECHNICAL REPORTS

Y. X. Tang, Yoshihiko Miyazaki and Takashi Tsuchida : Practices of Reused Dredgings by Cement Treatment

Takao Satoh, Takashi Tsuchida, Koji Mitsukuri and Z. Hong : Field Placing Test of Lightweight Treated Soil under Seawater in Kumamoto Port

To these contributors, the organizing committee of IS-Yokohama and TC30 extend their appreciation. It is hoped that this special issue together with the proceedings of IS-Yokohama 2000 will give an excellent source of information, data and ideas involved with the coastal geotechnical engineering for researchers and practicing engineers.

Keynote addresses

Coastal Geotechnical Engineering in Practice, Nakase & Tsuchida (eds)
© 2002 Swets & Zeitlinger, Lisse, ISBN 90 5809 151 1

Re-examination of established relations between index properties and soil parameters

H.Tanaka

Port and Airport Research Institute, Yokosuka, Japan

ABSTRACT: The well-known correlations of fundamental soil properties with index properties have been carefully examined, using data accumulated from the soil investigations at the Port and Harbour Research Institute (PHRI). All soil samples were retrieved by the Japanese standard sampler. No unique relation between ϕ' and I_p or (s_u/p) and I_p could be found. The distinguished characteristics of Japanese clays are large ϕ' and (s_u/p) in spite of having large I_p. It is found that most Japanese clays contain considerable amount of diatoms. The influence of diatom was studied using artificial mixture of clay with diatom. It is found that (s_u/p) and ϕ' increase remarkably, owing to even small contents of diatoms.

1 INTRODUCTION

Due to easy access of computer, various numerical models for predicting the ground behavior have been proposed, and some of them have already been used in practice. The soil parameters required in such numerical models, however, are seldom measured directly from laboratory or *in situ* tests, instead the parameters are usually estimated from empirical correlations based on the index properties. Especially, plasticity index (I_p) has been believed for long time to govern mechanical properties of soft clays and has been correlated with many soil parameters, such as strength or compressibility. As a typical example, Skempton's strength incremental ratio may be pointed out: $s_u/p =0.11+0.0037I_p$, where s_u is the undrained shear strength and p is the consolidation pressure. It should be kept in mind, however, that such well-known relations were established based on the limited data accumulated in relatively small-restricted areas such as the North Europe and the North America, where glaciers in the ice age have strongly affected its sediments.

Recently, owing to development of communication such as Internet, it has been easy to accumulate data from various sites in the world to establish useful correlations. Unlike the manufacturing industries such as automobile or computer, however, technical standards in geotechnical engineering are still not unified, and vary from country to country across the world. Inevitably, there still exist a lot of data with different quality in these proposed relations.

The author's research group has carried out the soil investigation in several sites in Japan as well as other parts of the world, using the Japanese standard sampling techniques (Tanaka & Tanaka 1999: Tanaka 2000: Tanaka *et al.* 2000a, b, c, d). All such soils samples were transported to Port and Harbour Research Institute (PHRI) and extensive laboratory tests were performed. Using test results from these investigations, this paper critically examines whether the I_p index is truly useful to predict the mechanical properties of clayey soils.

2 REGIONAL DIFFERENCE IN TESTING QUALITY

2.1 *Laboratory testing*

To determine geotechnical properties of a soil, there is no international standard for laboratory testing as well as field testing. In addition, a technician in charge of soil sampling and *in situ* testing may not be well trained, unlike the technician in other testing field, for example, medical field. Therefore, details in testing or sampling method vary with region. To make the matter worse, an engineer in-charge of site investigation does seldom go to the site to check how properly soil sample is recovered and how accurately *in situ* properties are measured in the field.

Among soil parameters, the density of soil particles (ρ_s) is the most fundamental parameters, and its test procedures seem to be strictly specified. As shown in Fig. 1, however, the values of ρ_s are completely different depending upon who performed the tests. The ρ_s indicated by solid marks in the figure was measured at a well-known institute in Southeast

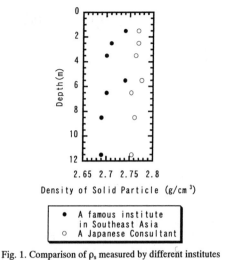

Fig. 1. Comparison of ρ_s measured by different institutes

Asia. It is clearly shown that the ρ_s value measured by this institute is smaller than that measured by a geotechnical consultant in Japan, following testing standard of the Japanese Geotechnical Society (JGS).

As shown in Fig. 2, the index properties of Bothkennar clay measured by both British and Japanese methods are in the same order, despite the existence of small difference. These differences may have been caused by the slight difference in the location of each boreholes (British data is referred from Hight *et al.* 1991). However, for the Louiseville clay, which is a Champlain sea clay widely distributed in Quebec of East Canada, slight differences in test re-

sults measured by Japanese and Canadian sides are recognized, as shown in Fig. 3. The liquid limit (w_L) measured by the Laval University using the fall cone test is slightly smaller than that obtained by the Japanese geotechnical consultants (denoted by A-JPN and B-JPN in the figure) using the Casagrande's cup method. The clay content (defined by particles < 2μm) of the Champlain clay as measured by the hydrometer test, is clearly different depending on where the test was performed. The clay content measured by the Laval University is much higher than that measured by the Japanese consultants who followed the laboratory testing standard of the JGS. As a result, the activity, which is defined as a ratio of I_p and the clay content, is 0.5 when calculated using data from Canadian side, and 1.0 from Japanese side. In case of the Bothkennar clay, British and Japanese sides obtained nearly the same order of clay content, as shown in Fig. 2. The great difference in the clay content of the Louiseville clay may be attributed to the difference in methods followed in dispersing the soil particles, such as the use of different type of chemical agents and variation in their amount. A cooperative research among institutes as well as among ISFMGE member countries is required to find reason for such difference in results, and to unify the testing procedure so that similar results could be obtained regardless of whoever performs such tests.

2.2 *Sample quality*

The above example well illustrates the difficulty in

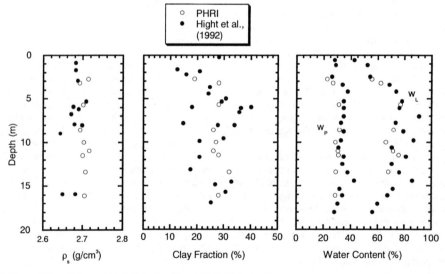

Fig.2. Comparison of index and gradation properties at the Bothkennar site

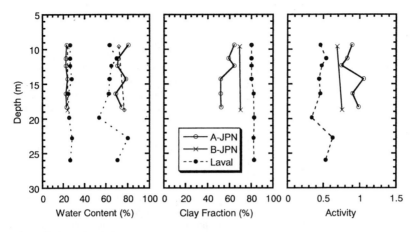

Fig. 3. Comparison of index and gradation properties at the Louiseville site

making database for establishing empirical relations in different regions. The problem of internationally unifying quality sampling is much worse than that of unifying laboratory testing. Geotechnical engineers have realized the importance of sample quality in obtaining soil parameters from laboratory tests and their interpretation. However, it is ironical that a geotechnical engineer seldom visits a sampling site to supervise the proper operation of drilling and sampling. In addition, there is no international standard for the sampling method. Tanaka (2000) performed comparative study on sample quality at the Ariake site, using six different samplers as indicated in Table 1. An example of the difference in sample quality is shown in Fig. 4. The unconfined compression strength (q_u) of the sample retrieved by the Shelby tube, which is widely used in many countries in practice, is 60% as small as that collected by the Sherbrooke sampler. The Sherbrooke and the Laval samplers, whose sample sizes are larger than that of the conventional tube sampler, have enjoyed the reputation for high quality sampling. It should be noted, however, that the Japanese sampler is also capable of getting almost similar quality sample as those two famous samplers types.

Another illustration of the importance of the sample quality can be demonstrated by the difference in the shape of e-log p curves, as shown in Fig. 5. For a sample collected by the Japanese sampler, the change in void ratio is very small at consolidation pressure (p') smaller than the yield consolidation pressure (p_y). In addition, for the high quality sample, the gradient of the e-log p curve after reaching the p_y becomes the largest just after p_y, and it gradually decreases to moderate value with further increase in p'. On the other hand, for a poor quality sample, the large volume change is observed even at the *in situ* effective vertical pressure (p'_{vo}). The

bending point of the e-log p curve at p_y pressure is not so clear as in the Japanese sampler, and also the relation between e and log p is almost linear when

Table 1. Main features of samplers used in the comparative study at the Ariake site (Tanaka 2000)

Sampler	Inside Diameter (mm)	Sampler Length (mm)	Thickness (mm)	Area Ratio (%)	Piston
JPN	75	1000	1.5	7.5	yes
LAVAL	208	660	4.0	7.3	no
Shelby	72	610	1.65	8.6	no
NGI54	54	768	13	54.4	yes
ELE100	101	500	1.7	6.4	yes
Sherbrooke	350*	250*	–	–	no

* Dimension of the soil sample

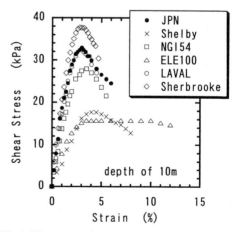

Fig. 4. UC test results for different sample quality (Tanaka 2000)

5

the consolidation pressure reaches the normally consolidation stage.

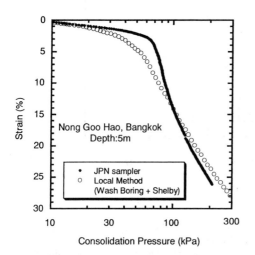

Fig. 5. Comparison of e-log p curves for samples with different quality

3 OBJECTIVE SOIL INVESTIGATION SITES

The author's research group has visited total of six different sites outside Japan to carry out site investigation. These sites are the Drammen of Norway, Pusan of Korea, Bothkennar of UK, Singapore, Louiseville of Canada and Bangkok of Thailand. The Japanese samplers were brought to these sites for soil sampling. All drilling and sampling operations were done under the author's supervision. The soil samples were transported by air cargo to PHRI, while keeping samples in the sampling tube. It has been confirmed, by comparing the strength measured at corresponding local institutes and PHRI, that the disturbance in the sample quality caused by the transportation can be ignored (Lunne, et al. 1997, Tanaka & Tanaka 1999). In addition to these non-

Japanese soils, PHRI has performed site investigation works in many soil deposits across Japan, among which the sites of Ariake, Hachirougata and Yamashita have been selected for comparative study in this paper. Therefore, in this paper, the investigation data from the six non-Japanese soils sites and the three Japanese soils sites are presented. The main features of these sites are summarized in Table 2.

4 CONSOLIDATION CHARACTERISTICS

4.1 C_c and liquid limit

Among the empirical correlations concerning consolidation of a soil, the correlation between compression index (C_c) and w_L, which is called Terzaghi's correlation, may be pointed out as follows:

$$C_c = 0.009(w_L\text{-}10) \quad (1)$$

Several researchers have tried to establish the correlation between C_c and w_L. As an example, relations obtained by Ogawa & Matsumoto (1978) for marine clays in coastal areas in Japan are shown in Fig. 6. As can be seen, a general trend exists between w_L and C_c: that is, w_L increases, as C_c increases. However, there are wide scatters in data. Before discussing the existence of a unique relation between w_L and C_c, a problem arises on sample quality and testing method for measuring the e-log p relation, as pointed out earlier. Sample quality does strongly affect the C_c value, especially after reaching p_y, as shown in Fig. 5. All data plotted in Fig. 6, however, are measured from the oedometer tests at PHRI, strictly following the testing standard of the JGS. The sample quality of each sample is also carefully examined and in case if the sample quality of a soil is judged to be not good, such data have been omitted from Fig. 6. Thus, it may be considered that the accuracy of the test data in Fig. 6 is quite reliable, and they are the representative relations for Japanese marine soils deposited in the Holocene era. It should be kept in mind, however, that the e-log p relations

Table 2. Main properties of clay carried out by this investigation

site	country	depth (m)	LL (%)	PL(%)	I_p	w_n(%)	CF* (%)	s_u**(kPa)	References
Ariake	Japan	1–18	55–178	27–62	28–116	42–200	40–68	8–40	Tanaka, H. (1994), Tanaka, H. et al. (2000a)
Yamashita	Japan	20–38	92–126	38–55	54–77	74–100	32–66	130–160	Tanaka, H. et al. (2000c)
Hachirogata	Japan	3–41	120–240	41–89	75–150	78–207	15–57	20–60	Tanaka, H. (1994)
Louiseville	Canada	9–19	69–81	21–23	46–57	64–76	49–64	45–66	Tanaka, H. et al. (2000c)
Bangkok	Thailand	6–17	46–101	19–27	41–73	41–81	33–60	20–40	Tanaka, H. et al. (2000a)
Singapore	Singapore	16–28	66–82	22–24	42–57	50–60	62–70	17–90	Tanaka, H. et al. (2000a)
Bothkennar	UK	3–16	55–77	22–32	32–45	51–68	19–32	20–60	Tanaka, H. (2000)
Drammen	Norway	5–19	34–48	18–22	16–27	30–43	36–48	20–40	Tanaka, H. (2000)
Pusan	Korea	6–22	54–73	22–26	30–47	46–65	29–64	22–38	Tanaka, H. et al. (2000d)

CF* Clay Fragment (<2μm)

s_u** Measured by Vane

based on C_c curve were measured by the conventional oedometer test, i.e., the incremental loading (IL) oedometer tests, in which the consolidation pressure at each step is increased to twice the previous step, and the duration of each step is 24 hours.

Except for the reconstituted soil at laboratory or for remolded soil, the e-log p curves do not consist of two straight lines joined at the yield stress (p_y), as

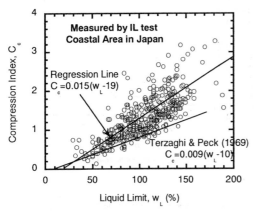

Fig. 6. Relation of C_c and w_L for clays in coastal areas in Japan, using conventional IL odedometer test (after Ogawa & Matsumoto 1978)

described in many textbooks. Instead, the gradient of the e-log p curve, that is, the C_c index is not constant, but changed non-linearity with the consolidation pressure (p) even at the normally consolidated state. Figure 7 shows the e-log p relations for the Louiseville and Yamashita clays, measured by Constant Rate of Strains (CRS) oedometer apparatus. Unlike the IL oedometer test, CRS test gives continuous e-log p relation where its non-linearly after p_y can be clearly observed.

As shown in Figure 7, the C_c index becomes the largest when p exceeds p_y and gradually decreases with increase in p. Then, it becomes constant when p reaches a certain stress level. In case of the Louiseville clay, which is well known cemented clay, the volume change is very small before p_y. When p exceeds p_y, however, a large volume change occurs. The Yamashita clay also shows tendency similar to the Louiseville clay, although the change in the C_c index according to the stress level is not as pronounced as that of the Louiseville clay. Figure 8 shows the normalized relation between C_c/C_{cl} and p/p_y for these soils, where C_{cl} is the C_c index at large enough p to yield constant C_c. It can be recognized that although the absolute values of $(C_c/C_{cl})_{max}$ for the Louiseville and Yamashita are completely different, both the curves have almost identical pattern of decrease or increase in C_c/C_{cl} with p/p_y. Evidently, the non-linearity of the e-log p curve becomes the most prominent when p/p_y reaches between 1 and 2.

According to the JGS standard on the oedometer testing, the C_c index is defined as the steepest gradient in the e-log p relation. The IL oedometer yields the e-log p relation only at some points, which results in large subjectivity in determination of C_c, especially when p is in the range of 1.0 p_y to 2.0 p_y, as shown in Fig. 8.

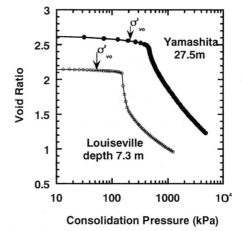

Fig. 7. e-log p curves for Louiseville and Yamashita clays, measured by CRS oedometer test

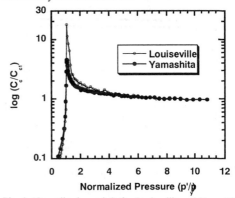

Fig. 8. Normalized p and C_c for Louiseville and Yamashita clays

When the C_{cl} is plotted against w_L, instead of the C_c index measured by IL oedometer test, a clear relation can be observed between these two indices, which follows correlation defined by the equation (1), as shown in Fig. 9. It has been confirmed from the previous studies that the equation (1) can be applied to re-constituted or remolded soil. This means that the C_c for intact soil becomes the same as that of the C_c of the reconstituted or remolded soil, when the consolidation pressure is significantly larger than p_y. On the other hand, as shown in Fig. 10, the correlation between C_{cmax} and w_L is poor, where C_{cmax} is the maximum value of C_c in the e-log p curve. These scatters between C_{cmax} and w_L indicate that C_{cmax} can

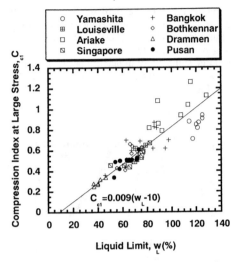

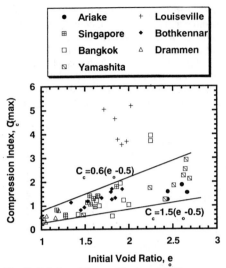

Fig. 9. Relation of C_{cl}, C_c at large enough pressure to be constant, and w_L

Fig. 11. Relation of C_{cmax} and *in situ* void ratio, e_o

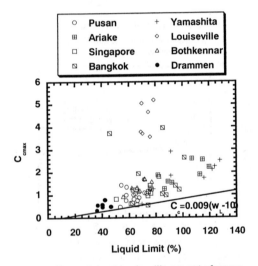

Some data of Louiseville are out of range

Fig. 10. Relation of C_{cmax}, maximum of C_c, and w_L

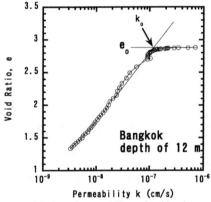

Fig. 12. Relation of permeability and void ratio, measured by CRS odedometer test

not be determined by measuring the soil properties in remolded condition such as index properties. Instead, it is strongly dependent on the sedimentation environment and the post-sedimentation history. Following the suggestion of Leroueil *et al.* (1983), the data in Figure 10 have been plotted again in Fig. 11 by replacing w_L with its *in situ* void ratio (e_o). Leroueil *et al.* (1983) have insisted that C_{cmax}/e_o is strongly correlated with sensitivity (S_r) of a soil. It should be kept in mind, however, that in their relation, they have used C_c measured by the IL test, which gives smaller C_c than that measured by CRS

test used in this research. As seen in Fig. 11, the relations between C_{cmax} and e_o are distributed in a relatively narrow range of $C_{cmax}=(0.6{\sim}1.5)(e_o - 0.5)$, except for the Louiseville and Bangkok clays. It can be recognized from both Figs. 11 and 7, that the consolidation characteristics of the Louiseville clay are completely different than that of other clays, which is caused by strong cementation.

4.2 *Permeability*

The coefficient of permeability (k) is an important parameter in considering the speed of consolidation. In addition, recently, k has also become a very important parameter in constructing waste deposal facilities. Previous studies by various researchers have established that k plotted in logarithm scale can be correlated to e. A typical example of this relation measured by the CRS oedometer test is demonstrated in Fig. 12. It is observed that k decreases with a decrease in e. The *in situ* permeability (k_o) may be defined as k corresponding to e_o in the extrapolated line of the log k and e. The k_o value is a rather complicated parameter because it may be related to e, the pore sizes and their continuity. In general, k for a sandy soil is much larger than that for a clayey soil, whereas e for the sandy soil is much lower than that for the clayey soil. The reason for large k of a sandy material, in spite of small e, is the large size of its pore.

Figure 13 shows the relation between e_o and k_o for various clays listed in Table 2. There is a general trend of decrease in k_o with a decrease in e_o, however, the Ariake clay has relatively small k_o in spite of its rather large e_o. Mesri et al.. (1994) assumed that k is a function of not only e, but also of Clay Fraction (CF) as well as of its Activity (A_c). Their explanation is that the pore diameter of soils containing larger CF is smaller than that of the coarse soils, even though they have the same e_0. Therefore, it is understandable that k for a fine material is much lower than that for a coarse material, even if both the soil types have the same e. In addition, in case of the same CF, k for the material with high A_c should be smaller than that for material with low A_c, since the size of soil particle with high A_c is smaller and inevitably its pore sizes are smaller. From these considerations, Mesri et al. (1994) proposed the following equation for k:

$$k = 6.54*10^{-9}(e/CF/(A_c+1))^4 \quad (2)$$

Figure 14 plots the relation between normalized void ratio using Eq.(2), $(e_o/CF/(A_c+1)$ and k_o. The equation proposed by Mesri et al, (1994) is also drawn in the figure. It is recognized that large scatter between k_o and e_o observed in Fig. 13, gets considerably reduced. However, still significant scatters exist between k_o and e_o. Especially, the data for the Pusan and Ariake clays are located above the line proposed by Mesri et al. (1994). Lapierre et al. (1990) have tried to establish relation between pore size and permeability for the Louiseville clay, using a mercury-intrusion porosimetry test. However, they could not found any unique relation between those two parameters. More researches are necessary to explain as to what parameters govern the permeability of a soil.

5 UNDRAINED SHEAR STRENGTH CHARACTERISTICS

5.1 *Stress-strain behavior of clays at various areas*

Characteristics of soils are not so simple as to be

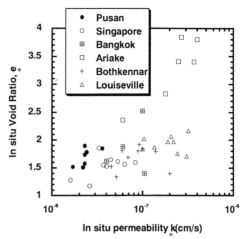

Fig. 13. Relation of *in situ* permeability and void ratio for various clays

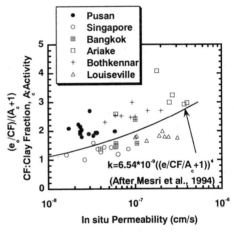

Fig. 14. Normalized *in situ* permeability proposed by Mesri, *et al.* (1994)

represented only by its I_p and OCR. Figures 15 and 16 show the stress-strain curves and stress paths for various soils, measured by the undrained triaxial compression test. All specimens were consolidated under the corresponding *in situ* effective stresses, following the Recompression method. Both the deviator as well as the mean principal stresses ((σ'_1-σ'_3)/2 and (σ'_1+σ'_3)/2, where σ'_1 and σ'_3 are the major and the minor principal stresses, respectively) are normalized by the *in situ* effective vertical stress (p'_{vo}). To make the comparison easily, test results are summarized into two figures; results for OCR less than 1.5 are shown in Fig. 15, while the results for OCR greater than 1.9 are plotted in Fig. 16. All the OCRs were measured by CRS tests at a strain rate of 0.02 %/min. The difference in OCRs of the

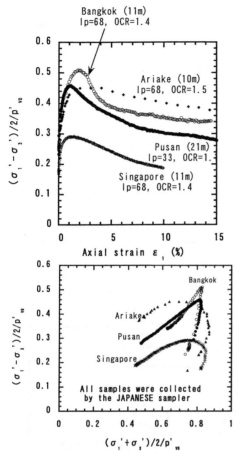

Fig. 15. Stress-strain curves and stress paths for clays with OCR(‚1.5), Pusan, Bongkok, Singapore and Ariake clays

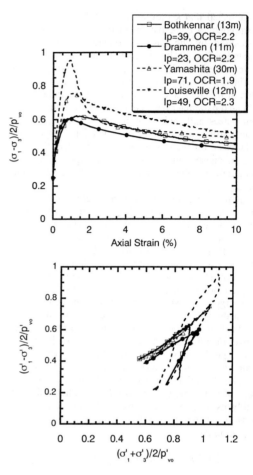

Fig. 16. Stress-strain curves and stress paths for clays with OCR(>2.0), Yamashita, Louiseville, Drammen and Bothkennar clays

soils is believed to have been caused by ageing or cementation effects rather than by the change in their stress history.

The shapes of the stress-strain and the stress path are somewhat different from that usually described in the conventional textbooks. For example, the w_n of the Ariake clay exceeds w_L, yielding the liquidity index (I_L) of over unity (I_L=1.36). It is usually believed that such soil should be highly sensitive and should show brittle behavior. As shown in Fig. 15, however, the strength reduction after the peak is relatively small, and the shape of the stress path is significantly ovular compared to that of the Pusan and Bangkok clays; the I_L for the Pusan and Bangkok clays being 0.94 and 0.66, respectively. The resulting internal effective friction angle (ϕ') at the residual state for the Ariake clay is quite high compared to other clays. On the other hand, although the shape of the stress path of the Singapore clay is

very similar to that of the Ariake clay, its ϕ' value is very small, which is its most distinguished characteristics compared to those of the other clays. The ϕ' for the Pusan and Bangkok clays fall in between them, and have moderate values. However, their deviator stresses increase almost vertically until the peak, generating only a small excess pore water pressure. After reaching the failure, their stress paths follow the failure envelope. Thus, these clays have relatively high normalized undrained strength, in spite of their moderate ϕ' values. These results confirm that the undrained shear strength is dependent of ϕ' as well as dilatancy characteristics of a soil, which could be strongly related to soil structure.

In case of clays with relatively high apparent OCR, the values of peak normalized deviator stresses are considerably different, although the residual φ' is nearly the same for all clays. In another words, these clays follow the same normalized stress path although their peak points are different. As a result, for example, the (s_u/p'_{vo}) ratio for the Louiseville and Drammen clays are 0.96 and 0.61, respectively. The ratio of the normalized undrained strength for these clays is as much as 1.6. It is of course impossible to explain these differences by either OCR or I_p.

5.2 The relation of φ' and I_p

It may be understood that there is no unique relation between I_p and φ' from the stress paths, as already shown in Figs. 15 and 16. However, several correlations between φ' and I_p have been proposed until recently. The relation shown in Fig. 17 is reported by Leroueil et al. (1990), indicating that the φ' decreases with an increase in I_p. It is especially apparent that relations for clays of the Scandinavian, Eastern Canada as well as the Southeast Asia follow well-known Kenney's (1959) or Bjerrum & Simon's (1960) empirical relations. The relation for the Japanese clays, which was examined by the author's research group, has also been plotted in Fig. 17. The φ' values for the Japanese clays were measured from the triaxial compression tests, by subjecting the specimen to consolidation pressure high enough to bring the specimen in the normally consolidated state. It is found that the φ' for the Japanese clays is relatively higher than that for non-Japanese clays, and its dependency on I_p cannot be recognized.

5.3 Relation of s_u/p and I_p

It has long been believed strength incremental ratio (s_u/p) of a soil increases with an increase in I_p. For example, Skempton's equation, $s_u/p = 0.11 + 0.0037I_p$, as mentioned earlier, is well known, and has been introduced in most of the text books as well as in many design standards. When Skempton proposed this relation, it was believed that p_y, which is still sometimes called preconsolidation pressure (p_c), is the same as the *in situ* effective burden pressure (p'_{vo}) in normally consolidated soil deposits (i.e., the maximum p'_{vo} in the past is the same as that at present). However, since then, it has been recognized that even in the mechanically normally consolidated soil deposit, p_y is greater than its p'_{vo}, therefore its OCR is apparently greater than 1.0. Since s_u increases with an increase in p_y, which in turn increases with time due to such factors as the secondary consolidation, clay mineralogy and depositional as well as post depositional history of a soil, s_u/p should strictly be defined as s_u/p_y, instead of as s_u/p'_{vo}. Figure 18 shows the relation between I_p and s_u/p_y measured in the present study, as well as the relation for the Scandinavian clays referred from Lars-

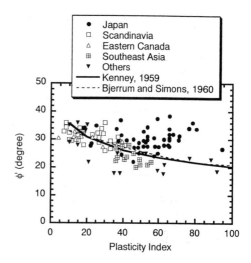

Fig. 17. Relation of internal friction angle and I_p. All data except for Japanese clays are referred from Leroueil, et al. (1990)

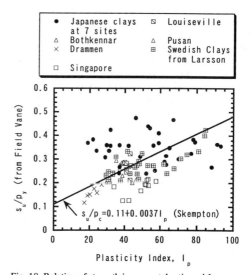

Fig. 18. Relation of strength incremental ratio and I_p

son (1991). The values of s_u for all clays were measured by the field vane tests. The values of s_u/p_y for the Singapore clays are considerably smaller compared to those of the other clays. It can be seen that in non-Japanese clays, in general, s_u/p_y increases with an increase in I_p. However, the Japanese clays do not follow this trend. Instead, s_u/p_y for the Japanese clays is almost constant with I_p, and the corresponding values are also relatively larger than those of non-Japanese clays.

11

5.4 Anisotropy and rate effect on the undrained shear strength

Many geotechnical engineers recognize that among many important factors that govern complicated behavior of a soil, anisotropy and rate effect on the shear strength are prominent. The testing method conventionally used in practice is a compression type, and the strain rates in such tests are generally faster than that encountered at failure in the field. To make the matter worse, these factors contribute to overestimation of shear strength, yielding unsafe and dangerous design parameters. To estimate the effects of shearing rate and strength anisotropy, many researchers have tried to correlate their influences with the simple index, such as I_p.

Figure 19 shows the relation between I_p and the strength anisotropy (s_{ue}/s_{uc}), where s_{ue} and s_{uc} are strengths measured by extension and compression tests following the recompression technique. For Japanese clays, including the Ariake and Yamashita clays, s_{ue}/s_{uc} increases with an increase in I_p, although the strength anisotropy is relatively small for the Yamashita clay compared to other Japanese clays having similar I_p. The Louiseville clay also follows the s_{ue}/s_{uc} trend of the Japanese clays. The Bangkok and Singapore clays show, however, relatively large values of s_{ue}/s_{uc}. Especially, for the Singapore clay, it is nearly unity, i.e., the Singapore clay behaves isotropically. It may be concluded from these observations that the anisotropy parameter is not so simple as to be correlated to only I_p.

shown in Fig. 20, Hanzawa & Tanaka (1992) have demonstrated that the rate effects on the undrained strength of Japanese clays are nearly independent of I_p. Figure 21 shows another experimental evidence that the rate effect is independent of I_p, where the rate effects on p_y measured by the CRS oedometer test at different strain rates are shown. As can be seen in the figure, any correlation between the rate effect and I_p cannot be recognized.

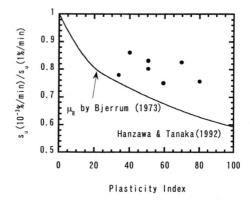

Fig. 20. Rate effect on undrained shear strength

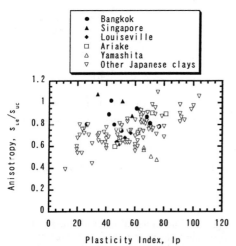

Fig. 19. Relation of anisotropy, defined by the ratio of extension and compression strengths, and I_p

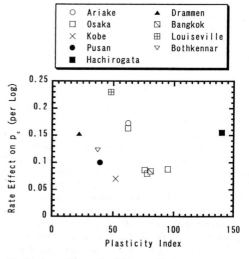

Fig. 21. Rate effect on the yield consolidation pressure

6 UNDRAINED SHEAR STRENGTH FOR DESIGNING

6.1 Bjerrum's correction factor

It has been believed that the rate effect can also be related to I_p. As a typical example, Bjerrum's correction factor is based on the idea that the rate effect increases with an increase in I_p, as mentioned later. As

For the determination of the undrained shear strength (s_u), the field vane shear (FVS) test is extensively used in most countries, except in Japan. It is usual that the s_u measured by the FVS test is corrected by Bjerrum's correction factor (μ) to get strength for use in design. This concept is believed to take into account of the strength anisotropy and the rate effect on the vane strength (Bjerrum 1973).

The vane used in general practice usually has the blade ratio of 1:2; the height of the vane blade is twice as much as its diameter. When the shear resistance in the horizontal plane is assumed to be fully mobilized, the torque of the vane (M) is given by the following equation:

$$M = (\pi/2)HD^2 s_{uv} + (\pi/6)D^3 s_{uh}$$
$$= (\pi/6)D^3(6s_{uv} + s_{uh}) \quad (3)$$

Where H and D are the height and the diameter of the vane: s_{uv} and s_{uh} are the shear strengths mobilized in the vertical and horizontal planes, respectively.

As indicated in Eq. (3), the strength measured by the FVS is mostly governed by s_u on the vertical plane; i.e., s_{uv}. Therefore, when the s_u measured by the FVS is used in a stability analysis, the strength anisotropy (s_{uv}/s_{uh}) effect should be considered (Bjerrum denoted this anisotropy effect as μ_A).

Although there is no international standard on the FVS test, most countries adopt the speed of the vane rotation as 6°/min. This rotation speed is considerably higher than the deformation speed at failure in the field. Since s_u, especially for soft clay, is affected by the strain rate as discussed previously, a factor should be considered for the rate effect (μ_R). Bjerrum found from extensive laboratory data that both μ_A and μ_R can be related to I_p and he finally proposed the correction factor to the strength measured by the FVS for design: $\mu = \mu_A\mu_R$, relating to I_p. He demonstrated that this correction factor could explain many failure records.

6.2 The q_u method

Japan is among a few countries, where s_u of a soft soil is determined only by unconfined compression (UC) test for design. Although field and laboratory tests such as the FVS and triaxial tests including UU, CIU or CK₀U, are also frequently carried out, such tests are primarily limited either to research purpose. The s_u used in design is calculated as $q_u/2$, which is obtained from the UC test, and is used without correction for I_p. The reason for the use of the q_u method is attributed to so-called "lucky harmony". It is well recognized that q_u value is strongly dependent on the sample quality. Because of loss of its in situ confining pressure, swelling of a sample leads to strength reduction. Therefore, UC test always underestimates the true strength, although the degree of the strength underestimation depends on the sample

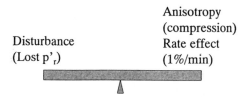

Fig. 22. Lucky harmony in the q_u method

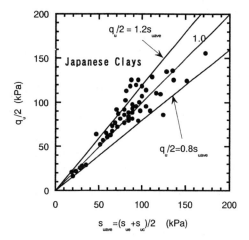

Fig. 23. Comparison of $q_u/2$ and the strength of $(s_{ue} + s_{uc})/2$, where s_{ue} and s_{uc} are extension and compression strength measured by recompression test, respectively

quality. On the other hand, the UC test is a kind of compression test, which yields rather high strength, as already shown in Fig. 19. Also the strain rate of the UC test is 1 %/min, which is considerably larger what is encountered in the actual failure in a field. These two factors, i.e. the anisotropy and the strain rate play roles in overestimating the strength of a soil. In the concept of "lucky harmony", under-and over-estimating factors such as the sample disturbance and the anisotropy as well as the strain rate effects are preserved in well balance with each other, as indicated in Fig. 22.

There is still controversy on how to determine s_u from laboratory test for design. Hanzawa (1982) has proposed that the s_u value for design be taken as the average strength from compression and extension strength (($s_{uc}+s_{ue}$)/2), measured by triaxial shearing at a strain rate of 0.001 %/min, and following the recompression technique in which specimens are consolidated under the in situ effective stress conditions. Figure 23 compares $q_u/2$ with ($s_{uc}+s_{ue}$)/2 for Japanese clays. It can be seen from the figure that these strengths are very closely related with each other, indicating that $q_u/2$ is a quite reasonable strength indicator for design. However, if the sample quality is not good enough to the hold balance shown in Fig. 22, then the $q_u/2$ measured from such sample underestimates the strength.

6.3 Comparison of the strengths estimated by UC and corrected FVS tests

A question arises whether the above two methods provide the same strength for design or not. Figure 24 compares strengths from the $q_u/2$ and the μ corrected FVS strength. This figure was originally prepared for Japanese clays (Tanaka, 1994) and recent data of non-Japanese clays obtained by the author's research group have been added in the figure. The following interesting results are found from the figure:

1) For the Japanese clays with I_p smaller than 40, the strengths from the q_u method are considerably lower those from the Bjerrum's method. This may be attributed to the fact that the Japanese clays with small I_p contain a large quantity of coarse materials such as silts or sand particles (Tanaka *et al*. 2000b). Therefore, these soils cannot hold residual effective stress large enough to maintain the lucky harmony balance, as shown in Fig. 22. It may be true that $q_u/2$ of the Japanese clays having small I_p might underestimate the design strength. On the other hand, for non-Japanese clay with small I_p such as that of the Drammen clay, the Bjerrum's correction method and the q_u method provide nearly the same strength.

2) For the Japanese clays with relatively large I_p, the strength ratio μ s_u *(vane)* / $q_u/2$ is less than unity. This means that if the Bjerrum's method is assumed to provide a proper design strength, the q_u method should overestimate the design strength. And if $q_u/2$ value is used for the design, then the foundation designed using $q_u/2$ should be failed. However, no such accident has ever been reported. There may be insistency that the q_u method luckily avoids failures because of the use of the safety factor. Indeed, in the design standard for the port construction using the q_u method, the safety factor of between 1.2 and 1.3 is adopted against the slip circle failure. This safety factor might be argued to compensate for the strength overestimated by q_u. This safety factor is, however, used to overcome uncertainties associated not only with the shear strength measurement, but also with other factors such as variability of the ground condition, simplifications of acting outer and inner forces in calculation method, and so on. It should also be noted that even if the design method proposed by Bjerrum is used, a certain safety factor is required to preserve the stability. Therefore, it may be reasonable to think that the Bjerrum's correction factor method gives considerably conservative strength values compared to that of the q_u method. This fact is applicable even to other clays such as the Bothkennar, Bangkok and Singapore clays.

For the Drammen clay, the Bjerrum's corrected vane strength and the $q_u/2$ values are nearly the same (Fig. 24). However, the q_u value for the Drammen clay is expected to underestimate its proper strength.

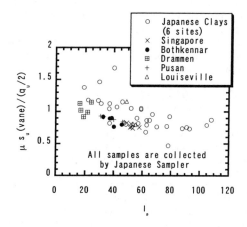

Fig. 24. Comparison of $q_u/2$ and vane strength corrected by Bjerrum's correction factor

Figure 25 shows the residual effective stress (p'_r) of the Drammen clay, which was measured before performing the UC test by placing the specimen on the ceramic disc with the air entry value of 200 kPa. As stated earlier, in order for UC test to be applicable, the degree of disturbance, as indicated by p'_r, should be compensated by the factors overestimating the strength such as anisotropy and rate effect. For the marine clays that the author's research group has conducted site investigation in the past, the order of the (p'_r/p'_{vo}) ratio is in the range of 1/4 and 1/6. For the Drammen clay, however, p'_r/p'_{vo} is considerably small. Especially at depths lower than 14 m, p'_r value is nearly zero. Therefore, it should be judged that the q_u method cannot be applied to the Drammen clay. Despite this, the μ corrected FVS strength obtained is equal to $q_u/2$. The μ factor is 1.0 for $I_p = 20$ so that the raw vane strength is the same as that corrected by μ. In Fig. 25, the average strengths of s_{ue} and s_{uc} are also plotted. It can be seen that the means strength is remarkably larger than $q_u/2$ or the Bjerrum's corrected vane strength. Therefore, it is confirmed that the vane shear strength underestimates the true strength, probably because of the disturbance caused in pushing the vane blade into the ground.

Another similar example from the Singapore site is presented in Fig. 26, which shows the strength evaluated by the average of $(s_{ue}+s_{uc})/2$, UU test and the direct shear test (Takada, 1993). It may be considered that the strength from the UU test is equivalent to the UC test. For determination of the design

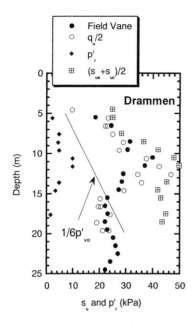

Fig. 25. Undrained shear strength evaluated by several methods and residual effective stress for Drammen clay

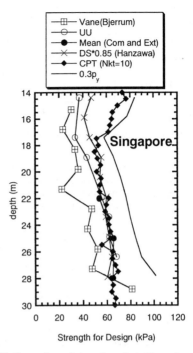

Fig. 26. Comparison of strength evaluated by various methods at the Singapore site

strength from the direct shear test, Hanzawa (1992) proposed to multiply the resulting strength by 0.85, to compensate for the rate effect. It can be recognized from the figure that only the vane strength corrected by the μ factor is considerably smaller than the strengths measured by other methods. Similar results have been observed in other sites as well. Therefore, it can be concluded that the concept of the Bjeruum's correction factor significantly underestimates the design strength. This conclusion is also supported by the fact that the rate effect does not increase with a increase in I_p, as shown in Figs. 16 and 17.

Based on the Bjerrum's relation with s_u/p'_{vo} and I_p, Mesri (1975) has proposed the incremental strength ratio (s_u/p) of 0.22 for young clays, regardless of I_p and OCR. Lets examine the validity of Mesri's proposal using a case history of the Kansai International airport. This airport was constructed in the Osaka Bay, the detail description of the construction, including soil conditions, are referred to Maeda (1988). Shown in Fig. 27 is a typical cross section of the sea wall, in which the step loading method was adopted. The soft clay beneath the reclaimed land was improved by the sand drain method. Each construction stage was carried out after getting strength owing to the weight of the structure placed in the previous stage.

Unfortunately, the sufficiency of strength gained for each succeeding states of constructions was confirmed by the UC test, and not by the FVS test, following the conventional Japanese method. As shown in Fig. 28, the vane strength (not corrected by Bjerrum's correction factor) is nearly the same as the $q_u/2$, provided that the I_p of the soil is not extremely low and the sample is retrieved by the proper sampling method. This fact is valid not only to Japanese clays, but also other clays that the author's research group has done in various parts of the world. To examine the validity of the Mesri's method, it is assumed that the raw FVS strength is equal to the $q_u/2$. That is, the design strength is $\mu(q_u/2)$ before construction. Following the Mesri's proposal, the strength increment (s_u/p) during the construction due to the weight of the sea wall is assumed to be 0.22. A stability analysis result showed the minimum safety factor (F_{smin}) of 0.953, which means the sea wall should have failed. However, no failure incidents were actually observed. For the comparison purpose, the F_{smin} calculated based on $q_u/2$ value is 1.107. Indeed, it is usual in Japan that s_u/p for the step loading is adopted as 0.3, and is sometime taken to as much as 1/3. It is easily understandable from these cases that, either the Bjerrum's correction factor or Mesri's proposal does not work for the Japanese clays.

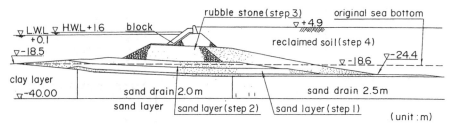

Fig. 27. Typical cross section of the sea wall at the Kansai International Airport (First phase)

Of course, it cannot be said that the q_u method is a perfect method for evaluating the undrained shear strength for design. The intensity of anisotropy and rate effect is not constant, but it considerably varies

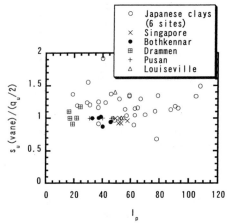

Fig. 28. Comparison of FVS strength (not corrected by Bjerrum's correction factor) and $q_u/2$

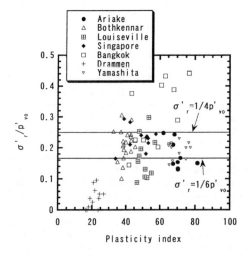

Plasticity index

Fig. 29. Ratio of residual effective stress and effective vertical pressure for various clays

with clay types as shown in Figs.21 to 23. Variability in sample quality adds to the degree of uncertainty. In the q_u method, it is assumed that the sample is recovered by an appropriate sampling method, such as the Japanese standard sampling method. Therefore, if a sample retrieved by a sampler is worse or better than the Japanese standard sampler, the q_u method cannot evaluate the proper strength value. Figure 29 shows relation between p'_r / p'_{vo} and I_p for various clays. It can be seen that p'_r/p'_{vo} is completely different for different clays, indicating that degree of reduction of strength due to the sample disturbance is different for different clays, although all samples were recovered by the Japanese standard sampling method. In addition, it should be remembered that the (p'_r/p'_{vo}) ratio is not dependent on I_p.

7 SMALL STRAIN STIFFNESS

Small strain stiffness (G_o) is gradually getting attention because of increasing requirements for the accurate prediction of a ground deformation caused by construction activities such as embankment construction or ground excavation. It is extensively recognized that the deformation modulus is quite dependent on strain level (for example, Jamiolkowski, et al.., 1991). However, extensive experimental facts show that the shear deformation modulus at strain smaller than a certain strain level, say 10^{-5}, can be considered to be practically constant. The stiffness in this strain level is called "small strain stiffness (G_o)".

Since Hardin & Black (1969) proposed an empirical relation for G_o of clays as indicated in Eq. (4), several researchers have tried to establish similar empirical equations for predicting G_o using fundamental soil parameters (for example, Shibata & Soelarno, 1975: Shibuya & Tanaka, 1996).

$$G_o = 3,270\{(2.97-e)^2/(1+e)\}p'^{0.5} \text{ (in kPa)} \quad (4)$$

Most of the proposed equations of G_o are the functions of the void ratio and a power of the confining pressure (p'), which is usually smaller than 1.0 and is typically 0.5. On the other hand, the undrained shear strength (s_u) linearly increases with an increase in p'.

$$s_u = (s_u/p)_{nc}OCR^m p' \quad (5)$$

Where $(s_u/p)_{nc}$ is an incremental strength ratio at normally consolidated state, and its typical order is 0.3 for the Japanese clays, as described earlier. Unlike G_o in Eq. (4), magnitude of void ratio is not considered in the undrained shear strength. Indeed, for the normally consolidated state, e and p' have strong relation as can be seen in the e-log p curve, implying that e and p' are not independent parameters. It should also be noted that, in practice, Young's modulus is usually correlated to the s_u value such as $E = 210s_u$. If the governing equations for the shear strength and shear modulus are different, as indicated in Eqs. (4) and (5), then the ratio of (G_o/s_u) should be dependent on e and p'.

Figure 30 shows the relation between the rigidity index (G_o/s_u) and e_o, where G_o and s_u have been measured by the seismic cone and the FVS test, and e_o is the *in situ* void ratio. In this figure, data for the Scandinavian, Bothkennar and Louiseville clays are obtained from the following references: Larsson (1991) for Scandinavian clays, Nash et al. (1992) for Bothkennar clay, and Laval University for Louiseville clay. It is found from the figure that G_o/s_u increases with a decrease in e_o. It is also apparent that the G_o/s_u for the Japanese clays is not dependent with I_p, while it increases with a decrease in I_p for non-Japanese clays. Also the rigidity index is significantly larger for the non-Japanese clays than that for Japanese soils. The value of e_o may be correlated with w_L, because these samples have nearly the similar order of the liquidity index (I_L), except for large depths. It should also be kept in mind that there is a strong relation between w_L and I_p, because these values for most marine clays are distributed along A line in the plastic chart. Therefore, the e_o can be roughly replaced by the index I_p. Indeed, instead of e_o in Fig. 30, certain correlation between G_o/s_u and I_p can be recognized in Fig. 31, although this correlation is not so significant as that of G_o/s_u and e_o.

There may be two reasons for dependency of G_o/s_u on e_o: one reason is that s_u may decrease with a decrease in e_o or I_p. Indeed, as already shown in Fig. 18, s_u for many non-Japanese clays decreases with a decrease in I_p, or a decrease in e_o. Another reason is that G_o increases with a decrease in e_o, as suggested by Eq. (4). Figure 32 plots G_o/p'_{vo} ratio against e_o for various clays. The relation between (G_o/p'_{vo}) ratio and e_o in Fig. 32 involves various clays with different OCR, which may have been caused by ageing effect, not by stress change, as mentioned already. The relation is re-plotted again in Fig. 33, using the relations in Fig. 32, with parameter of OCR. It is clear that with increasing OCR, G_o/p'_{vo} increases at a given e_o. Thus, at a given OCR, it is recognized that G_o/p'_{vo} is strongly affected by e_o. From these observations, it may be concluded that G_o is a function of e_o (or I_p), unlike the undrained strength.

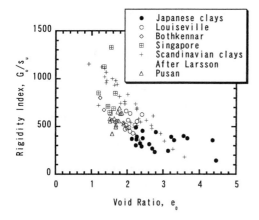

Fig. 30. Relation of rigidity index G_o/s_u and e_o, where G_o and s_u are measured by seismic cone and FVS test, respectively

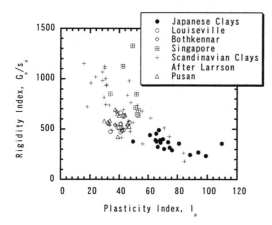

Fig. 31. Relation of G_o/s_u and I_p, instead of e_o in Fig. 30

8 PECULIARITY IN CHARACTERISTICS OF JAPANESE SOILS

As shown in the preceding sections, I_p related correlations for many Japanese as well as non-Japanese soils are not applicable. More outstanding observations, however, are that Japanese soils are somewhat peculiar compared to European and Canadian, as well as some Asian clays. Briefly reviewing the properties of Japanese soils here again, their index properties have large values, in spite of the presence of relatively small clay fraction. Figure 34 compares

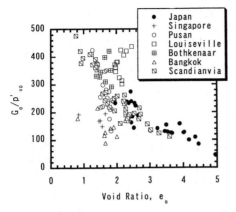

Fig. 32. Relation of G_o/p'_{vo} and e_o for various clays

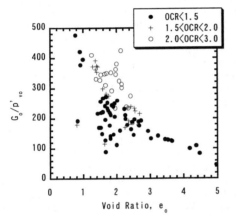

Fig. 33. Relation of G_o/p'_{vo} and e_o in Fig. 32, with OCR's parameter

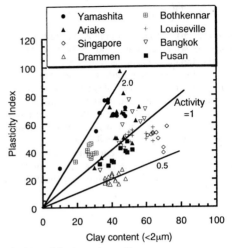

Fig. 34. Activity for various soils

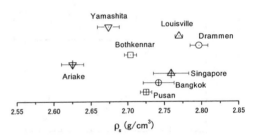

Fig. 35. Distribution of ρ_s for various clays in the world

activity, for various clays in the world. Indeed, it is found that the activities for the Ariake and Yamashita clays are quite high. However, although they have high I_p, the ϕ' is quite large, as shown in Fig. 17. It is not rare to observe ϕ' exceeding 40°, as shown in Fig. 15 for the Ariake clay. Also, the strength incremental ratio (s_u/p) is quite large and it has already been mentioned that the s_u/p of 0.3 is usually taken in the step loading in Japan. Also, it should be remembered that the unit weight of solid particle (ρ_s) for the Japanese clays are relatively small. Figure 35 shows the values of ρ_s for the investigated sites by the author's research group, and it is noted that the values of ρ_s for the Yamashita and Ariake clay are less than 2.7, while ρ_s for most of the non-Japanese soils, the values are beyond 2.7. For the Drammen clay, it is as high as 2.8.

One of reasons for such peculiar features of the Japanese clays may be attributed to its clay minerals. Figure 36 shows X-ray diffraction charts for the Ari-

ake and Hachirougata clays (as Japanese clays) as well as for the Singapore and Pusan clays (as non-Japanese clays). The examined grain sizes of all specimens were less than 2 μm. For clear identification of clay minerals, the specimens were prepared in natural state (denoted by N in the figure), Glycolation (G) treated state and heated state (at 550 C°, indicated by H). It can be seen that the Ariake and Hachirougata clays might contain a lot of smectite, whose activity is very high. Locat et al. (1996) have also shown that main clay minerals in Japanese clays are smectite. On the other hand, main clay mineral of the Singapore clay is kaolin; no smectite was detected in it. It is very interesting to note that smectite was also not found in the Pusan clay, although it is located just opposite to the Kyushu Island, which is separated by relatively narrow channel, called Tsushima Straight. The large index properties and high activity of the Japanese clays may be attributed to the presence of large amount of smectite. It is reported, however, that the ϕ' values for pure smectite are smaller compared to that of illite or kaolin whose I_p are rather small (see for example, Mitchell 1976).

18

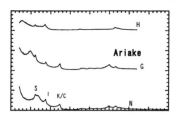

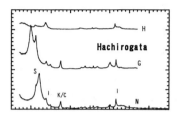

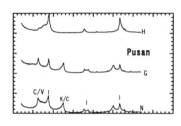

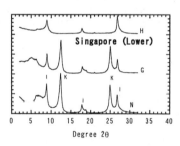

Fig. 36. X-ray diffraction chart for various clays

N: Natural G: Glycolated H: Heated
Clay mineral
C: chlorite I: illite K: kaolin S: smectite V: vemiculite

These studies have created an illusion that φ' decreases with an increase in I_p.

Another reason for strange behaviors of the Japanese clays may be due to existence of microfossils, especially, the presence of diatoms (Tanaka & Locat 1999). Figure 37 shows pictures taken by Scanning Electronic Microscope (SEM). Large numbers of diatom skeletons with shape of a hollow cylinder can be seen in these pictures. Most of them are diatoms.

It can be seen in the picture of Osaka bay clay that some diatoms have retained their shape even at a great depth of 400m. It is anticipated that if a soil contains a lot of diatoms, its large sized pores in skeletons lead the high water content. It is found from the figures that diatoms have dominantly silt sized skeletons. Generally speaking, it is believed that activity of a clay is defined by the water captured by clay particles. In clays containing diatoms, however, a large part of water is entrapped in pores of their skeletons, which may be classified into silt by grading test. Therefore, it may be anticipated that the Atterberg limits are strongly influenced by the presence of diatoms. In addition, diatom may affect mechanical properties such as φ' value because of its rough and angular shape.

9 STUDY ON CHARACTERISTICS OF DIATOMS MIXED SOIL

9.1 Preparation of mixtures

In order to examine the influences of diatoms on physical as well as mechanical properties of a soil, the artificial mixture of diatom and clay was prepared in laboratory (Shiwakoti, et al. 1999). Diatomite was recovered from an industrial quarry at Hi-

(a) Hachirogata clay

(b) Osaka Bay clay

Fig. 37. Scanning electronic microscope pictures for typical Japanese clays

19

Table 3. Properties of artificial mixture soil

Soil Type	w_n (%)	w_L (%)	I_p	ρ_s (g/cm3)	Silt size (%)
Diatomite	240–280	NP	NP	2.26–2.37	58
Kaolin	–	68.8	33.9	2.78	20
Singapore	56	82.5	59.8	2.77	28

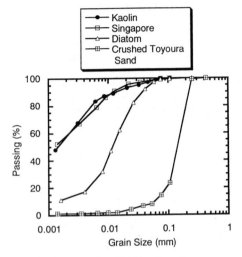

Fig. 38. Grading curves for artificial mixture clay

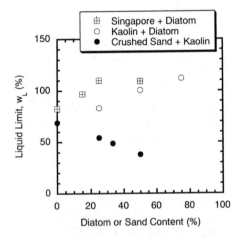

Fig. 39. Change in liquid limits with increasing in diatomite and crushed Toyoura sand

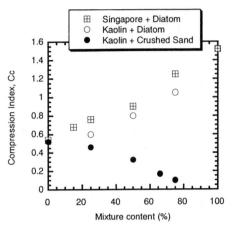

Fig. 40. Change in C_c with content of diatom and crushed Toyoura sand

ruzenbara, Okayama prefecture, Japan. The main properties of diatomite used in this study are shown in Table 3, and its grain size distribution is given in Fig. 38. It can be observed from the grading curve that the main grain size of diatomite is in the range of silt particles. It should also be noted that the ρ_s of diatomite is very small compared to that of a soil. This fact also supports that Japanese soils contain a large amount of diatom. To differentiate the properties of diatomite with ordinary soil, a soil material consisting of solid particles (without pore insides) was also prepared as a mixture. It was desired that the grain size of this material be similar to that of the diatom used in this study. Therefore, the Toyoura sand was crushed using a mill machine. Unfortunately, the grain size of the crushed Toyoura sand is about ten times greater than that of diatom, as shown in Figure 38.

Two clays were mixed with diatom: kaolin available in the commercial market and the Singapore clay. The reason for selecting these two clays as the mixture component is that there is no diatom in these clays. The main properties of these clays are indicated in Table 3.

9.2 *Change of physical properties due to mixture of diatomite*

Figure 39 shows the change in w_L due to the increase in content of diatom and the crushed Toyoura sand. It is very interesting that w_L increases with an increase in content of diatomite, while for the mixture of the crushed sand, w_L decreases with an increase in the content of the crushed Toyoura sand. These facts show that a large volume of water is trapped by pores of diatom skeletons, which contributes to large value of w_L. On the other hand, because the crushed sand does not have the similar capacity for holding water, the w_L decreases with a decrease in its content. Therefore, it should be kept in mind that even if the content of silt particles is the same, the index properties can be completely different, depending on whether a soil contains diatoms or usual silt sized soil particles.

20

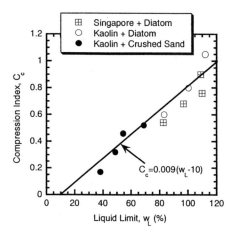

Fig. 41. C_c and w_L relation for diatom and crushed Toyoura sand mixtures

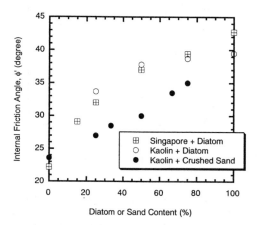

Fig. 42. Change in ϕ' with increasing in diatom and crushed Toyoura sand

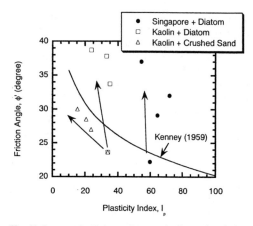

Fig. 43. Increase in ϕ' due to increase in diatom in relation of I_p and ϕ'

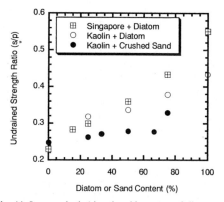

Fig. 44. Increase in (s_u/p) ratio with content of diatom and crushed Toyoura sand

9.3 Consolidation properties

The values of the Cc index for the mixtures were measured by the conventional IL oedometer. Test results are shown in Fig. 40. For the diatomite mixture, C_c increase with an increase in diatomite content, while it decreases with an equivalent increase in the crushed Toyoura sand. Thus the effect of diatom microfossils on C_c of a soil is completely opposite to what is generally observed for an ordinary soil. It is interesting to note that the relation of the C_c and w_L follow the Terzaghi's relation: $C_c=0.009(w_L-10)$ for the diatomite mixture as well as the crushed sand mixture, as shown in Fig. 41. This relation is the same as that for natural clays, if C_c is defined at very large consolidation pressure, which is not influenced by effect of structure, as descried earlier.

9.4 Strength properties

The internal effective friction angles (ϕ') for these mixtures were studied using a direct shear test with constant volume during shearing. The ϕ' for both mixtures of diatom and the crushed Toyoura sand increase with increase in their content, as shown in Fig. 42. However, for the mixture of the crushed Toyoura sand, ϕ' increases proportionally with the sand content, while ϕ' for the mixture of diatom increases at much larger rate even for a small proportion of diatomite content, and the ϕ' becomes almost constant at a certain amount of diatomite, say, 50 %. It is also interesting to note the fact that ϕ' for mixtures with both kaolin and Singapore clays is in the nearly same order. As pointed out earlier, ϕ' value for natural Singapore clay is very small compared to many other clays, Fig. 15. However, as shown in Fig. 43, if ϕ' is plotted against I_p, the relation between ϕ' and I_p for Singapore clay mixtures is similar to that of the Japanese natural clays. An interesting assumption can be made that Singapore clay would have the same order of φ' if diatoms were breed in Singaporean sea, as in Japan Japanese clays.

The strength incremental ratio (s_u/p) measured by the direct shear test also increases with diatom content, as shown in Fig. 44. It is very interesting to note that (s_u/p) for the mixture of the crushed Toyoura sand is not much affected by its content. The influence of diatom on the (s_u/p) of a soil, however, is significant. For example, by adding only 25 % of diatomite, (s_u/p) for both kaolin and Singapore clay increased by as much as 20 %.

It is found from these investigations on the artificial mixture of diatom that the physical properties can be drastically changed due to the presence of diatom. The value of ρ_s become small and w_L increases with diatomite content. The value of C_c also increases with an increase in diatom content and the relation between C_c and w_L follows the Terzaghi's relation: $C_c=0.009(w_L-10)$. The strength properties such as φ' and (s_u/p) are also significantly influenced by the presence of diatom. Therefore, it is likely that the peculiar properties of Japanese clays are caused by diatom. A question has arisen as to what is the proportion of diatoms present in the Japanese clays. This issue will be discussed in the succeeding section. Also the influence of diatoms in a natural clay will also be examined.

10 QUANTITATIVE ANALYSIS ON THE AMOUNT OF DIATOM AND ITS INFLUENCE ON THE PROPERTIES OF NATURAL SOILS

10.1 *Estimation of the amount of diatom in natural clay*

The ingredient of diatom is mainly the silica (SiO_2), which is sometimes called biogenic or amorphous silica because of its non-crystallized nature. There are still controversies on how to quantify biogenic silica, since some parts of non-biogenic silica derived from quartz etc., is also inevitably measured in the process of quantifying total quantity of biogenic silica (see, Kamatani & Oku 1999).

A preliminary method has been developed for analyzing the amount of diatoms. The specimen, weighing 0.6 g by dry weight, was treated by hydrogen peroxide and hydrochloric acid for removing organic matter and dispersion of diatoms cell from soil particles. Part of the solution was taken with a pipette and mounted on the glass sheet. The specimen, thus prepared, was observed under the optical microscope under the 400 times magnification. The numbers of diatoms in the restricted area were counted, from which, the total number of diatoms present per unit dry weight of soil were calculated.

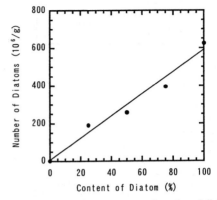

Fig. 45. Relation of diatom content and number of diatom counted by the present study

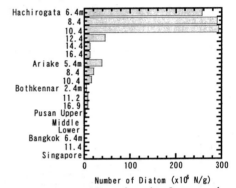

Fig. 46. Histograms of diatom for various Japanese and non-Japanese clays

Of course, there are many problems in evaluation of diatom numbers by this method. The size of diatom is not the same but varies with species. And also, as shown in Fig. 37, some diatoms are not in the original shape, but are found broken into smaller pieces. In this test procedure, such a fragment is counted as one. Therefore, the number of diatom determined by this method does not directly indicate the volume of diatoms present in the soil specimen, nonetheless, it provides at least a rough idea on whether a soil contains plenty of diatoms or not. Following this method, Figure 45 shows the relation between diatomite content and the counted diatoms for the mixture of diatomite and kaolin. It can be seen that there is a linear relation between them, indicating that this method has a capability of quantifying diatom.

Figure 46 shows the number of diatoms in various natural clays, for both the Japanese as well as non-Japanese clays. It is found that Hachirougata clay, which will be described in more detail, contains a large amount of diatom. On the other hand, diatom cannot be found in Pusan, Singapore and Bangkok clays. Although the Bothkennar clay also contains

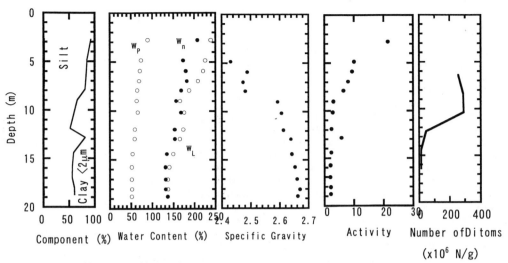

Fig. 47. Soil profiles at the Hachirogata site

small amount of diatoms, content of diatoms in Japanese clays is much larger than that in overseas clays.

10.2 A case study on influence of diatom in real soil

As already discussed, it is found from the study on the soil artificially mixed with diatomite that influence of diatom is important in mechanical as well as physical properties of a soil. And also it is found that Japanese soils contain a lot of diatoms compared with overseas soils. In this section, let us consider the influence of diatom for a real soil, the Hachirougata clay.

The Hachirougata site is located in Akira prefecture, which is in the northern part of Honshu Island of Japan. It used to be the second largest lake in Japan and was reclaimed in 1950's for rice field. The investigated site was at the center of the reclaimed lake, where the soft clay with about 50 m thickness is deposited.

Figure 47 shows the soil profile until 20 m depth of this site. As a general tendency, Atterberg limits for Hachirougata clay are extremely high even among Japanese clays whose index properties are rather high compared with clays in other areas, as already discussed. It should also be noted that the proportion of clay particle content in Hachirougata is very small in spite of its high w_L and w_P. As a result, its activity is more than 10 at shallow depths. The solid particles density (ρ_s) is also extremely low near the ground surface, which increases with depth in inverse proportion with the activity. The main reason for such high index properties is smectite content. Clear shift of diffraction angle (2θ) after glycolation

treatment, indicates the existence of smectite in this soil, Fig. 36. Another reason for such strange physical properties of Hachirougata clay may be attributed to the existence of diatom. As shown in the SEM picture of Fig. 37, a large number of diatoms exist in clay specimen. The right hand side of Fig. 47 indi-

Table 4. Change of the Atterber limits due to application of consolidation pressure (1MPa)

	w_L (%)	w_P (%)	I_p
Before (intact)	243.9	67	176.9
After (10MPa)	179.8	55.5	124.3

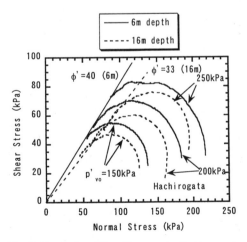

Fig. 48. Results from Direct Shear test for Hachirogata clay at 6 m and 16 m depth

23

cates the number of diatoms measured by the method described above. With the aid of these test results, high activity and low ρ_s for Hachirougata can be easily explained: that is, a large part of silt size particles are anticipated to consists of diatom, which has relatively small ρ_s and can hold a large amount of water in its skeleton.

There are some indirect evidences of the existence of diatoms in the soil near the ground surface. Table 4 compares the Atterberg limits before and after consolidation at 10 MPa. The w_L for a consolidated specimen also decreased because the skeleton was crushed so that same quantity of water could not be hold in its pore. Another evidence is the strength. Figure 48 shows a test result from the direct shear test for specimens at depths of 6 m and 16 m. From Fig. 47, it is known that a specimen at a depth of 16 m contains considerably small amount of diatom compared to that at 6 m. The peak undrained strength normalized by the p'_{vo} as well as ϕ' for a specimen at 6 m depth is much higher than those at 16 m. This test result implies that the existence of diatom creates high ϕ' and the undrained shear strength, which is exactly the same result as that obtained in the artificially prepared diatomite mixture.

11 CONCLUSIONS

Due to locality in geotechnical engineering in laboratory testing and sample quality, it has long been difficult to establish unified database of soil parameters from all over the world. In addition to the Japanese clays, soil data for clays overseas were accumulated by laboratory testing at Port and Harbour Research institute, on soil samples retrieved by unified sampling technique using the Japanese standard sampler. The well-established correlations with plasticity index (I_p) were carefully examined. No unique relation is found either between I_p and the internal effective friction angel (ϕ'), or between I_p and the strength incremental ratio (s_u/p). It is also revealed that the vane strength corrected by Bjerrum's correction factor gives considerably conservative design strength. However, the compression index (C_c) at very large consolidation pressure can be correlated with liquid limit (w_L), and the relation can be expressed by Terzaghi's equation: $C_c=0.009(w_L-10)$.

Characteristics of Japanese clays are very peculiar compared with clays in other parts of the world. They have: 1) large I_p in spite of small proportion of clay particles, i.e., large activity, 2) compressibility is high, 3) large ϕ' and (s_u/p). The reason for large activity for Japanese clays may be attributed to clay mineral: in general, Japanese clays contain a large amount of smectite. It is known, however, that the ϕ' of pure smectite is quite small. This characteristic is completely opposite to that of Japanese clays. It is found from the observation of SEM that Japanese

clays contain a lot of diatom. The influence of diatom is very significant on mechanical properties; i.e., as diatom content increases, the ϕ' and (s_u/p) increase significantly. It may be concluded that the existence of diatom gives large ϕ' and (s_u/p) ratio, despite having large I_p.

REFERENCES

Bjerrum, L. & Simons, N. E. 1960. Comaprison of shear strength characteristics of normally consolidated clay. Proc. of Research conference of shear strength of cohesive soils, *ASCE*:711-726.

Bjerrum, L. 1973. Problems of soil mechanics in unstable soils. Proc. of 8th ICSMFE. 3:111-159.

Hanzawa, H. 1982. Undrained shear strength characteristics of Alluvial marine clays and their application to short term stability problems. PhD thesis, Tokyo University, Japan.

Hanzawa, H. 1992. A new approach to determine soil parameters free from regional variations in soil behavior and technical quality. *Soils and Foundations*. 32(1):71-84.

Hanzawa, H. & Tanaka, H. 1992. Normalized undrained strength of clay in the normally consolidated state and in the filed. *Soils and Foundations*. 32(1):132-148.

Hardin, B. O. & Black, W. L. 1969. Vibration modulus of normally consolidated clay. Jour. of the SMF Div., Proc. *ASCE*. 95(6):1531-1537.

Hight, D. W., Bond, A. J. & Legge, J. D. 1992 Characterization of the Bothkennar clay: an overview. *Geotechnique*. 42(2):303-347.

Jamiolkowski, M., Leroueil, S. & Lo Presti, D.C.F. 1991. Design parameters from theory to practice. Proc. of the international conference on geotechnical engineering for coastal development (Geo-Coast '91). 2:877-917.

Kamatani, A. & Oku, O. 1999. Measuring biogenic silica in marine sediments. *Marine Chemistry*, (in print).

Kenney, T.C. 1959. Discussion. Proc. *ASCE*. 85(SM3):67-79.

Lapierre, C., Leroueil, S. & Locat, J. 1990. Mercury instruction and permeability of Louiseville clay. *Canadian Geotechnical Journal*. 27:761-773.

Larsson, R. 1991. Shear moduli in Scandinavian clays. Report of Swedish Geotechnical Institute, No. 40.

Leroueil, S., Tavenas, F. & Le Bihan, J.P. 1983. Proprietes caracteristiques des argiles de l'est du Canada. *Canadian Geotechnical Journal*. 20(4):681-705.

Leroueil, S., Magnan, J. P. & Tavenas, F. 1990. Embankments on soft clays. Translated by Wood, D. M. Ellis Horwood.

Locat, J., Tremblay, H., Leroueil, S., Tanaka, H. & Oka, F. 1996. Japan and Quebec clays: Their nature and related environmental issues. Proc. of the second International Congress on Environmental Geotechnics. 1:127-132.

Lunne, T., Berre, T. & Strandvik, S. 1997. Sample disturbance effects in soft low plastic Norwegian clay. Proc. of the International Symposium on Recent Developments in Soil and Pavement Mechanics:81-102.

Maeda, S. 1988. Observational control system on the ground improved by sand drain to construction of the large size artificial island. PhD thesis, Kyushu University, (in Japanese).

Mesri, G. 1975. New design procedure for stability calculation of embankments and foundations on sot clay. Discusion, *ASCE*. 101(GT4):409-412.

Mesri, G., Kwan Lo, D. O. & Feng, T. W. 1994. Settlement of embankments on soft clays. Proc. of Settlement '94. *ASCE*. Geotechnical Special Publication No. 40:8-56.

Mitchell, J. K. 1976. Fundamentals of soil behavior. Series in soil engineering. John Wiley & Sons, Inc.

Nash, D.F.T., Powell, J.J.M. & Lloyd, I.M. 1992. Initial investigations of the soft clay test site at Bothkennar. *Geotechnique*. 42(2):163-181.

Ogawa, F. & Matsumoto, K. 1978 Correlation of the mechanical and index properties of soils in harbour districts. Report of the Port and Harbour Research Institute. 17(3):3-89 (in Japanese).

Shibata, T. & Soelarno, D. S. 1978. Stress-strain characteristics of clays under cyclic loading. Jour. of JSCE. 276:101-110 (in Japanese).

Shibuya, S. & Tanaka, H. 1996. Estimate of elastic shear modulus in Holocene soil deposits. *Soils and Foundations*. 36(4):pp.45-55.

Shiwakoti, D. R., Tanaka, H., Locat, J. & Goulet, C. 1999. Influence of microfossils on the behaviour of cohesive soil. Proc. of the 11th Asian Regional Conference, Seoul, South Korea: 23-26.

Takada, N. 1993. Mikasa's direct shear apparatus, test procedures and results. Geotechnical Testing Journal, *GTJODJ*. 16(3):314-322.

Tanaka, H. 1994. Vane shear strength of Japanese marine clays and application of Bjerrum's correction factor. *Soils and Foundations*. 34(3):39-48.

Tanaka, H. & Tanaka, M. 1999. Key factors governing sample quality. Characterization of soft marine clays: 57-82. Balkema.

Tanaka, H. & Locat, J. 1999. A microstructural investigation of Osaka Bay clay: the impact of microfossiles on its mechanical behaviour. *Canadian Geotechnical Journal*. 36:493-508.

Tanaka, H. 2000. Sample quality of cohesive soils: Lessons form three sites, Ariake, Bothkennar and Drammen, Soils and Foundations. 40(4). (in press)

Tanaka, H., Locat, J., Shibuya, S., Tan, T. S. & Shiwakoti, D. R. 2000a. Characterization of Singapore, Bangkok and Ariake clays. *Canadian Geotechnical Journal* (accepted)

Tanaka, H., Tanaka, M. & Shiwakoti, D. R. 2000b. Characterization of low plastic soils: intermediate soil from Ishinomaki, Japan and low plastic clay form Drammen Norway. *Soils and Foundation*. (accepted).

Tanaka, H. Shiwakoti, D. R., Mishima, O., Watabe, Y. & Tanaka, M. 2000c. Comparison of mechanical behavior of two overconsolidated clays: Yamashita and Louiseville clays. *Soil and Foundations*. (submitted)

Tanaka, H., Mishima, O. Tanaka, M., Park, S.Z. Jeong, G. H. & Locat, J. 2000d. Characterization of Yangsan clay, Pusan, Korea. *Soils and Foundations*, (accepted).

Coastal Geotechnical Engineering in Practice, Nakase & Tsuchida (eds)
© 2002 Swets & Zeitlinger, Lisse, ISBN 90 5809 151 1

Use of direct shear and cone penetration tests in soft ground engineering

H.Hanzawa
Toa Corporation, Yokohama, Japan

ABSTRACT: A new procedure for soil investigation, design and construction control is presented for soft ground engineering. The proposed method, consisting of direct shear and cone penetration tests, DST and CPT, focuses to do soft ground engineering minimizing the regional variations both in soil behaviour and technical quality with the approach as simplified as possible. Design parameters to be necessary in soft ground engineering are firstly summarized together with various characteristics of natural clay deposits investigated up to the present. Theoretical backgrounds of the proposed method based on the various advantages of DST and CPT are then presented. Validity of the method are also demonstrated by extensive field and laboratory tests results on marine clays in the world as well as lots of case studies of stability on actual earth structures constructed on soft clays at and near failure.

1 INTRODUCTION

Very soft and highly compressible marine clay is widely found in the huge area from the East to the Southeast Asia. The shear strength of this material for design use, designated with $S_{u(mob)}$ in this paper, has been determined by different methods. Unconfined compression test, UCT, is very common in Japan, while field vane test, FVT and unconsolidated triaxial compression test, UTCT are much more popular in other countries. The shear strengths from these tests, $S_{u(UCT)}$, $S_{u(FVT)}$ and $S_{u(UTCT)}$, however, are quite unclear, and their applicability as $S_{u(mob)}$ strongly depends on regional experiences.

In order to overcome their experience dependency, various methods to decide $S_{u(mob)}$ have been proposed and practiced (Bjerrum, 1972; Bjerrum, 1973; Ladd and Foott, 1974; Mesri, 1975; Hanzawa, 1977; Trak, et al., 1980; Ohta, et al., 1989; Hanzawa, 1989; Tsuchida, 1989). Most of these methods are, however, based on different concepts, which give different $S_{u(mob)}$. In addition, some of them are difficult to use in routine work because of their highly qualified and complicated techniques.

It is an important task for geotechnical engineer in this region to find out a practical procedure to give reliable $S_{u(mob)}$ minimizing the regional experience dependency with the approach as simplified as possible. In this paper, a new procedure consisting of direct shear and cone penetration tests, DST and CPT, is presented for soft ground engineering with supports of extensive field and laboratory tests on marine clays in the world, and lots of case studies on actual structures constructed on soft clays at and near failure.

2 PARAMETERS REQUIRED IN DESIGN OF SOFT SOILS

Undrained shear strength of natural clay deposit since sedimentation to the present, designated wit the symbol, S_{uf}, in this paper, is as shown in Fig. 1 when plotted versus effective vertical stress, σ'_v. Since natural clay possesses an additional shear strength developed by aging effect, S_{uf} is always greater than the shear strength in the normally consolidated state (hereafter called N.C. state), S_{un} even there has been no release of the effective overburden stress, σ'_{vo} since sedimentation.

$$S_{uf} > S_{un} = S_{un}/\sigma'_v \times \sigma'_{vo}$$

where S_{un}/σ'_v = strength increment ratio in the N.C. state

When the clay at point B is subjected to loading like embankment, S_{uf} does little change until σ'_v reaches σ'_v at point C at which clay starts to yield changing to the N.C. clay. The σ'_v at point C is therefore the consolidation yield stress, σ'_y which is also an important design parameter given by Eq. (2).

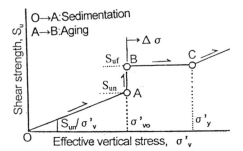

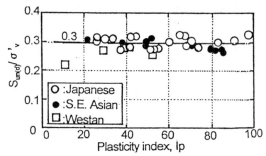

Fig. 1 S_u vs. $\acute{o}'_v$ correlation for natural clay deposit

Fig. 2 $S_{un(d)}/\acute{o}'_v$ ratios in the N.C. State of various marine clay in the world measured by DST

$$\sigma'_y = S_{uf} \div S_{un}/\acute{o}'_v \qquad (2)$$

Appropriate determination of S_{uf} and S_{un}/σ'_v is therefore very important for design purpose. In determining these parameter, two requirements must be satisfied:- 1) strength anisotropy should be taken into account, and 2) simple and easy testing procedures should be employed. When triaxial test is adopted, both compression and extension tests must be done under K_o-consolidated state. It is evident that such complicated test with high quality technique is definitely not recommended in routine work.

3 SHEAR STRENGTH INCREMENTAL RATIO AND STRENGTH ANISOTROPY IN THE NORMALLY CONSOLIDATED STATE

There are lots of advantages in direct shear test, DST, when compared with triaxial test:- 1) much easier and faster in operations, 2) automatic achievement of K_o-consolidation and swelling, 3) similar deformation state to the actual one in shear stage, 4) easier choice of highly homogeneous specimen from actual soil sample and so on. In order to make the best use of such marked advantages of DST, Mikasa (1960) successfully developed a new apparatus which eliminates various defects of the ordinary device and makes constant volume shear test possible. This is the most important advantage of this apparatus.

Lots of DST with Mikasa's apparatus were carried out on various clays in the world in the N.C. state together with K_o-consolidated triaxial compression and extension tests in order to investigate direct shear strength increment ratio in the N.C. state, $S_{un(d)}/\sigma'_v$ ratio and the strength anisotropy among the direct shear, compression and extension strengths in the N.C. state. Strength increment ratio, $S_{un(d)}/\sigma'_v$ ratio, and strength anisotropy evaluated by $S_{un(c)}/S_{un(d)}$ and $S_{un(e)}/S_{un(d)}$ ratios, obtained from the tests are plotted versus plasticity index, Ip in Figs. 2 and 3. It can be said that $S_{un(d)}/\sigma'_v$ ratios of Asian marine clays are approximately constant showing a value of 0.3 irrespective of Ip, and $S_{un(d)}$ is also about the same as the average shear strength between $S_{un(c)}$ and $S_{un(e)}$.

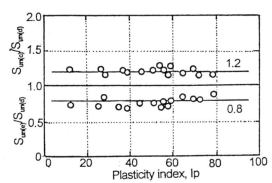

Fig. 3 Strength anisotropy evaluated by $S_{un(c)}/S_{un(d)}$ and $S_{un(e)}/S_{un(d)}$ ratios in the N.C. State

Hanzawa and Tanaka (1992), on the other hand, found that the strain rate effect is also constant irrespective of Ip for Asian marine clay based on result of K_o-consolidated triaxial compression test with different strain rate as shown in Fig. 4.

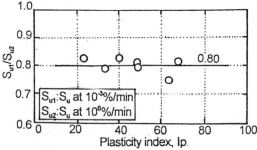

Fig. 4 Change of S_u at different strain rate plotted versus Ip

These three findings, being different from knowledge obtained for clays found in the regions covered with glacier, are very important.

$$S_{un(d)}/\sigma'_v = const. \fallingdotseq 0.3 \neq F(Ip) \qquad (3)$$

$$S_{un(d)} = (S_{un(c)} + S_{un(e)})/2 \qquad (4)$$

Strain rate effect = constant $\neq$ F (Ip) $\qquad (5)$

28

4 SHEAR STRENGTH CHARACTERISTICS OF NATURAL CLAY DEPOSITS INVESTIGATED BY DIRECT SHEAR TEST

4.1 Procedure to Determine Shear Strength of Overconsolidated Clays

Natural clay is in an overconsolidated state even there has been no release of the σ'_{vo} because of the additional shear strength developed by aging effect. Such clay is termed to be the normally consolidated aged clay, N.C.A. clay, while the overconsolidated clay formed by release of the σ'_{vo} is termed to be the overconsolidated young clay, O.C.Y. clay which possesses no aged structure according to Bjerrum's classification (1973). Shear strength in-situ, S_{uf} of two types of overconsolidated clays are given by Eq. (6) as schematically presented in Fig. 5.

$$S_{uf} = S_{un}/\sigma'_v \times \sigma'_{y1} \text{ for N.C.A. clay} \qquad (6.1)$$

$$= S_{un}/\sigma'_v \times \sigma'_{y2} \times \alpha \text{ for O.C.Y. clay} \qquad (6.2)$$

where $\sigma'_{y1} = \sigma'_y$ of the clay with the path of O→A →B, $\sigma'_{y2} = \sigma'_y$ of the clay with the path of O→A' →B and α = correction factor for strength reduction through swelling from A' to B.

Two methods were proposed to determine S_{uf} of natural clay deposits in seventies, 1) the recompression method (Bjerrum, 1973; Hanzawa, 1977) and 2) the SHANSEP method (Ladd and Foott, 1974), which are based on quite different concepts. The recompression method considers the effect of aging on S_{uf} to be important, while aging effect on S_{uf} is considered to be the same as that of the man-made overconsolidated clay in the SHANSEP method. S_{uf} in the SHANSEP method is therefore given by Eq. (6.2). It should also be pointed out that S_{uf} from Eq. (6.2) is always smaller than S_{uf} from Eq. (6.1) when same σ'_y is used.

This subject was studied in detail by Hanzawa and Kishida (1981), and they made clear that

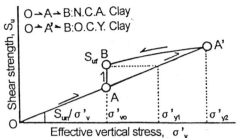

Fig. 5 S_u vs. $\acute{o}'_v$ path of N.C. Aged and O.C. Yong clays

strength-deformation behaviour of natural clays should be investigatede by the recompression method as demonstrated in Fig. 6. In this figure, differences in K_o values, strains at failure and stress-displacement behaviours of N.C.A. and O.C.Y clays measured by the recompression and the SHANSEP methods are presented.

4.2 Testing Procedure to Determine $S_{uf(d)}$ of Natural Clay Deposits

Although K_o-consolidated triaxial compression and extension tests had been carried out until beginning of eighties, DST with Mikasa's apparatus has been used in order to obtain $S_{uf(d)}$ up to the present because of lots of advantages of DST:- mainly two, 1) several 10 times faster working rate than triaxial test and 2) compensate the shear strength anisotropy.. In the DST, the specimen with diameter of 60mm and height of 20mm prepared from undisturbed sample is consolidated at σ'_{vo} until primary consolidation has been achieved, which generally takes 5 to 10 minutes, and then sheared under the constant volume condition. Since $S_{uf(d)}$ compensates the strength anisotropy effect, $S_{u(mob)}$ is given as a simple function of $S_{uf(d)}$ as expressed by Eq. (7).

$$S_{u(mob)} = S_{uf} \times \mu_A \times \mu_R = S_{uf(d)} \times \mu_R \qquad (7)$$

where $ì_A$ and $ì_R$ are correction factors for strength anisotropy and strain rate effect

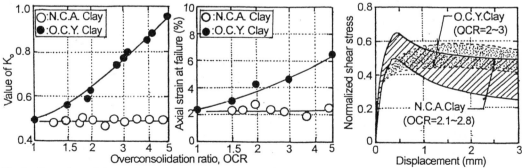

Fig. 6 Comparison of K_o, strain at failure and atress-displacement behaviours of N.C.A. and O.C.Y. clays measured by the recompression method (for N.C.A. clay) and the SHANSEP method (for O.C.Y clay)

4.3 Distribution Pattern of Suf of Natural Clay Deposits Related to Aging Effect

Because of quick operation of DST, extensive data on $S_{uf(d)}$ values of natural clay deposits have been collected. Hanzawa (1995) found that distribution pattern of $S_{uf(d)}$ with σ'_{vo} is fundamentally classified into the two patterns as indicated by Eq. (8) as also schematically shown in Fig. 7.

Pattern I $\quad S_{uf} = S_{ufo} + S_{un}/\sigma'_v \times \sigma'_{vo}$ (8.1)

Pattern II $\quad = k \times S_{un}/\sigma'_v \times \sigma'_{vo}$ (8.2)

where S_{ufo} = shear strength at the ground surface and k = index to show the degree of secondary compression being equivalent to overconsolidation ratio, OCR, $k > 1.0$

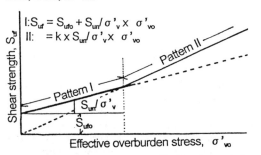

Fig. 7 Distribution patterns of S_{uf} with $\acute{o}'_{vo}$ of natural clay deposits investigated by DST with the recompression method

From the distribution pattern given by Eq. (8), Hanzawa and Adachi (1983) already discussed the interaction of chemical bonding and secondary compression, which are the two major aging effect, to the structure of the clay. They suggested that the case, where chemical bonding has first taken place

with secondary compression occuring thereafter, is most probable case. In this case, development of secondary compression will be prevented because of the structure already formed by chemical bonding. However, when secondary compression progresses with time even should take this in a small amount, it could damage the chemically bonded structure. Since amount of secondary compression is proportional to σ'_{vo}, it is suggested that secondarily compressed structure completely replaces the chemically bonded structure when σ'_{vo} exceeds a certain value. The structure of the clay formed through this process will be as presented in Fig. 8 which well explains the distribution pattern of S_{uf} given by Eq. (8) when the shear strength is used instead of the structure.

The values of $S_{uf(d)}$ measured for Ariake (Japan), Bothkennar (United Kingdom), Champlain (Canada) and Drammen (Norway) clays are plotted vs. σ'_{vo} in Fig. 9. (Dam, et al 1999). Although four clays are located at quite different regions and significantly differ in properties, it is clear that $S_{uf(d)}$ versus $\acute{o}'_{vo}$ exactly follow the patterns given by Eq. (8). Values of k obtained for Japanese, Indonesian and above 4 clays except Champlain clay are plotted versus Ip, elapsed time since sedimentation and thickness of the clays in Fig. 10. It is suggested that the value of k is related to these three factors.

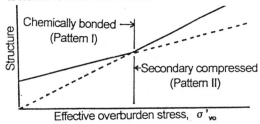

Fig. 8 Distribution pattern of aged structure of natural clay deposits suggested from distribution patterns of S_{uf} vs. $\acute{o}'_{vo}$

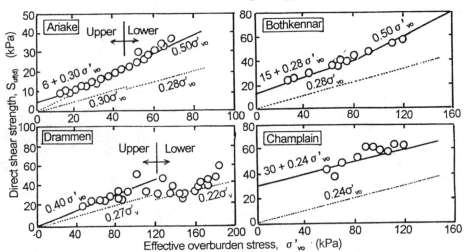

Fig. 9 $S_{uf(d)}$ with the recompression method plotted vs. $\acute{o}'_{vo}$ for Ariake Bothkennar , champlain and Bothkennar clays

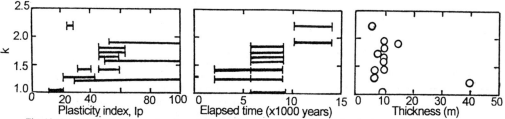

Fig. 10 Volues of K of Asian marine clays plotted vs. Ip , elapsed times since sedimentation and the thickness of each clay

4.4 Correlation between $S_{uf(d)}$ and Other Shear Strength Parameters

Unconfined compression, field vane and cone penetration tests, UCT, FVT and CPT have also been carried out together with DST. Shear strengths, $S_{u(UCT)}$ and $S_{u(FVT)}$, and point resistance, $(q_T - \sigma_{vo})$ measured in these tests are plotted versus $S_{uf(d)}$ values in Fig. 11. The followings can be said to characterize the correlation between each shear strength parameter and $S_{uf(d)}$.

1) Correlation between each shear strength parameter and $S_{uf(d)}$ are as follows:-

$$S_{u(UCT)} = (0.4\text{~}1.0)S_{uf(d)} \qquad (9.1)$$

$$S_{u(FVT)} = (0.7\text{~}1.0)S_{uf(d)} \text{ for clean clay} \qquad (9.2)$$

$$= (1.0\text{~}1.4)S_{uf(d)} \text{ for clay with high sand} \qquad (9.2')$$

$$(q_T - \sigma_{vo}) = (8\text{~}13)S_{uf(d)} \qquad (9.3)$$

2) Scatter between $S_{u(UCT)}$ and $S_{uf(d)}$ was the largest among the three parameters. $S_{u(UCT)}$ is influenced by sample quality and the effective stress at the testing stage, $\sigma'_{(UCT)}$, which can not be controlled. For an example, ó'$_{(UCT)}$ of Drammen clay (Ip = 15) is only 1/8 of ó'$_{(UCT)}$ of Ariake clay (Ip = 40~60) even highly quality sampling technique is used. This extremely low ó'$_{(UCT)}$ is the reason for very low low $S_{u(UCT)}$ of Drammen clay as reported

by Tanaka and Tanaka (1995). Since $S_{u(UCT)}$ is influenced not only by sample quality but also by $\sigma'_{(UCT)}$ which can not be controlled technically, UCT is not recommended to be used for any practical purpose in author's opinion.

3) Correlation given by $S_{u(FVT)} = (0.7\text{~}1.0)S_{uf(d)}$ to clean clay is reasonable when mechanism of shear in FVT and DST is examined. $S_{u(FVT)} = (1.0\text{~}1.4)$ $S_{uf(d)}$ to clay with high content of sand seam or shell is also reasonable because $S_{u(FVT)}$ on vertical plane is higher than $S_{u(FVT)}$ on horizontal plane of this type of clay. Use of $S_{u(FVT)}$ of this type of clay to practice should therefore be careful. It should also be pointed out that $S_{u(FVT)}$ values in Fig. 11 were obtained by sheathed FVT which gives much more accurate and less scatter $S_{u(FVT)}$ values than the values measured by drilling type FVT.

4) Correlation of $(q_T - \sigma_{vo}) = (8\text{~}13)S_{uf(d)}$ irrespective of clay properties should be an important advantage of CPT together with its quick working rate and much less influenced feature by technical quality when compared with FVT and UCT. Because of these advantages, CPT is the most recommendable in-situ test for soft ground engineering. It should be informed that double piped portable cone penetration device played an important role in evaluating shear strength of clay and stability of excavated slope in an power plant project in a country in Southeast Asia.

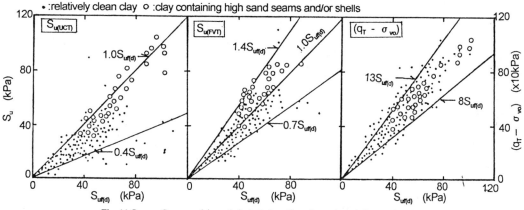

Fig. 11 $S_{u(UCT)}$, $S_{u(FVT)}$ and $(q_T - ó_{vo})$ plotted vs. $S_{uf(d)}$ for various Asian marine clays

31

Appropriate determination of the consolidation yield stress, σ'_y is also an important subject in practice. Standard oedometer test does not give clear answer to this subject, while strain control oedometer test gives much clear yield point and compressibility as indicated in Fig. 12.

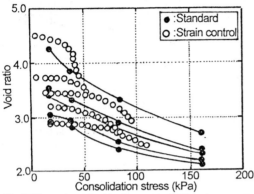

Fig. 12 Comparison of void ratio vs. consolidation stress curves from standard and strain control oedometer tests

Although the yield point obtained from the strain control test is not known to give the proper σ'_y, in actual project, lots of strain control oedometer tests were also carried out to get the clear σ'_y at a strain rate of 0.02%/min. The values of σ'_y obtained are plotted versus $(q_T - \sigma_{vo})$ in Fig. 13, and a correlation of $(q_T - \sigma_{vo}) \fallingdotseq 3\sigma'_y$ was derived. On the other hand, $(q_T - \sigma_{vo}) = (8 \sim 13)S_{uf}$ was also obtained as givens by Eq. (9.3) previously, while S_{uf}/σ'_y ratio of natural clay should be the same as S_{un}/σ'_v as conceptually given by Eq. (2). The relationship of $(q_T - \sigma_{vo}) = 3$ $\sigma'_y = (8 \sim 13)S_{uf(d)} = 10.5S_{uf(d)}$ in average leads $S_{uf(d)}/\sigma'_y = 0.29$. This is approximately the same as the $S_{un(d)}/\sigma'_v$ ratio of 0.3 for Asian marine clays in the N.C. state shown in Fig. 2.

$$(q_T - \sigma_{vo}) = 3\sigma'_y = 10.5S_{uf(d)} \tag{10}$$

$$S_{uf(d)}/\sigma'_y = 0.29 = S_{un(d)}/\sigma'_v \tag{11}$$

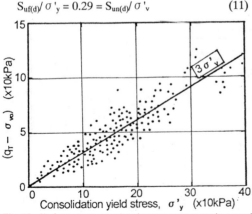

Fig. 13 σ'_y from strain cotrol oedometer tests at strain rate of 0.02%/min plotted vs. ($q_T - \sigma_{vo}$) for Asian marine clays

4.5 Shear Strength Reduction Induced by Unloading

S_{uf} decreases when subjected to unloading like excavation as shown in Fig. 14. The shear strength reduction ratio, α given by S_{uo}/S_{uf}, where S_{uo} is the shear strength after swelling induced by unloading, is obtained using DST as explained in the following:- Undisturbed sample is firstly consolidated at OCR = 1.0 (at $\sigma'_{v(at\ B')}$) , then swelled at corresponding OCR to $\sigma'_{vo}/\sigma'_{uo}$ ($\sigma'_{v(at\ B')}/\sigma'_{v(at\ C')}$) and sheared to obtain α (= $S_{u(at\ C')}/S_{u(at\ B')}$) as schematically shown in the figure. Repeating this tests at different OCR, values of α are plotted versus OCR as presented in Fig. 15 which indicate α vs. OCR for Ariake, Banjarmasin (Indonesian) Bothkennar and Drammen clays. This technique gives a little bit conservative S_{uo} value than actual one as suggested from Fig. 14.

$$S_{uo(d)} = S_{uf(d)} \times \alpha \tag{12}$$

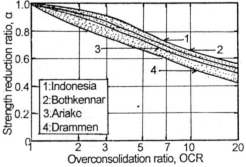

Fig. 14 Schematic presentation to show the S_{uf} reduction induced by unloading and its determining method by DST

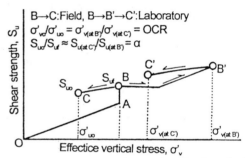

Fig. 15 Correction factor, α vs. OCR for 4 marine clays obtained by DFT

4.6 Pre and Post Failure Behaviour

Since Miki et al.(1994) indicated that apparent shear modulus, $G_{50(ap)}$ from DST given by Eq. (13) as also shown in Fig. 15 is approximately the same as G_{50} from the direct simple shear test, $G_{50(ap)}$ have been used to evaluate the deformation of earth structures as reported by Tsuji(2,000).

$$G_{50(ap)} = S_{uf(d)}/2 \div D_{50}/H_o \qquad (13)$$

where D_{50} = displacement at $S_{uf(d)}/2$ and H_o= initial height of the specimen.

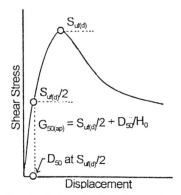

Fig. 16 Procedure to determine $D_{50(ap)}$ from DST

Since displacement at failure, D_f depends both on clay properties like Ip and structure, which is evaluated by $S_{uf(d)}/S_{un(d)}$ ratio, Dam et al (1999) recommended to express $D_{50(ap)}$ in normalized form as shown in Fig. 16 in which $G_{50(ap)}/S_{uf(d)}$ values of Ariake, Bothkennar, Champlain and Drammen clays are plotted versus Ip. $G_{50(ap)}/S_{uf(d)}$ values of Ariake, Bothkennar and Drammen clays look to be located on a unique line when plotted versus. Ip. Higher $G_{50(ap)}/S_{uf(d)}$ value of Champlain clay could be because of relatively higher cemented structure of this clay among 4 clays.

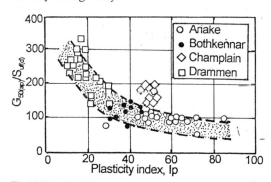

Fig. 17 Normalized $G_{50(ap)}$ values with $S_{uf(d)}$ plotted vs. Ip for 老 world famous 4 marine clays measured by DST

Brittleness or strain softening should be used in characterizing the post-failure behaviour of the clay where destruction of the structure is progressed. Dam et al (1997) proposed the three parameters to present the brittleness of natural clay property as schematically shown in Fig. 17. More detailed discussions on this subject are presented in the references by Dam et al (1997;1999), Suzuki et al (2,000), and Choa and Hanzawa (2,000) to application in practice.

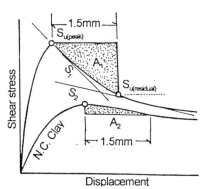

Fig. 18 Idealized presentation for 3 parameters to evaluate post-failure behaviour

5 PROPOSED METHOD TO USE DST AND CPT IN SOFT GROUND ENGINEERING

5.1 Practical Sequences of the Proposed Method

From the discussions given in the previous sections, it is evident that DST and CPT are the best tools in soft ground engineering from both practical and theoretical points of view. In this section, actual procedures of the proposed method, which was so named ACCESS (Advanced Construction Control for Earthwork on Soft Soils) System, focusing to do investigation, design and construction control with the approach as simplified as possible is firstly described.

Actual procedures of the ACCESS System is as explained below:-
1) CPT are firstly carried out all over the project site in order to understand ground and drainage conditions as well as $(q_T - U_{vo})$ values.
2) Undisturbed sampling with hydraulic piston sampler is done at the locations showing representative ground conditions. The reason to use hydraulic piston sampler is its high working rate with reasonable sample quality.
3) DST to determine $S_{uf(d)}$ are carried out on all the undisturbed samples with the recompression method, while the tests to decide $S_{un(d)}$ (and α if necessary) are also conducted on representative clay samples.
4) Correlation between $(q_T - \sigma_{vo})$ and $S_{uf(d)}$, and 3 design parameters are determined. The value of 0.85 to $S_{u(mob)}$ is the correction factor for strain rate effect.

$$(q_T - \sigma_{vo}) = k \times S_{uf(d)} \rightarrow \text{correlation}(=\text{cone factor})$$

①$S_{u(mob)} = 0.85 S_{uf(d)}$, ② $S_{un(d)}/\sigma'_v$ and

③ $\sigma'_y = S_{uf(d)} \div S_{un(d)}/\acute{o}'_v \rightarrow$ 3 design parameters

33

5) Design is done using parameters obtained at stage 4 and construction is started.
6) During construction, CPT are conducted in order to evaluate $S_{uf(d)}$, σ'_v and the degree of consoli dation , U which are predicted by Eq. (14).

$$S_{uf(d)} = (q_T - \sigma_{vo} - \Delta\sigma)/k$$

$$\sigma'_v = S_{uf(d)} \div S_{un(d)}/\sigma'_v \qquad (14)$$

$$U = (\sigma'_v - \sigma'_y) \div (\sigma'_{vo} + \Delta\sigma - \sigma'_y)$$

The processes described above are presented as the flow chart in Fig. 19.

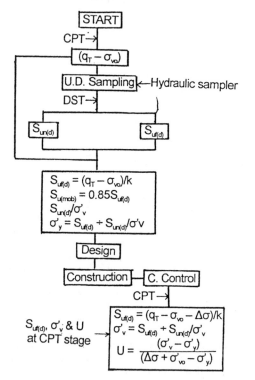

Fig. 19 Flow chart to show the actual sequences of the proposed method

5.2 Case Studies for Stability of Actual Structures

5.2.1 Outline of Case Studies

Locations of the projects are shown in Fig. 20. Although they are also presented in the papers by Hanzawa et at and Choa and Hanzawa in this Symposium, outlines of the projects are briefly described in the followings.
1) Fao Steel Jetty, Iraq (1976):Fao Steel Jetty was constructed at the river mouth of the Arave river which is the river after meeting the Tigris and the Euphrates rivers. Severe stability problem was found immediately after commencement of construction when $F.S._{min}$ was calculated using $S_{u(UCT)}$ ($F.S._{min}$ = 0.6). A detailed field and laboratory tests was carried out parallel with construction and a special property of Fao clay was brought to light through this investigations. The recompression method to determine S_{uf} was actually developed through this investigation. $F.S._{min}$ value obtained using $S_{u(mob)}$ determined by K_o-consolidated triaxial compression and extension tests with the recompression method was indeed 2.4times of $F.S._{min}$ from $S_{u(UCT)}$, and the jetty was successfully constructed on time.
2) Al-Zubair Embankment, Iraq (1978): 5 embankments for preloading were rapidly constructed on Al-Zubair clay which is found at the old river channel of the Arab River . In order to shorten the preloading time, embankment was designed with the minimum possible safety of $F.S._{min}$ = 1.05, which was determined using $S_{u(mob)}$ from the recompression method with K_o-consolidated triaxial compression and extension tests. Because of a slight difference between designed and constructed embankments, one embankment failed immediately after completion and two others were probably on the verge of failure.
3) Daikokucho Daike, Tokyo Bay (1981):A temporary dike for reclaimed land was constructed on a high plastic marine clay. Immediately after completion, it failed and sunk into the sea. The author was given the chance to investigate the cause of failure.
4) Banjarmasin Revetment, Kalimantan, Indonesia (1989):The revetment was initially designed to be constructed under multi-loading stages with support of soil improvement by vertical drain because of inadequate $F.S._{min}$ obtained by $S_{u(UCT)}$. Immediately after the contract, a series of DST, FVT and CPT was carried out, and $S_{u(mob)}$ was newly determined by DST. Sine the value of $F.S._{min}$ from $S_{uf(d)}$ was a little bit greater than 1.0, the revetment was then constructed in one stage.
5) Kameda Embankment, Niigata (1993): Embankment for an expressway was constructed on peaty subsoil with multi-loading stages. When the height of the second embankment was rapidly increased from 4.5m to 6.3m, a large deformation took place together with tensile crack and heave. $S_{u(mob)}$ was determined by DST, FVT and UCT.
6) Vungtau Revetment, Mekong Delta, Vietnam (1996): Vungtau Revetment is a fisherman's port along Mekong Delta. $F.S._{min}$ values from $S_{u(UCT)}$ was 0.41, while the value from $S_{u(FVT)}$ was 2.85 at the construction commencement stage. In order to evaluate more reliable $F.S._{min}$ value, CPT were carried out mainly in front of the revetment par allel with construction and the revetment was safely constructed.

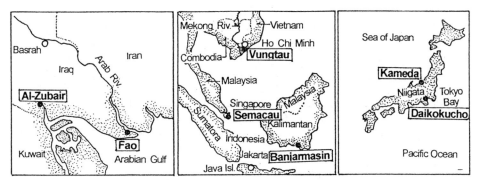

Fig. 20 Locations of 7 projects of case studies for stability problem

7) Semakau dike, Singapore (1995): Some of the reclamation projects in Singapore have been conducted on coral sand islands which are commonly located on slopes. In order to make a early work base along the coral sand island, a higher dike was temporarily constructed at the initial stage of the project. During construction of the dike, a large scale slip failure suddenly took place. A inspecter who was working at the top of the dike witnessed throwing into the sea just in a few seconds.

5.2.2 *Field and Laboratory Tests Conducted*
Characteristics of each clay, and field and laboratory tests carried out to determine $S_{u(mob)}$ are presented in the followings.
1) Fao clay with Ip = 10~30 is characterized by indicating significant different in $ó'_y$ when subjected to different stress incremental ratio in oedometer test. $ó'_y$ from the standard oedometertest were 70% of $ó'_{vo}$, while $ó'_y$ from the special oedometer test with small stress increment ratio were about the same to $ó'_{vo}$. Noticing this feature, a practical technique of the recompression method was actually developed. In addition to K_o-consolidated compression and extension tests, a series of consolidated (OCR = 1.0), swelled under different OCR and undrained triaxial compression test (C.S.U-T) was also car ried out to investigate the reduction of S_{uf} induced by excavation in front of the Jetty.
2) Al-Zubair clay with Ip = 30~35 found about 60km west of Fao is much older than Fao clay and has been subjected to complicated aging ef fects such as desiccation, cementation and sec ondary compression. K_o-consolidated compres sion and extension tests with the recompression and the SHANSEP methods were carried out to decide $S_{u(mob)}$ together with $S_{u(FVT)}$.
3) Daikokucho clay with Ip = 40~60 is a typical marine clay in Tokyo Bay but contains relatively higher sand contents. K_o-consolidated triaxial compression and extension tests with the Re

compression and the SHANSEP methods were done together with UCT, FVT and CPT. CPT was however executed to investigate the failure plane alone.
4) Banjarmasin clay with Ip = 40~110 is classified into the upper and the lower clays bounded by the desiccated clay at a depth of 20m, which was formed when the sea level was lowered about 10,000 years ago. UCT was only done at the design stage, but CPT, DST and FVT were carried out to determine new $S_{u(mob)}$ after the contract..
5) Kameda cohesive soils consist of peat with w_N = 100%~300%, and sandy clay with w_N = 50%~70%. DST, CPT, FVT and UCT were done. Correlation between $S_{uf(d)}$ and $(q_T - ó_{vo})$ and $S_{un(d)}/ó'v$ were as folloes.

$$S_{uf(d)} = (q_T - ó_{vo})/10$$

$$S_{un(d)}/ó'_v = 0.45$$

6) Vungtau clay is divided into the upper clay with Ip = 20~40 and the lower clay with Ip = 20 divided by a thin sandy layer. About 5m excavation is made in front of the revetment, while 2m reclamation to the behind. UCT and FVT were executed at the design stage giving $F.S._{min}$ = 0.41 for $S_{u(UCT)}$ and 2.85 for $S_{u(FVT)}$. In order to evaluate more reliable $F.S._{min}$ value, portable CPT with sectional area of 6.45cm2 and point angle of 30 were carried out at 10 points in front and behind the revetment. On the other hand, Cone factor was determined by laboratory tests on a Japanese marine clay as shown below.

$$S_{uf(d)} = (q_c - ó_{vo})/7$$

Using this relationship, $F.S._{min}$ 1.25 was obtained and the revetment was safely constructed on time.
7) Semacau clay with Ip = 30~60 classified into the upper and the lower clays bounded by the desiccated clay formed by lowering the sea level at 10,000 years ago. It is one of the typical marine clay formed in coastal zone in Singapore

35

Table 1 Field and laboratory tests carried out in each project

Project	Area	CPT	KoTCT	KoTET	DST	FVT	UCT	Investigation time
Fao	Iraq		O	O		O	·O	During construction
Al-Zubair	Iraq	O-1	O	O		O		Design stage
Daikokucho	Tokyo Bay	O-1	O	O		O	O	After failure
Banjarmasin	Indonesia	O-1			O	O	O	After contract
Kameda	Niigata	O-2			O	O	O	After contract
Vungtau	Vietonamu	O-3			O	O	O	Desgn & contract
Semakau	Singapore	O-2			O	O	O	Investigation stage

CPT-1:Mechanical cone with A= 10cm^2 and θ = 60° , CPT-2:Electric piesocone with A =10cm^2 and θ = 60° . CPT-3:Portable cone with A= 6.45cm^2 and θ = 30°

but much more brittle than the marine clay found at Changi area. Lots of CPT, DST and FVT were carried out in order to understand the ground conditions and mechanical properties of the upper and the lower clays. At the initial stage of reclamation, large scale of slip failure took place.It coul not be explained by circular arcfail ure plane and then residual shear strength wasused instead of peak shear strength in the analysis as presented later.

All the tests carried out in each project are summarized in Table 1.

5.2.3 Results of Stability Analysis

S_{uf} of the clay was determined by the recompression method in all the projects. In addition, $S_{u(FVT)}$, corrected $S_{u(FVT)}$ by Bjerrum's method, $S_{u(UCT)}$ and the SHANSEP method were also used to determine $S_{u(mob)}$. K_o-consolidated triaxial compression and extension tests were used in Fao, Al-Zubair and Daikokucho projects, that is, until the beginning of eighties, while DST replaced the position to determine S_{uf} after it was proved that $S_{uf(d)}$ compensates the strength anisotropy. In the projects in which S_{uf} increases by drainage under loading (in the projects of Daikokucho and Kameda) or decreases by swelling under unloading (in the projects of Fao and Vungtau), S_{uf} values were modified. Both peak and residual direct shear strengths, on the other hand, were used to explain the actual failure plane at Semakau Dike as described in detail by Choa and Hanzawa in this Symposium.

$S_{u(mob)}$ used for stability analysis is fundamentally Presented in Eq. (15) and summarized for each of the projects in Table 2.

$$S_{u(mob)} = 0.85 S_{uf(d)} \times \alpha \times \beta \tag{15}$$

where 0.85 = correction for strain rate effect, α = correction factor for change of S_{uf} during construction, $\alpha \geqq 1.0$ (loading) and < 1.0 (unloading), and β = correction factor to progressive failure, β = 1.0 in the first 6 projects and β <1.0 in the project in Singapore

$S_{u(mob)}$ of Fao, Al-Zubair and Daikokucho projects in which Suf was determined by K_o-consolidated triaxial compression and extension tests, was mechanically determined to be $0.85(S_{uf(c)} + S_{uf(e)})$. This $S_{u(mob)}$ gives somewhat different F.S.$_{min}$ values from the original F.S.$_{min}$ values given by Hanzawa (1983).

Minimum factor of safety, F.S.$_{min}$ values obtained from the analyses with the use of $S_{u(mob)}$ determined by DST, FVT, UCT, corrected $S_{u(FVT)}$ (Bjerrum's method) and the SHANSEP method are summarized in Table 2 together with behaviour of each structure. F.S.$_{min}$ values obtained using $S_{u(mob)}$ determined by DST given by Eq. (15) well explain the actual behaviour of structures. Calculated failure planes to give F.S.$_{min}$ are compared with actual and predicted failure planes in Fig. 21. Calculated failure planes to give F.S.$_{min}$ also show good agreement with the actual ones.

Table 2 F.S.$_{min}$ values of each of structure obtained using $S_{u(mob)}$ values determined by various methods

Project	Location	Behaviour	$S_{u(FVT)}$	$S_{u(UCT)}$	$S_{u(FVT)}$·μ	$0.85 S_{uf(d)}$	SHANSEP	α	β
Fao	Arab Riv. Mouth	Stable	1.24	0.68	1.24	1.23		<1.0	1.0
Al-Zubair	Arab. Riv. Delta	Failed	0.88		0.83	0.96	0.71	1.0	1.0
Daikokucho	Tokyo Bay	Failed	1.26	1.07	1.03	0.97	0.91	>1.0	1.0
Banjarmasin	Indonesia	Stable	1.06	0.85	0.80	1.06		1.0	1.0
Kameda	Niigata Pref.	Kameda	1.44	0.69		0.98		>1.0	1.0
Vungtau	Mekong Delta	Stable	2.85	0.41		>1.25		<1.0	1.0
Semakau	Singapore	Failed				0.92~1.05		1.0	<1.0

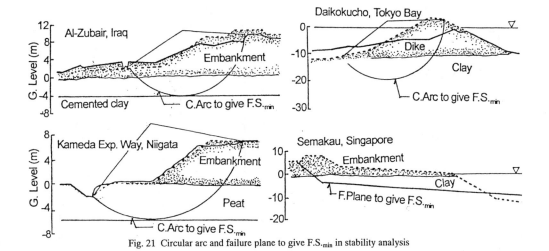

Fig. 21 Circular arc and failure plane to give F.S._min in stability analysis

5.3 Application to Construction Control

The proposed method has been adopted in actual projects in Japan, Singapore and Philippine as well as in Vietnam. Among these projects, construction control to trial embankments for bypath(Site A) and airport (Site B) on marine clays in Japan is briefly presented here because construction control in Singapore and Philippine are on the way.

Correlation between $(q_T - \sigma_{vo})$ vs. $S_{uf(v)}$ obtained at investigation stage are indicated in Fig. 22. On the other hand, $S_{un(d)}/\sigma'v$ ratio were 0.30 for both marine clays as show in Fig. 23.

$$(q_T - \sigma_{vo}) = 10S_{uf(d)} \text{ (Site A)}$$

$$= 11.5S_{uf(d)} \text{ (Site B)}$$

$$S_{un(d)}/\sigma'_v = 0.30 \text{ (Sites A and B)}$$

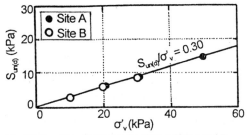

Fig. 23 $S_{un(d)}/\sigma'v$ ratios for marine clays at Sites A and B

CPT were carried out at 400 days at site A and 470 days at site B after commencement of embankment as indicated in Fig. 24 in which settlement with final one, S_f predicted from settlement measurement are also shown. Values of $S_{uf(d)}$, σ'_v and U at the time of CPT calculated by Eq. (14), in which k = 10 (site A) and 11.5 (Site B), and $S_{un(d)}/\sigma'_v = 0.30$

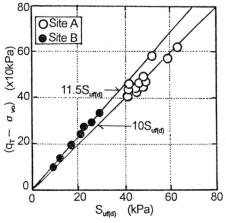

Fig. 22 $(q_T - \sigma_{vo})$ vs. $S_{uf(d)}$ at Sites A and B

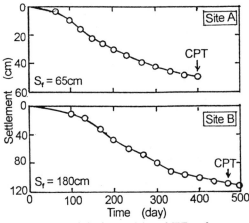

Fig. 24 Settlement behaviours and time of CPT performance

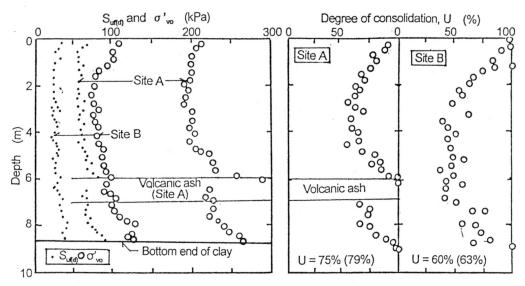

Fig. 25 $S_{uf(d)}$, $\acute{o}'_v$ and U at the time of CPT performance evaluated by the proposed method

(sites A and B) are used, are presented in Fig. 25. Values of U from the proposed method were 75% (site A) and 60% (site B), while the values from settlement measurement were 79% for site A and 63% for site B, respectively. The values of U obtained by the two methods showed good agreement with each other. Such accuracies in $S_{uf(d)}$, $\acute{o}'_v$ and U are adequate from practical point ov view.

6 CONCLUSING REMARKS

In order to overcome regional variations both in soil behaviour and technical quality, a new procedure, so named ACCESS (Advanced Construction Control for Earthwork on Soft Soils) System was developed for soft ground engineering. This system consists of DST and CPT because of their various technical advantages, and easy and quick working operations. From the studies and experiences up to the present, the following conclusions can be drawn.

1) Direct shear strength compensates the strength anisotropy and can directly be used as $S_{u(mob)}$ after correction for strain rate effect, which is the most important advantage of DST.

2) Strength increment ratio of Asian marine clays in the N.C. is approximately constant irrespective of Ip, showing a value of $S_{un(d)}/\sigma'_v \fallingdotseq 0.3$

3) Shear strength of natural clay deposit, S_{uf}, must be decided by the recompression method, and its distribution pattern with σ'_{vo} can be classified into the two patterns:-

$$S_{uf} = S_{ufo} + S_{un}/\sigma'_v \text{ x } \sigma'_{vo} \text{ (Pattern I)}$$

$$= k \text{ x } S_{un}/\sigma'_v \text{ x } \sigma'_{vo} \text{ (Pattern II)}$$

S_{ufo} = shear strength at the ground surface and

k = index to show the degree of secondary compression being equivalent to OCR

Pattern I, in which chemical bonding is main aging effect, is predominant until σ'_{vo} reaches a certain value, while Pattern II in which secondary compressed structure replaces the chemical bonded structure, is predominant after σ'_{vo} exceeds the above σ'_{vo}.

4) Correlation of (q_T σ_{vo}) to $S_{uf(d)}$ is much less scattered and less affected by technical quality and clay properties when compared with those of $S_{u(FVT)}$ and $S_{u(UCT)}$ to $S_{uf(d)}$. In addition, because of its easy and quick working rate, CPT is the most recommendable in-situ test in soft ground engineering.

5) Stability analysis of actual structures constructed at and near failure strongly demonstrated the validity of $S_{u(mob)}$ determined by the proposed method, that is, $S_{u(mob)} = 0.85 S_{uf(d)}$ (x α).

6) It can be concluded from the experiences up to the present that the ACCESS System is a good tool for soil investigation, design and construction control.

It should finally be emphasized that there are lots of regions and projects in the worlds where such simplified procedure like the ACCESS System is strongly desired.

REFFERENCES

Bjerrum, L. 1972. Embankment on soft soil. Proc. ASCE Conf., Performance on Earth and Earth-Supported Structures. 2 1-54.

Bjerrum, L. 1973. Problems of soil mechanics and construction on soft clay and structurally unstable soils. Proc. 8th ICSMFE. 3. 111-159.

Choa, V. and Hanzawa, H. 2000. Case study of a failed embankment with consideration of progressive failure. Proc., Coastal Geotechnical Engineering in Practice. IS-Yokohama.

Dam, T. K. L., Yamane, N., Hanzawa, H. and Porhaba, A. 1997. Evaluation of progressive failure potential of natural clay deposit. Proc., Deformation & Progressive Failure in Geomechanics. IS-Nagoya 199-204.

Dam, T. K. L., Yamane, N. and Hanzawa, H. 1999. Direct shear test results on four soft marine clays. Characterization of soft marine clay. Balkema. 229-240.

Hanzawa, H. 1977. Geotechnical property of normally consolidated Faoclay, Iraq. Soils and Foundations. 17 (4): 1-15.

Hanzawa, H. 1983. Three case studies for short term stability of soft clay deposits. Soils and Foundations, 23 (2):140-154

Hanzawa, H. 1995. In-situ shear strength of marine clay related to aging effect. Proc. 11th European Conf. ICSMFE. 141-146.

Hanzawa, H. and Kishida, T. 1981. Fundamental consideration on undrained strength characteristics of alluvial marine clay. Soils and Foundations, 21(1): 39-50

Hanzawa, H. and Adachi, K. 1983. Overconsolidation of alluvial clays. Soils and Foundations. 23 (4): 106-118.

Hanzawa, H. and Tanaka, H. 1992. Normalized undrained strength of clay in the normally consolidated state and in the field. Soils and Foundations. 32 (1): 132-148.

Hanzawa, H., Kishida, T., Fukasawa, T. and Suzuki, K. 2,000 Case studies of six earth structures constructed on soft clay deposits. Proc., Coastal Geotechnical Engineering in Practice. IS-Yokohama.

Ladd, C. C. and Foott, R. 1974. New design procedure for stability of soft clay. ASCE 100 (GT100): 763-786.

Mesri, G. 1975. Discussion on new design procedure for stability of soft clay. ASCE. 101 (GT4): 409-411.

Mikasa, M. 1960. Direct shear device newly developed. Proc. 15th JSCE: 45−48(in Japanese)

Miki, H., Kohashi, H., Asada, H. and Tsuji, K. 1994. Deformation-pore pressure behaviour measured and analyzed for trial embankment. Proc. Pre-failure Deformation Characteristics of Geomaterials. IS-Sapporo. 1. 547-552.

Ohta, H., Nishihara, A., Iizuka, A., Morita, Y., Fukagawa, R. and Arai, K. 1989. Unconfined compression strength of soft aged clay. Proc., 12th. ICSMFE. 1. 71-74.

Suzuki, K., Fukasawa, K., Hirabayashi, H., Asada, H. and Kamata, R. 2,000 Stress−displacement characteristics of natural clay deposits investigated by direct shear test. Proc., Coastal Geotechnical Engineering in Practice. IS − Yokohama

Tanaka, H. and Tanaka, M. 1997. Application of UC test for two European clays. Proc. 14th ICSMFE. Hamburg. 209-212.

Trak, B., LaRochelle, P., Tavenas, F., Leroueil, S. and Roy, M. 1980. A new approach to the stability analysis of embankment on sensitive clays. Canadian Geotechnical Journal, 17 (4): 526-544.

Tsuchida, T. 1989. New method for determining undrained strength of cohesive soils by means of triaxial tests. Doctorate thesis. University of Tokyo.

Tsuji, K. 2000. Mechanism of direct shear test on marine clay under constant volume condition, and its application to evaluating failure and deformation of soft ground. Doctorate Thesis. Tokyo Institute Technology (in Japanese).

Special lectures

Coastal Geotechnical Engineering in Practice, Nakase & Tsuchida (eds)
© 2002 Swets & Zeitlinger, Lisse, ISBN 90 5809 151 1

Mechanical properties of Pleistocene clay and evaluation of structure due to aging

T.Tsuchida, Y.Watabe & M.S.Kang
Port and Airport Research Institute, Yokosuka, Japan

ABSTRACT: In recent years, as the scale of structures has been enlarged and the construction works have been carried out in deeper sea areas, the loads due to constructions has become close or larger than the consolidation yield stress, p_c, of the Pleistocene clay layer at large depth. The author's research group has been carried out a series of studies on the mechanical properties of aged clays having structures that were formed during the process of long-term sedimentation and consolidation. In the former part of this report, the compression and shear characteristics of aged clay, at first, are reviewed with related to the case study on Kansai International Airport project. In the latter part, the method for evaluating the degree of structure is newly proposed, and a unique consolidation behavior of structured clay is shown.

1 INTRODUCTION

In the research of soil mechanics, remolded and re-constituted clays have been used in laboratory test to avoid scatter of experimental results. However, it was clarified that mechanical properties of remolded clays are quite different from those of natural clays, because natural clay has a structure due to secondary compression and/or cementation effects during sedimentation. (Tavenas and Leroueil, 1977; Jami-olkowski et al., 1985). Although these differences are observed in Holocene clays, they have raised important engineering problems in the construction projects on Pleistocene clays layers.

In coastal regions of Japan, Pleistocene clay layer is often located beneath a Pleistocene gravel layer, which lies under a 5 to 30 m thick Holocene clay layer. In most construction projects before 1970, the Pleistocene gravel layer had been considered as a stable layer that could support the superstructures, and the engineers seldom consider the behavior of Pleistocene clay layers beneath the gravel layer. In recent years, however, as the scale of structures has been enlarged and the construction works have been carried out in deeper sea areas, the loads due to constructions has become close to or larger than the consolidation yield stress, p_c, of the Pleistocene clay layer at large depth. Thus, the evaluation on engineering properties of the Pleistocene clay layer is one of the most important issues for design and construction in the coastal areas.

A typical example is shown in Figure 1, which are time-settlement curves obtained from 4 corners of a ventilation tower for undersea tunnel constructed at a reclaimed land in Osaka Port (Nakashima et al., 1995). The ventilation tower was constructed at the depth of 33.5m trenched down to the top of Pleistocene gravel layer. However, as shown in Figure 2, the soil profiles in the site, below the gravel layer, there exist three Pleistocene clay layers whose over-consolidation ratios were 1.8 – 2.3 before construction. As the reclamation continued after the installa-

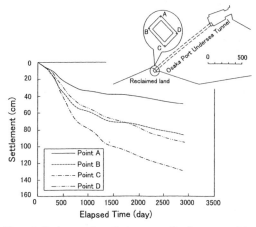

Figure 1 Settlement of ventilation tower of under sea tunnel in Osaka Port

tion of the ventilation tower, the overburden stress on Pleistocene clay layers was increased consequently decreasing the over-consolidation ratios. Figure 3 is OCR profiles of Pleistocene clay layers after the completion of reclamation, showing that all the layers are still in the over-consolidated regions. Although the overburden stress did not exceed the consolidation yield stress, the settlements of 50-130cm have taken place continuously for 2500 days. According to Nakashima et al., the predicted settlements with coefficient of volume compressibility, m_v and coefficient of consolidation, c_v, obtained by conventional step loading consolidation tests underestimated the observed settlements, and to explain the observed settlement, it was necessary to use larger compressibility m_v and smaller c_v in the normally consolidation region and to consider the secondary settlements with the coefficient of secondary compression, C_α.

Another example is the Kansai International Airport (KIA) construction project, which was constructed on an artificial island in the Osaka Bay, which is 5 km off the southwest of Osaka City (Arai,1991;Arrai et al.,1991; Endo et al.,1991). The airport was inaugurated in September 1994, when the first phase of construction was completed. As the airport is being operated with only one runway, the second phase construction project commenced in July 1999, aiming the completion of the second runway parallel to existing one up to 2007.

Figure 4 is the soil profile of the site, showing that the Pleistocene gravel-sand layers and clay layers are accumulated alternately up to 400m depth, and the over-consolidation ratio before the reclamation are 1.0-1.5. When the project was planned in 1970s, most of engineers hardly expected that the settlements of Pleistocene clay layers might make an important problem, because they had no experiences of large settlements of Pleistocene layer at that time.

Figure 5 shows measured settlement with time at an observation point in KIA, where the largest settlement was observed (Kobayashi,1994). As shown in Figure 5, the consolidation settlement of Holocene clay layer improved by sand drain method ended up within 6 months after reclamation, while the settlement of Pleistocene clay layer commenced when the level of reclaimed land reached almost the sea level, and has continued after the airport was opened. Recently the total settlements in the island have become larger than 11m and about 5m out of the total settlement is by the settlement of Pleistocene layer. The reclaimed area in the second phase has an average water depth of 19.5 m, in comparison with 18.0 m in the first phase. The average thickness of the Holocene clay layer in new site location is 24 m (18 m in the first phase), which will increase the volume of reclaimed sand or gravel by 40%, compared to the first phase. The prediction of settlement in the 2nd

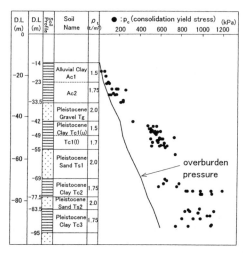

Figure 2   Soil profiles at the site (Osaka Port)

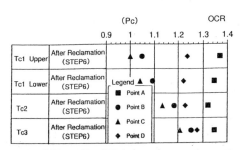

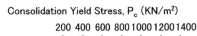

Figure 3   OCR profile of Pleistocene layers under ventilation Tower

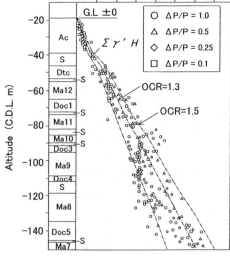

Figure 4   Soil profiles at the site (KIA)

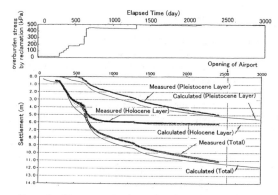

Figure 5   Measure settlement with time in KIA island (Kobayashi,1994)

phase construction has been carried out based on the experiences of the 1st phase, and the mean predicted total settlement in the airport island is 18m, which will be the largest consolidation settlement that geotechnical engineers have experienced. In the 2nd phase project, both the observation of settlement and the version-up of settlement prediction has become the most important work as the construction control.

The problems of settlement prediction of Pleistocene clay layer in Osaka Bay area are summarized as follows;

1) Pleistocene clay layers often show much larger compressibility than Holocene clay layers. The settlement of Pleistocene clay commences sharply when the overburden stress exceeds a critical value, which is not necessarily equal to the consolidation yield stress obtained in laboratory test.

2) Even if the overburden stress is less than consolidation yield stress, large settlements can take place in Pleistocene clay layers. In this case, it seems that the settlement includes both the primary settlement and the secondary settlement, although the distinction is usually difficult.

3) Because Pleistocene clay layer exists at large depth, counter-measures such as ground improvement methods are practically not available and the correct prediction of the behavior is inevitable to design.

The author's research group in Port and Harbour Research Institute has been carried out a series of studies on the mechanical properties of aged clays having structures that were formed during the process of long-term sedimentation and consolidation. In the former part of this report, the compression and shear characteristics of aged clay, are reviewed with related to the case study on KIA project. In the latter part, the method for evaluating the degree of structure is newly proposed, and a unique consolidation behavior of structured clay is introduced.

2   COMPRESSIBILITY OF AGED CLAY AND STRENGTH INCREASE DUE TO CEMENTATION

2.1   Compression index ratio

The importance of structure and aging effect on mechanical properties of clay deposits was firstly pointed out by Bjerrum (1967, 1973). Figure 6 shows a schematic e-log p relationship proposed by Bjerrum, where the difference between young clay and aged clay was explained by the secondary or delayed consolidation, i.e.; the aged clay was characterized by the increase of consolidation yield stress p_c and the decrease of void ratio due to the secondary consolidation during a long-term sedimentation age.

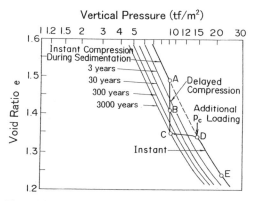

Figure 6 Schematic e-log p relationship during the sedimentation process (Bjerrum,1967)

Another type of the difference in e-log p curve between natural aged clays and reconstituted young clays is frequently seen as shown in Figure 7, where e-log p curve has a sharp concave downward for natural clay, while young clay has a gentle e-log p curve. This type of change on e-log p curve cannot be explained by Bjerrum's concept in Figure 6. Tsuchida et al (1991) defined a compression index ratio, r_c by the following equation and measured the values of r_c for Japanese marine clays;

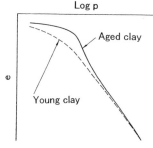

Figure 7   Difference in e-log p curve between aged clay and reconstituted clay

45

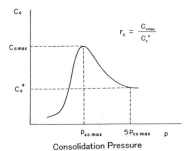

Figure 8   Compression index ratio

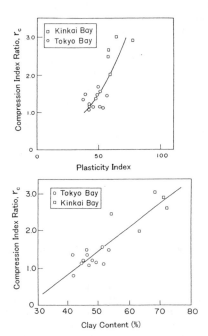

Figure 9(a)   Compression index ratio, r_c, with I_p and fine content

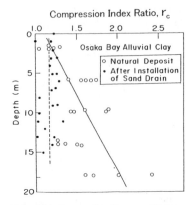

Figure 9(b) Change of r_c after consolidation

$$r_c = C_{cmax}/C_c^* \qquad (1)$$

where, C_{cmax} is the peak value of C_c in Figure 8 and C_c^* is the C_c at which consolidation pressure is 5 times larger than p_c. Compression index ratio, r_c of Holocene clays are plotted with the plasticity index I_p and clay content in Figure 9(a). As shown in figure, r_c clearly shows the positive correlation with I_p and clay content. Figure 9(b) shows r_c with depth at a site in Osaka Bay, where the landfill was carried out after constructing sand drain. The sampling and soil tests were conducted before the installation of drain and during the landfill. As shown in Figure 9(b), r_c of Osaka Bay Holocene clay increased with depth before the installation of drain, however, during the consolidation with landfill, the r_c became almost constant value near 1.0. According to Tsuchida et al. (1991), the value of r_c of Holocene clays and Pleistocene clays in Japan are 1.1-3.0 and 1.1-6.0, respectively, increasing with depth, while r_c of reconstituted young clays are almost 1.0.

Figure 10 illustrates a schematic one-dimensional compression curve explaining the effects of cementation and secondary compression. If the consolidation pressure is increased up to p_0, and ceased at point A for a long time, the void ratio will reduce along the path AB, accompanied with secondary compression as explained in Figure 6. At the same time, formation change and chemical alteration take place among soil particles under conditions not related with volumetric compression. These changes have the effect of strengthening particle interlocking, namely bonding and will certainly make a contribution to strength gain. The component of strength gain not associated with volumetric compression is classified as cementation, in a very broad sense in this paper. The cementation stated here includes various factors, such as flocculation, thixotropy, leaching and so on. When the sample at point B is subjected to additional pressure increase, the natural clay generally shows compression characteristics as shown by curve BCDE.

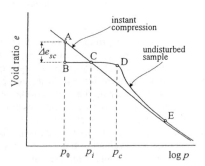

Figure 10 Effect of cementation and secondary compression on e-log p curve

Here, the author define the difference $(p_i - p_0)$ as the result of secondary compression and $(p_c - p_i)$ as the result of cementation. Based on the distinction, as suggested in Figure 10, the compression index ratio, r_c, is an index of the effect of cementation $(p_c - p_i)$ and not related to the secondary consolidation $(p_i - p_0)$.

2.2 Duplication of structure by high temperature consolidation

Tsuchida et al. (1991) proposed the high temperature consolidation method for restoring the structure of aged clay in laboratory, by consolidating remolded clayey slurry under 75° C. Tokyo Bay clay was thoroughly remolded at a water content more than twice the liquid limit. The slurry was put into a 20 cm diameter consolidation cell and consolidated one-dimensionally. The consolidation cell was surrounded by hot water with the temperature of constant 75° C adjusted by an electric heater as shown in Figure 11. Clay was first consolidated under the weight of loading plate and 4 incremental pressures (10, 20, 40, 80 kPa) were applied subsequently by the air cylinder. After the completion of consolidation, the sample was unloaded and cooled down at the room temperature (25° C).

Figure 12 shows the typical time-settlement relationship of sample consolidated under high temperature (HTC), and conventional sample consolidated under room temperature (RTC), for each pressure increment. As shown in Figure 12, the settlement of HTC under the weight of loading plate was smaller than that of RTC. However, under subsequent loading, the settlement of HTC was larger than that of RTC. Comparing the water contents of both samples after the completion of consolidation, the water content of HTC sample was slightly larger than RTC sample. The consolidation settlement of HTC is obviously accelerated by high temperature condition.

Figure 13 shows the e-log p curves obtained for both HTC and RTC samples with the e-log p curve of undisturbed Tokyo Bay. As shown in the figure, e-log p curve of HTC sample resembles that of undisturbed clay which must have undergone aging for thousands years, showing distinctive yielding and larger compressibility after yielding. As the compression index ratio r_c of HTC sample were 1.9-2.0, it can be said that the procedure of high temperature consolidation is a useful technique to reproduce the structure as Holocene clay deposits have in the laboratory. As the water content of HTC sample was slightly larger than RTC sample, it is considered based on the concept shown in Figure 10 that the acceleration of cementation action will be the main effect of high temperature consolidation.

Kitazume and Terashi (1994) used a procedure of high temperature consolidation to make a model round in the centrifuge model test of slope stability.

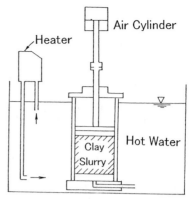

Figure 11 Apparatus of high temperature consolidation

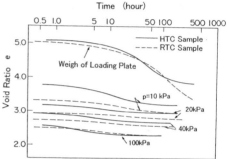

Figure 12   Time settlement relationship under high temperature consolidation

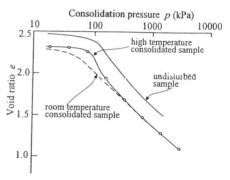

Fig.13   e-log p curves of HTC sample and RTC Sample (Tsuchida et al.,1991)

Figure 14 shows the comparison of deformation in the model slope when the failure took place. As shown in the figure, the model slope of HTC clay failed suddenly and the deformation was concentrated around a sliding surface, while the model slope of RTC clay failed after the large deformation of whole model slope. It can be said that the slope of HTC clay shows a brittle behavior similar to usual natural slopes. The procedure of high temperature consolidation seems to be a useful technique to duplicate aged clay ground in laboratory

2.3 Increase of strength due to the cementation effect

As mentioned in Figure 10, the strength gain with time can be divided into two components. One is associated with the volumetric compression, while the other is attributed to the cementation effect. When the shear strength gain from point A to point B in Figure 10 is shown as Δs_u, it can be assumed that the strength gain due to the volume decrease will be given as $(s_u/p)(p_i - p_0)$, where (s_u/p) is a strength increment ratio of normally consolidation condition. Accordingly, the strength increment due to the cementation is obtained as $(\Delta s_u - (s_u/p)(p_i - p_0))$.

Using the above assumption, Tan and Tsuchida (1999) investigated the shear s̶t̶r̶ ̶ ̶ ̶ ̶ ̶ ̶ ̶ ̶ due to cementation effect experimentally, and ̶ ̶owed that the strength increment by cementation is proportional to the logarithm of elapsed time with the increment related to the effective overburden stress.

Figure 15 shows the relationship between the normalized strength increment per one log-scale of elapsed time, $(\Delta s_u/p)/\Delta(\log t)$, and the effective overburden pressure p. As shown in the figure, the strength increment can be expressed by the following simple equation;

$$(\Delta s_u/p)/\Delta(\log t) = 0.3/\sqrt{p} \qquad (2)$$

or $\quad \Delta s_u = 0.3\sqrt{p}\ \Delta(\log t)$ $\qquad$ (2)'

where, p is the effective overburden pressure (unit: kN/m^2), and t is elapsed time after the end of primary consolidation.

Equation (2) and (2)' mean that the strength increment by cementation is proportional to $\sqrt{p}$ and $\Delta(\log t)$. Accordingly, the smaller is the overburden stress, the larger is the contribution of cementation in strength increase with time. Generally, when the overburden stress is less than $100kN/m^2$, the cementation effect is lagcr than that of the secondary compression, while when the overburden stress is larger than $100kN/m^2$.the secondary compression becomes the main component of strength gain with time.

3 STRENGTH OF OSAKA BAY PLEISTOCENE CLAY, – CASE OF KIA PROJECT –

3.1 Importance of shear strength of Pleistocene Clays in KIA second phase project

In second-phase project in Kansai International Airport, the shear strength of Pleistocene clay layer at the large depth has become an important design problem. In the first-phase project, the shear strength of Pleistocene clay was not studied in detail because

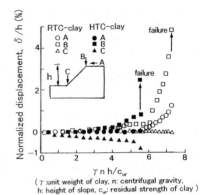

(γ :unit weight of clay, n: centrifugal gravity, h: height of slope, c_{ur}: residual strength of clay)

Figure 14 Deformation of slope at failure in centrifuge model test (Kitazume and Terashi, 1994)

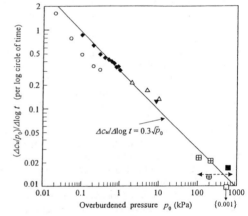

◆ Sedimentary (vane, Kumamoto clay)
△ Oedometer (fall cone, Honmoku clay)
▼ Large oedometer (vane, Honmoku clay)
○ Zreik (1997, fall cone, Boston blue clay)
■ Mitachi et al (1987, triaxial, Hayakita clay)
□   "   ("   Ohnegai clay)
⊞ Yasuhara el al (1983, direct shear, Ariake clay)
⊕ Ue el al (1997, triaxial, Okayama clay)
◄--→ Mesri (1995)

Figure 15 Normalized strength increment due to cementation (Tan and Tsuchida, 1999)

the stability analysis of the seawall structures indicated that the safety factors against deep-seated slip circles passing through the Pleistocene clay layer were much greater than the required design safety factor of 1.3. However, as the water depth and the load of reclaimation become larger in the second-phase project, the slip circle that gives the minimum safety factor possibly passes through the Pleistocene clay layer.

Undisturbed samples used in this study were obtained from Osaka bay in 1995 during the in-situ investigation for second construction phase of the Kansai international airport. The location of boring are shown in Figure 16(a). The borings carried out as deep as 400m indicated that the average depth of water was 19.2m and the depth of alluvial soft clay was 25m. The sampling method and the characteristics of soils collected from Osaka bay were reported in detail by Horie et al. (1984), Ishii et al. (1984) and Tsuchida et al. (1984).

Figure 16(b) shows the consistency profiles including liquid limit, w_L, plastic limit, w_P and natural water content, w_0. Although clay and sand layers have been deposited alternatively, only the clay layers were studied, because the shear strength data of clay is significant for the stability analysis.

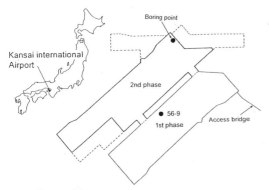

Figure 16(a)   Location of undisturbed sampling

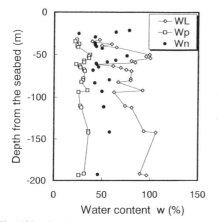

Fig.16(b)   Consistency profile of Pleistocene clays

3.2 Measurement of residual effective stress of undisturbed sample

Evaluation of sample quality by measuring the suction, which corresponds to the residual effective stress of sample, was firstly attempt by Okumura (1974). Shimizu and Tabuchi (1993), Shogaki et al. (1995) and Mitachi and Kudo (1996) measured the residual effective stress of sample and corrected the unconfined compression strength. As most of these researches studied on the saturated Holocene clays, the measured data of suction were less than vacuum (98.1kN/m^2). On the other hand, Mitachi and Kudo(1996) measured suction larger than 98.1kPa in Kaolin clay, with applying a back air pressure.

To evaluate the quality of Pleistocene clay samples, the measurement of suction was carried out. The suction measuring system used in this study is schematically illustrated in Figure 17. The system can measure the suction, by placing the specimen under a certain high air pressure u_a (back air pressure). The specimen trimmed as 35 mm in diameter and 80mm in height is placed on the ceramic filter which has an air entry value of 200 kN/m^2 and is saturated by de-aired water previously. After setting the cell, air pressure was applied up to a certain value u_a, and the valve, which is placed between inside of ceramic filter and inside of cell, was opened for a moment to balance both pressures in the pedestal and in the cell.

Figure 18 shows an example of data sets measured by suction tests, which were conducted on specimens taken from the depth of 93m. The measured suction under atmospheric condition was approaching to the vacuum, however much smaller than the suction when the back air pressure was applied. When the back air pressure was larger than 100kN/m^2, the measured residual stresses became constant with not dependent on the back air pressure. The constant value of the suction is considered to be a true residual stress of the specimen.

Figure 19 shows the variation of residual effective stresses based on the measured suction Osaka Bay Pleistocene clays taken from different depths. As shown ion the figure, the residual effective stresses in Pleistocene clays collected from large depth were equal to or more than one fifth of effective overbur-

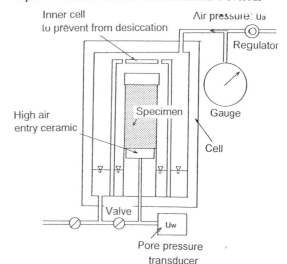

Figure 17 Measurement of suction under applying back air pressure

49

den stresses. These values are consistent with the test results of Tanaka, M. et al. (1995), who reported that residual effective stresses of Japanese Holocene clays are between one forth and one sixth of effective overburden stress. According to Okumura (1974), when the residual effective stress is equal to or more than one fifth of effective overburden stress, the disturbance ratio R, an index of sample quality, ranges from 1.5 to 3, which means that the sample quality is classified to "very good". It can be said that quality of undisturbed samples of Osaka Bay Pleistocene clay are better than anticipated, although they were collected from large depths.

3.3 Recompression method

The recompression method is aimed to reduce the effect of disturbance, in which the shear test is conducted after the specimen is consolidated by the in-situ effective stress conditions in triaxial cell (Berre and Bjerrum, 1973). In this method, a specimen from undisturbed sample is first consolidated under K_0 condition at the same effective stresses in the field, and after the completion of consolidation, it is subjected to shear under undrained condition. Hanzawa and Kishida(1982) and Hanzawa(1982) applied the shear strength obtained by the recompression method for stability analyses of several practical projects and reported excellent agreements between the analyses and the field behaviors. In testing procedures of Hanzawa, K_0-consolidation was carried out by the end of primary consolidation, and the average shear strength of compression and extension loading at strain rate of 0.01%/min were used for considering the effects of strength anisotropy and time to failure.

In this study, the recompression method similar to the procedure by Hanzawa was used. It must be noted that, although the shear strain rate of 0.01 %/min is recommended, the rate of 0.1 %/min was adopted to complete a number of tests within a limited time, and no correction on the shear strain rate was carried out. K_0 value was first determined by K_0-consolidation test in the triaxial test (Tsuchida et al., 1989). For the recompression method, the in-situ stress conditions must be applied without destroying the structure as much as possible by K_0 consolidation. Anisotropic consolidation was carried out by increasing the stresses linearly until $\sigma'_1 = \sigma'_{v0}$ and $\sigma'_3 = K_0 \cdot \sigma'_{v0}$ in 720 min and this condition was maintained until the excess pore pressure was completely dissipated. After that, the specimen was sheared in undrained condition for compression (CAUC) or extension (CAUE) loading with the axial strain rate of 0.1%/min.

As a simplified procedure of recompression, the Simple CU (consolidated undrained) test was proposed by Tsuchida et al. (1989) for a practical purpose. The Simple CU test is a CIU (isotropically consolidated undrained) compression triaxial test, in

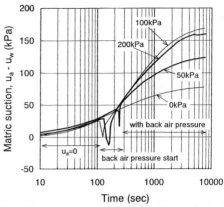

Figure 18   Measured suctions under different back air pressure

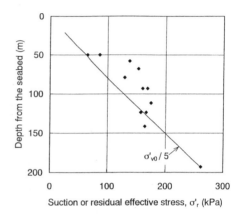

Figure 19   Residual effective stress of sample with depth

which an isotropic consolidation under the mean in-situ effective confining stress of $(1+2K_0)/3 \cdot \sigma'_{v0}$ is carried out, instead of the K_0 consolidation. For the stain rate, 0.1 %/min is recommended. They also proposed to take 75 % of the shear strength measured from the simplified CU test as the design strength, to account for the strength anisotropy and time effect.

Both in Bjerrum's method and Hanzawa's method of recompression (Berre and Bjerrum,1973; Hanzawa and Kishida,1982), the undrained shear strength is defined as the average of compression strength and extension strength. The problem is how to determine the extension strength. One definition of the extension strength is the maximum strength within the 15 % axial strain.

A typical stress-strain relationship for Japanese Holocene clay observed in CAUC and CAUE is shown in Figure 20. The peak strength appeared at a small strain in compression side, while the peak strength in extension side is usually mobilized with

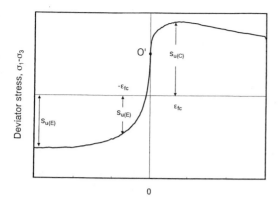

Figure 20 Typical stress-strain curve of Holocene clay

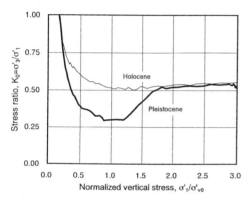

Figure 21 Measured K_0 value with axial consolidation pressure

peak strength is much larger than that of compression strength. In addition, extension shear strength is also defined as the strength at the strain when the peak strength is mobilized in compression test. The reason for the latter definition is that the average strength resulted from compression and extension tests should be considered at the same strain level. The extension strengths obtained by the two definitions were used for design shear strength, and the influences on stability analysis were discussed.

3.4 Test result and discussions

Figure 21 shows the observed variations of K_0-value ($= \sigma'_3/\sigma'_1$) with consolidation stress σ'_1 normalized with the effective overburden stress σ'_{v0} for an Holocene clay obtained from a depth of 21m and a Pleistocene clay from a depth of 53m below the seabed. For the Holocene clay, the K_0-value decreases during over-consolidation range, after that K_0-value reaches a plateau at 0.5 after σ'_{v0}. On the other hand, for the Pleistocene clay, which is relatively an old deposit, the K_0-value decreases remarkably during over-consolidation range to the bottom level around σ'_{v0}. After that K_0-value increases, it finally settles in a constant value of 0.5, that is almost the same as that for the Holocene clay. The difference between the alluvial clay and the Pleistocene clay is probably due to the well-developed structures in the Pleistocene clay. In this study, the K_0 value of 0.5 observed for normally consolidation stage is adopted to recover the in situ stresses by anisotropic consolidation for CAUC and CAUE tests.

Figure 22(a) shows the stress-strain relationships of Pleistocene clay sample of 55 m depth, in which one specimen was consolidated one dimensionally up to the normally consolidated stage of $3\sigma'_{v0}$ and for the other one, the recompression specimen was reconsolidated by σ'_{v0} with K_0=0.5. Figure 22(b) shows the effective stress paths of the tests. In com-

larger strain or sometimes no peak strength is shown within the maximum axial strain of 15%. When extension strength is defined as the maximum strength within the 15% axial strain, the axial strain at the

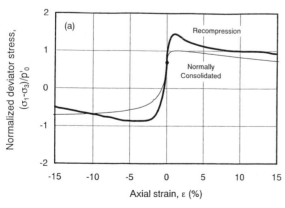

Figure 22(a) Stress-strain curves

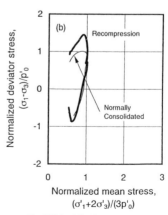

Fig.22(b) Effective stress path

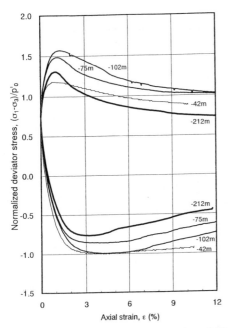

Figure 23 Stress-strain curves of Pleistocene clays of different depths

sion specimen were gradually approaching those of the normally consolidated clay. In extension loading, the normally consolidated clay did not have a maximum deviator stress, and the strain hardening behavior were shown even though the axial strain was as much as 15%. However, the specimen of recompression had a peak strength with a strain of 2 to 5 % and the strain softening behavior were shown after the yielding.

Figure 23 shows the stress-strain curves of Pleistocene clays of different depths. As indicated in the figure, the samples from the larger depth seem to show more brittle behaviors especially in extension loading. The differences of shear behaviors are related to the modes of failure. The aged Pleistocene clay failed with a clearly defined slip surface for compression and with necking for extension, while, the normally consolidated clay collapses uniformly without local failure. Therefore, it is able to say that characteristics of shearing behavior for a clay changes from ductile to brittle by developing structure due to aging effects.

Profiles of shear strength; s_u, determined by CAUC and CAUE tests with depth are shown in Figure 24(a) and Figure 24(b). The extension shear strengths were determined by the maximum strength within the 15% axial strain (definition A, Figure 24(a)), and the strength at the axial strain when compression strength showed the peak (definition B, Figure 24(b)), respectively. In the figure, the shear strengths of the Modified Bjerrum's method $s_{u(M.B.)}$ is defined by the mean of the compression strength $s_{u(C)}$ and the extension strength $s_{u(E)}$, and the $s_{u(SCU)}$ is a strength determined by the simplified CU test with correction factor of 0.75. The strength increases with depth linearly, and this means that, although the geological evidences of the periodical sea level

ression loading, the normally consolidated clay yielded just at the beginning of shear, shifting from the elastic stage to the plastic stage with strain softening behavior. On the other hand, the recompression specimen showed larger yield stress with larger secant modulus, while in the post-yielding region, tress-strain relation and stress path of the recompres-

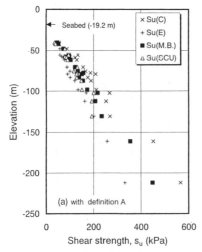

Figure 24(a) Shear strength with depth
(extension strength with definition A)

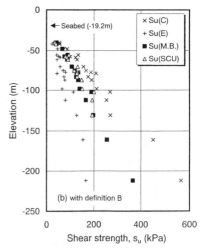

Figure 24(b) Shear strength with depth
(extension strength with definition B)

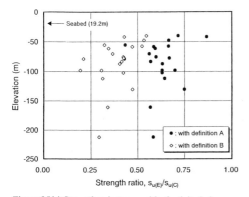

Figure 25(a) Strength anisotropy with plasticity index

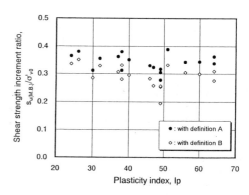

Figure 25(b) Strength increment ratio with plasticity index

changes are observed, the ground has not experienced major tectonic movements after the deposit.

The strength anisotropy, expressed as the ratio of extension strength to compression strength of Pleistocene clays are plotted in Figure 25(a). As shown in the figure, the ratio is about 0.6 when the maximum extension strength is used, while the ratio is about 0.4 when extension strength is at the strain equivalent to the peak compression strength, respectively. Tsuchida and Tanaka (1995) showed that the strength ratio of typical Holocene marine clays, in which the definition A was used, ranged from 0.6 to 0.8. It can be said that the strength anisotropy of Osaka Bay Pleistocene Clay is large, compared with other marine clays in Japan.

Figure 25(b) shows the relationship between $s_{u(M.B.)} / \sigma'_{v0}$ and the plasticity index I_p. The values of $s_{u(M.B.)} / \sigma'_{v0}$ are $0.3 \sim 0.5$ when the extension strength with definition A is used, and $0.25 \sim 0.40$ with definition B. According to the conventional

consolidation tests, over-consolidation ratios of Pleistocene clay are between 1.25 and 1.30, therefore by dividing the $s_{u(M.B.)} / \sigma'_{v0}$ with 1.30, strength increment ratio, s_u/p, in normally consolidation region, is obtained. The s_u/p of the Pleistocene clays were 0.29 when the extension strength with definition A was used, and 0.25 with definition B.

Mesri (1975) pointed out that s_u/p in normally consolidated region of clays in North America and Europe are 0.22, while Tanaka (1994) showed that the s_u/p of Japanese marine clays ranges from 0.25 to 0.35. The results of Osaka Bay Pleistocene clay confirm the results reported by Tanaka (1994). Taking average value with depth in Figures 24, $s_{u(M.B.)} / \sigma'_{v0}$, were 0.34 with the definition A of extension strength, and 0.29 with the definition B. The average normalized strength of the simplified CU test was 0.31.

Figure 26 shows the profiles of strain at failure for the recompression specimens (CAU tests). Strains at

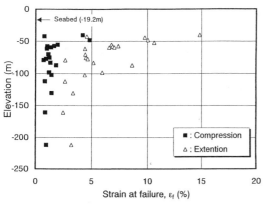

Figure 26 Strain at failure with depth

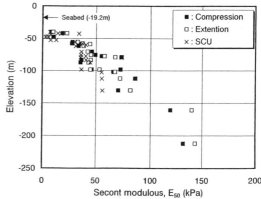

Figure 27 E_{50} obtained by triaxial tests with depth

failure of compression tests; ε_{fC}, ranges from 2 % to 6% for Holocene clays, while in Pleistocene clays, most of them are as small as 1%. Strains at failure of extension tests; ε_{fE}, were more than 15% in Holocene clays; while in Pleistocene clays, ε_{fE} were decreased with depth and ranged from 3 to 4% for the depths lower than 80m.

Figure 27 shows the secant modulus, E_{50}, obtained by CAU tests and the simplified CU test. As shown in the figure, the modulus determined from CAUC and CAUE tests were almost the same, implying that the secant modulus of Pleistocene clay is isotropic. The values of E_{50} obtained from the simplified CU test were smaller by 10-20% than those obtained from CAU tests, suggesting that the simplified CU may underestimate E_{50}, because of the difference of consolidation condition.

3.5 Effect of Shear Strength on the stability of sea-wall structure

The stability analysis was carried out for the typical sea-wall structure of Kansai International Airport in the second phase of construction. The assumed cross section is shown in Figure 28. The safety factor for the base slide was calculated by the slip circle analysis with modified Fellenius method. Table 1 shows the assumed material properties used in the stability analyses.

The layer number 11 represents the Pleistocene clay stratum in which the undrained shear strength increasing with the depth is varied corresponding to the triaxial test results. As shown in Figures 25(a) and 25(b), the strength increase ratio; s_u/σ'_{v0} is 0.34 for the average of compression and the maximum extension strengths (definition A), 0.29 for the average strength based on definition B of extension strength, and 0.31 for the simplified CU test with the correction factor 0.75. Considering these test results, for the range of strength increase ratio, $s_u/\sigma'_{v0} = 0.22 \sim 0.36$, the slip circle analysis was carried out

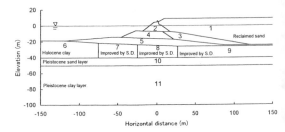

Figure 28 Typical cross section of seawall structure in KIA construction project

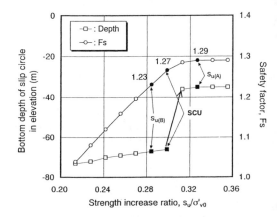

Figure 29 Safety factor and strength of Pleistocene clay

to understand the effect of strength on the safety factor.

Figure 29 shows the calculated variations of the safety factor; Fs, and the bottom depth of slip circle with strength increase ratio; s_u/σ'_{v0}. As shown in the figure, Fs reached a constant value at about 1.29, where s_u/σ'_{v0} was larger than 0.32. This is because the critical slip circle is within the Holocene clay layer and did not pass through Pleistocene de-

Table 1. Parameters for stability analysis

Layers	Soil type	Unit weight (kN/m³)	Strength parameters
1	Reclaimed sand	19.6	$\phi = 30$
2	Concrete caison	21.6	$\phi = 30$
3	Sand	19.6	$\phi = 30$
4	Sand	19.6	$\phi = 30$
5	Sand	19.6	$\phi = 30$
6	Holocene clay, not improved	16.9	$s_u = 1.5z$ (kN/m²)
7	Holocene clay improved by sand drain	16.9	$s_u = 49 + 1.5z$ (kN/m²)
8,9	Holocene clay improved by sand drain	16.9	$s_u = 118$ (kN/m²)
10	Pleistocene sand	19.6	$\phi = 35°$
11	Pleistocene clay	16.9	strength by recompression

z: Initial bottom of sea (DL −19m)

posit layer. When s_u/σ'_{v0} was less than 0.32, F_S was smaller than 1.27, increasing linearly with s_u/σ'_{v0}, because the critical slip circle reached the Pleistocene clay deposit and the strength of Pleistocene clay determined the stability.

Because the shear strength of Pleistocene sand layer between the Holocene clay layer and the Pleistocene clay layer was larger than the strengths of both clay layers, the depth of bottom of critical slip circle varied discontinuously at s_u/σ'_{v0} of 0.32.

Figure 30 shows the critical slip circles giving the minimum safety factor in the stability analysis. The $s_{u(peak)}$ corresponds to the average strength with the extension strength of definition A, and the critical circle passes in the Holocene clay. The $s_{u(strain)}$ corresponds to the average strength with the extension strength of definition B, in which the extension strength equivalent to the same strain level as compression, and $s_{u(SCU)}$ corresponds to the strength determined from the simplified CU test. In these 2 cases, the critical slip circles pass through the Pleistocene clay.

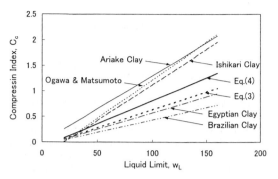

Figure 31 Comparison of empirical equations on C_c-w_L relation

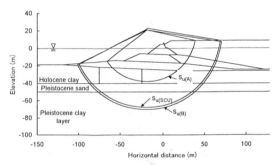

Figure 30 Critical slip circle with different strengths

As discussed above, depending on the definition of design shear strength, the critical slip circle passes in the Holocene clay layer or in the Pleistocene clay. The results of the stability analysis have shown that, for all the cases of different strength definitions, the minimum safety factor of the seawall structure is larger than 1.20, however the margin for the safety is not so much. The cross section used for stability analysis in Figure 28 is an example of the seawall whose total length is as much as 10km. When the slip circle passing though the Pleistocene clay layer at great depth may take place, the influenced area of the failure will be quite large and possibly give the serious damage to the whole construction project. Consequently, it is concluded that the accurate estimation of shear strength and the careful stability analysis covering all the construction sites is necessary. Further, the field observations and the strict construction control will be definitely important during the reclamation.

3.6 Summary

The undrained shear strengths and deformation characteristics of Pleistocene clay in Osaka Bay were studied by the recompression method with triaxial tests. The conclusions derived from this study are as follows;

1) The shear strength of Pleistocene clay increase linearly with depth, implying that a particular stress history has not been involved.

2) Residual effective stresses in Pleistocene clays collected from large depth were about one fifth of effective overburden stresses. The quality of undisturbed samples of Pleistocene clay are better than the anticipated, although they were collected from large depth.

3) Pleistocene clay at large depths showed brittle stress-strain behavior and it shifts very rapidly from elastic deformation to failure. The clays taken from the depth lower than 80m showed peak extension strength at axial strains of $3 \sim 4$ %.

4) When the undrained shear strength for design is defined as average of compression and extension strengths, the strength increase ratio; s_u/σ'_{v0}, depends on the definition of the extension strength.The value of s_u/σ'_{v0} was 0.34 if the extension strength is defined as a peak strength, while the value of s_u/σ'_{v0} was 0.28 if extension strength is defined as strength at the same strain as the compression peak strength. The value of s_u/σ'_{v0} determined by the simplified CU test was 0.31.

5) The critical slip circle in stability analysis passed thorough the Pleistocene layer of the depth of 70m, when the extension strength was defined as strength at the same strain as the compression peak strength. Although the safety factors obtained were larger than 1.20, the margin of the safety were not so much. The accurate estimation of the shear strength of Pleistocene clays and the

55

field observations and strict construction control will be necessary to complete the projects safely.

4 EVALUATION METHODS OF STRUC-TURE OF AGED MARINE DEPOSITS

4.1 Ultimate Standard Compression Curve (USC)

In the previous sections, the mechanical properties of aged clay have been examined experimentally. In the practical sense, it seems that the method to evaluate the degree of the structure quantitatively is necessary. Recently the author has proposed a concept of standard compression curves (SCC) for marine clays to explain the various aspects of e-log p relationship of clays in a unified manner. Here the SCC concept is introduced and it is shown that the degree of structure of marine deposits can be easily evaluated, using the SCC concept.

A number of studies have been carried out on the e-log p relationship of clays. An early study by Skempton (1944) showed the following fundamental findings;

a) Compressibility index, C_c has a strong relationship with the liquid limit w_L.

b) Sample disturbance of natural clay and the initial water content of laboratory-prepared clay have important effects on the e-log p relationship.

c) There exists some discrepancy between the laboratory compression curve and the sedimentation compression curve, which is obtained as the relationship between in-situ void ratio and the overburden stress in geologically normally consolidated layer.

The empirical relationship between C_c and w_L has been studied by a lot of researchers. Terzaghi and Peck (1969) proposed the well-known equations based on the experimental data by Skempton as follows;

$$C_c = 0.007 (w_L - 10) \quad \text{for remolded soil} \quad (3)$$

$$C_c = 0.009 (w_L - 10) \quad \text{for undisturbed soil} \quad (4)$$

However, as the above equations do not always give good estimation of C_c especially for undisturbed soils, various empirical equations for different clays have been proposed as follows:

$C_c = 0.015(w_L - 19)$, for Japanese marine clays
(Ogawa and Ogawa, 1978)

$C_c = 0.014(w_L - 20)$, for Ishikari clay
(JSSMFE, 1966)

$C_c = 0.013 w_L$, for Ariake clay, (JSSMFE, 1966)

$C_c = 0.0063(w_L - 10)$, for Egyptian clay
(Abdrabbo and Mahmoud, 1990)

$C_c = 0.0048(w_L - 10)$, for Brazilian clays,
(Bowles, 1979)

Figure 31 shows the comparison of above equations. Looking at the figure, a question arises such as "why the compressibility of natural soil is so much different among the soil types?". The concept of standard compression curve is to provide a unified model to explain the compressibility of different clays with a few fundamental parameters, such as, liquid limit, sensitivity and an initial void ratio at the sedimentation.

Let us consider some typical characteristics of e-log p curves of clays. Figure 32(a) is e-log p curve of reconstituted Osaka Bay clay, showing a typical linear relationship between void ratio e and log p, while Figure 32(b) is e-log p curve of undisturbed sample of Osaka Bay Pleistocene clay. In cases of these structured clays, e and the log p are not linear in normally consolidation region.

Figure 32(c) shows the change of e-log p curves due to the sample disturbance (Okumura, 1974). As shown in the figure, when the clay sample is seriously disturbed or is given some shear strain before the consolidation, the bending of e-log p curve at consolidation yield stress becomes obscure and the void ratio decreases as a whole by the disturbance. Figure 32(d) is e-log p curve of Kumamoto Port clay, which was sedimented from the initial water content of 400% and was consolidated carefully to the final pressure of 1,000 kN/m². As shown in the figure, the e-log p curves from p=0.01 kN/m² to 1,000 kN/m² is not linear but shows a convex to the bottom. It has been known that e-log p curve is not linear when the clay is consolidated from extremely high water content (Imai,1982).

In Figures 33(a),(b),(c) and (d), the same data in Figures 32(a),(b),(c) and (d) are plotted in ln f –log p system, respectively, where f is a specific volume of soil and equal to 1+ e, and ln and log mean natural and common logarithm, respectively. In Figure 33(d), the linearity between ln f and log p is recognized as a whole, and in Figures 33(a),(b) and (c), ln f - log p seems to be close to a straight line as the consolidation pressure increases.

Summarizing Figures 32 and Figures 33, the following suggestions are obtained;

· The linearity between e and log p is not commonly recognized, depending on the existence of structure, sample disturbance and the initial water contents of consolidation.
· However, by focusing on the e-log p relationships when the consolidation pressure is large enough, a linearity between ln f and log p is observed. This seems to be because the effects of structure, disturbance and initial void ratio will disappear with the increase of consolidation pressure.

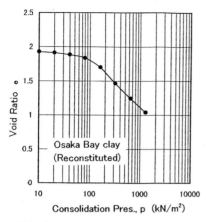

Figure 32(a) *e*-log *p* curve of reconstituted Osaka Bay Clay

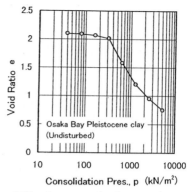

Figure 32(b) *e*-log *p* curve of undisturbed Pleistocene clay

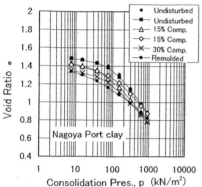

Figure 32(c) Change of *e*-log *p* curve due to sample disturbance

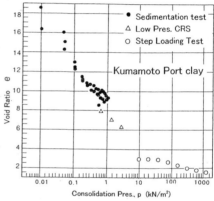

Figure 32(d) *e*-log *p* curve of clay with high water content

Based on the above suggestions, the following assumption on the existence of Ultimate Standard Compression Curve (USC) was made;

a) When clay is consolidated one-dimensionally from slurry with a large initial void ratio, it shows the linear relationship between $\ln f (=1+e)$ and $\log p$. This curve is named Ultimate Standard Compression Curve (USC), and it can be determined mainly by the liquid limit of clay.

b) In case that natural sedimentary clay has a structure due to aging, the void ratio e_0 can be larger than the value given by USC. When natural clay sample is disturbed before the consolidation, its void ratio becomes smaller than the value given by USC. When the sedimentation and the consolidation of natural clay starts from a small initial void ratio, the e-log p relationship locates USC.

c) By consolidating clay with the pressure much larger than the consolidation yield stress, the effects of structure due to aging, sample disturbance or small initial void ratio come to disappear, and finally the

e-log *p* curves converge into USC.

Figure 34 illustrates the USC concept in the log *f* - log *p* system. USC can be described by the following equation;

$$\ln f = -C (\log p - 1) + \ln f_{10} \tag{5}$$

where, f_{10} is a specific volume when $p=10$ kN/m² on the USC, and. C is a gradient of USC, given by the compression index C_c and e as follows;

$$C = C_c / (1 + e) \tag{6}$$

$C_c / (1+ e)$ is a parameter called compression ratio (Terzaghi and Peck,1967).

Butterfield (1979) firstly proposed the linearity between $\ln f$ and $\log p$ shown in Eq.(5). By the differentiation of Eq.(5), the following is obtained;

$$d f / f = -0.434 C (d p / p) \tag{7}$$

57

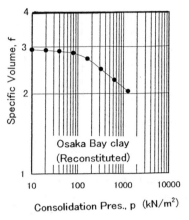

Fig.33(a) ln f-log p curve of reconstituted Osaka Bay clay

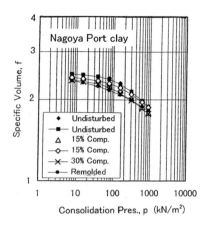

Figure 33(c) ln f-log p curve of disturbed sample

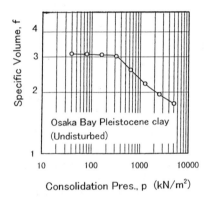

Figure 33(b) ln f-log p curve of undisturbed Pleistocene clay

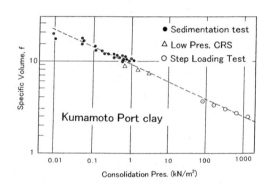

Figure 33(d) ln f-log p curve of clay with high water content

d f /f is a natural strain of soil, while dp/p was named *natural stress* by Butterfield. According to , Butterfield, Eq.(5) means the linear relationship between natural strain and natural stress, and it is a fundamental law of geotechnical materials.

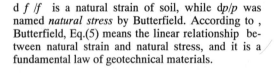

Figure 34 Ultimate standard compression curve, USC

4.2 Existence of a unique USC determined by liquid limit

To study the validity of the existence of USC, 725 consolidation test data of 18 marine deposits listed in Table 2 were analyzed by the following procedure:

1) To plot the specific volume f and the consolidation pressure p in ln f–log p system.
2) To ascertain the linearity between ln f and log p in the region that p is much larger than the consolidation yield stress p_c, and to determine the straight line (USC) for ln f - log p relation.
3) To calculate the compression ratio, C, and the specific volume on USC when p=10kN/m^2, f_{10}.

Figure 35(a) and (b) are the ln f – log p relationships of reconstituted Yokohama clay and undisturbed Osaka Bay Pleistocene clay, which were obtained by the constant rate strain consolidation tests, respectively. In both figures, when the consolidation pressures are 3 - 4 times larger than the consolidation yield stress, the linearity of ln f and log p were observed. The determination of USC was made with the regression analysis of data of these range. In Yokohama Clay, the specific volume f was smaller than USC in all the range of consolidation pressure, however, in Osaka Bay Pleistocene Clay, the specific volume was larger than USC when the p was close to p_c, and with increase of the consolidation pressure, the ln f- log p relationship con-verged to the USC.

Table 2 List of Data of Consolidation Tests

Clay	Number of Data	Liquid limit w_L (%)
1. Haneda,Japan	270	31.1 - 125.0
2. Yokohama Port, Japan	67	54.7 - 129.9
3. Kasumigaura,Japan	51	66.9 - 205.0
4. Sakata Port, Japan	14	63.1 - 102.8
5. Osaka Bay Holocene, Japan	56	48.8 - 117.7
6. Osaka Bay Pleistocene, Japan	58	40.4 - 138.0
7. Hachirogata, Japan	12	94.6 - 225.6
8. Kuwana, Japan	16	50.3 - 92.1
9. Maizuru, Japan	27	33.0 - 91.6
10.Kinkai Bay Japan	17	53.5 - 112.9
11.Tamano, Japan	7	52.0 - 92.5
12.Izumo, Japan	12	61.6 - 138.0
13.Ariake, Japan	38	54.5 - 137.4
14.Banjarmasin, Indonesia	27	76.8 - 137.4
15.Singapore	19	50.5 - 82.3
16.Bankok, Thailand	10	45.8 - 99.6
17.Drammen, Norway	14	31.6 - 55.2
18.Bothkenar, United Kingdom	10	55.2 - 76.9

Figure 36(a) and 36(b) are the relationships between C and w_L, f_{10} and w_L of Osaka Bay Holocene and Pleistocene clays, respectively. As shown in the figures, both C and f_{10} increase with the liquid limit and the correlation is fairly good. These results are seen in other marine clay. Figure 37(a) and 36(b) are

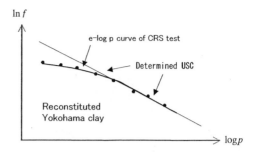

Figure 35(a) ln f – log p relationship (reconstituted Yokohama Clay)

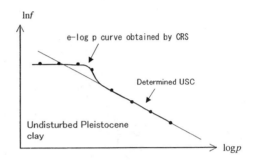

Figure 35(b) ln f – log p relationship f USC (undisturbed Osaka Bay Pleistocene Clay)

C - w_L and f_{10} - w_L relationships of Haneda clay and Yokohama clay, located in Tokyo Bay area. The relations are similar to those of Osaka Bay clay.

Figure 38(a) and 38(b) are the same figures of Hachirogata clay, Sakata clay, Maizuru clay, Izumo clay and Kasumigaura clay, which are located at the

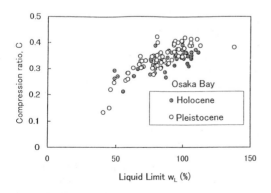

Figure 36(a) C - w_L relationship (Osaka Bay clay)

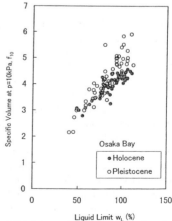

Figure 36(b) f_{10} - w_L relationship (Osaka Bay clay)

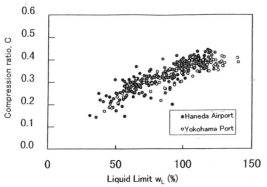

Figure 37(a) C - w_L relationship (Haneda clay and Yoko-hama clay)

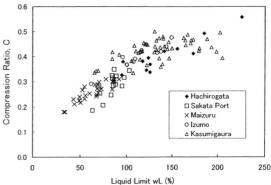

Figure 38(a) C-w_L relationship (Hachirogata, Sakata Port, Kasumigaura, Maizuru Port and Izumo clays)

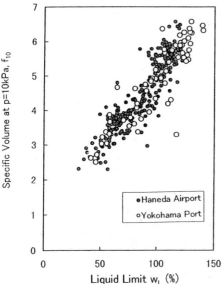

Figure 37(b) f_{10}-w_L relationship (Haneda clay and Yoko-hama clay)

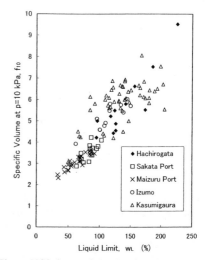

Figure 38(b) f_{10}-w_L relationship (Hachirogata, Sakata Port, Kasumigaura, Maizuru Port and Izumo clays)

coastal areas facing to Japan Sea or Kasumigaura Lake. Figure 39(a) and 39(b) are the same figures of Kinkai clay, Tamano clay, Ariake clay and Kuwano clay, which are located at coastal area in Western Japan. The results for non-Japanese clays, such as Indonesia, Scotland, Norway, Singapore and Thailand, are shown in Figure 40(a) and (b). In all these figures, common relationships are observed.

Figure 41 and Figure 42 are C-w_L relation and f_{10}-w_L relation of all of 18 marine clays, respectively. As shown in the figures, both C and f_{10} correlated with w_L fairly well, and the assumption that the USC is determined mainly by the liquid limit of soil seems to be valid for these marine clays. The regression analyses gave the following equations for C and f_{10}:

$$C = 0.0027\, w_L + 0.1 \tag{8}$$

$$f_{10} = 0.042\, w_L + 0.55 \tag{9}$$

USC is the e-p relationship, where the e-p curves converge ultimately with the increase of consolidation pressure. Using Eqs.(8) and (9) with Eq.(5), USC is given by w_L as follows:

$$\ln f = -(0.0027 w_L + 0.1)\,(\log p - 1) \\ + \ln(0.042 w_L + 0.55) \tag{10}$$

Here we consider the interrelationship between USC concept and the C_c.-w_L relations. Conventionally, C_c is determined as the maximum value of $\Delta e / \Delta(\log p)$ in normally consolidation region.

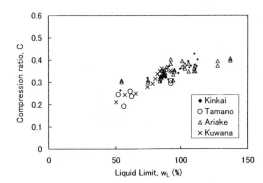

Figure 39(a) C-w_L relationship (Kinkai Bay, Tamano, Ariake and Kuwana clays)

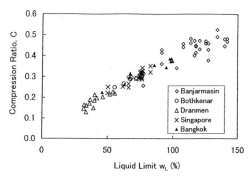

Figure 40(a) C-w_L relationship (Banjarmasin, Bothkenar, Drammen, Singapore and Bangkok clays)

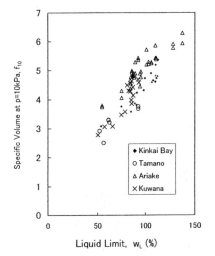

Figure 39(b) f_{10}-w_L relationship (Kinkai Bay, Tamano, Ariake and Kuwana clays)

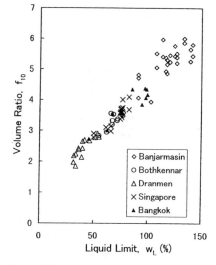

Figure 40(b) f_{10}-w_L relationship (Banjarmasin, Bothkenar, Drammen, Singapore and Bangkok clays)

On $C_c - w_L$ relationship of undisturbed clays, Eq.(4) by Terzaghi and Peck (1967) is well-known, while the following equation by Ogawa and Matsumoto(1978) is often referred for Japanese marine clays;

$$C_c = 0.015(w_L - 19) \qquad (11)$$

Figure 43(a) presents $C_c - w_L$ relationship of Osaka Bay Clay and Haneda Clay, C and f_{10} of which are plotted in Figures 36 and Figures 37, respectively.

As shown in the figures, Eq.(11) is fitting the data well and Eq.(4) is giving a lower boundary of C_c. Instead of C_c as the maximum value of $\Delta e/\Delta (\log p)$, C_{c1000}, which is $\Delta e/\Delta (\log p)$ at the consolidation pressure $p = 1000$ kPa, were calculated and plotted with w_L in Figure 43(b). The relationship between

C_{c1000} and w_L is approximated well by Eq.(4).Using the concept of USC, C_{c1000} is the compressibility on USC at $p=1000$kN/m^2, which does not include the effect of structure of aged clay. On the other hand, when the clay has a structure due to the aging, C_c obtained as the maximum value of $\Delta e/\Delta (\log p)$ becomes larger than C_{c1000}. Accordingly, the difference between Eq.(4) and Eq.(11) can be explained as follows:

$C_c = 0.009 (w_L - 10)$: for undisturbed soil *without* structure due to aging

$C_c = 0.015(w_L - 19)$:for undisturbed soil *with* structure due to aging

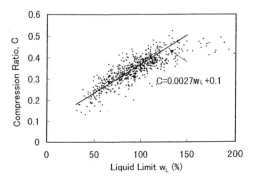

Figure 41 C-w_L relationship (all data)

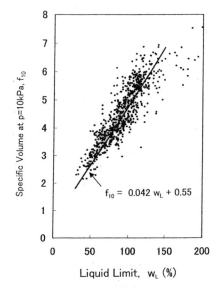

$f_{10} = 0.042\ w_L + 0.55$

Figure 42 f_{10}-w_L relationship (all data)

4.3 Normalization of USC on liquid limit of clays

In Figure 44 and Figure 45, instead of water content at liquid limit, w_L, C and f_{10} are plotted to the specific volume at liquid limit, f_L, respectively. As shown in Figures, the following equations are obtained by the regression analysis:

$$C = 0.27 \ln f_L \qquad (12)$$

$$\ln f_{10} = 1.20 \ln f_L \qquad (13)$$

Using Eqs(12) and (13) with Eq.(5):

$$\ln f = -0.27(\ln f_L)(\log p - 1) + 1.20 \ln f_L \qquad (14)$$

Dividing both terms by $\ln f_L$:

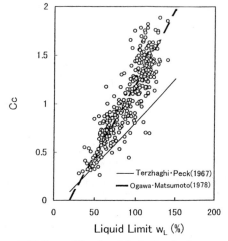

Figure 43(a) $C_c - w_L$ (Haneda and Osaka Bay clays)

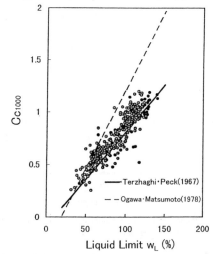

Figure 43(b) $C_{c1000} - w_L$ (Haneda and Osaka Bay clays)

$$\frac{\ln f}{\ln f_L} = -0.27 \log p + 1.47 \qquad (15)$$

Here a specific volume index I_{sv} is newly introduced to normalize specific volume of different soils on the liquid limit. I_{sv} is defined as follows;

$$I_{sv} = \ln f / \ln f_L \qquad (16)$$

Using the specific volume index, USC is simply given by;

$$I_{sv} = -0.27 \log p + 1.47 \qquad (17)$$

Although it has been known empirically that the

liquid limit of clay is a useful index for the compressibility of clay, there seems to be no theoretical explanation why the compressibility of clay is determined mainly by liquid limit of clay. Using the USC concept with Eq.(17), the interpretation can be given as follows:

Figure 46 shows USC by Eq.(17) in I_{sv}- log p system. As shown in the figure, I_{sv} is equal to 0, i.e. void ratio of clay is equal to 0, when $p=280MN/m^2$ on USC (point X). The pressure of $280MN/m^2$ is close to the rupture strength of soil particle, such as quarts, feldspar and limestone (Yashima,1986). Therefore, $I_{sv}= 0$ at $p=280$MPa on USC means the ultimate state that voids cannot exist in soil because of rupture of soil skeleton. This consideration is supported by some experimental data. The ln f-log p curves in high-pressure range are shown in Figure 47, which are the data of Osaka Bay Pleistocene clay in CRS (constant rate of strain) consolidation test up to the maximum pressure from $30MN/m^2$ to $50MN/m^2$ of consolidation pressure. By extrapolating ln f-log p curves of high pressure range, it seems that the void

ratios finally become near zero when the consolidation pressure reaches $200-400MN/m^2$.

In Figure 46, the consolidation pressure $p_S = 55N/m^2$ on USC, when $f =f_L$ and $I_{sv}=1$ (point S in Figure 46). At a void ratio of liquid limit, the shear strength in thoroughly remolded (point Q in Figure 46), $(s_u)_{LL}$, ranges from 1.0 to 2.0kN/m² (JSSMFE, 1992). As the shear strength of point S is given by $(s_u/p)_{USC} p_S$,where $(s_u/p)_{USC}$ is strength increment ratio on USC, the sensitivity, s_t , at the liquid limit is given :

$$s_t = (s_u/p)_{USC} \cdot p_S \ /(s_u)_{LL} \qquad (18)$$

Using typical values, for example, $(s_u/p)_{USC} = 0.25 -0.35$ and $(s_u)_{LL} =1.0-2.0$kPa, the calculated sensitivity s_t for $p_S=55$kPa ranges from 7 to 19, which agrees well with measured values of typical marine clays. When the sensitivity at liquid limit and strength increment ratio on USC are given, p_s at the

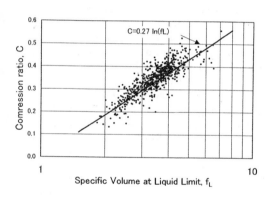

Figure 44 $C-\ln f_l$ relationship (all data)

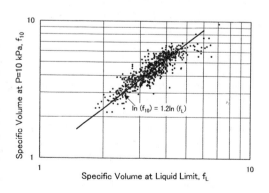

Figure 45 $\ln f_{10} - \ln f_L$ relationship (all data)

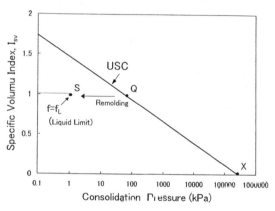

Figure 46   USC in $I_{sv} - \log p$ system

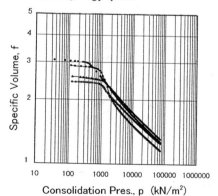

Figure 47 f-log p relationship of Osaka Bay clays obtained by high pressure consolidation test

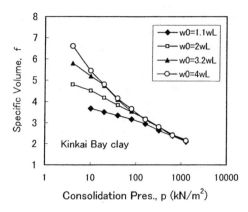
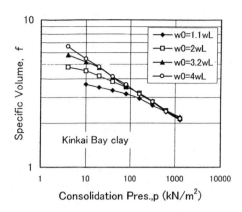

Figure 48 Typical e-log p curve of remolded clay of different e_0

point S is determined by:

$$p_S = (s_u)_{LL} \; s_t \; / \; (s_u/p)_{USC} \qquad (18)'$$

USC is given by Point X and Point S in Figure 46, and the former is related to the rupture strength of soil particles, and the latter is related to the sensitivity at liquid limit and strength increment ratio on USC. If we assume that both parameters are almost constant in different marine clays, it can be understood that USC is determined mainly by the liquid limit of soil.

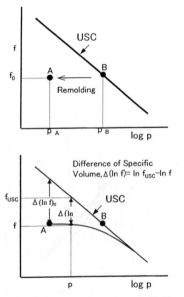

Figure 49 Standard compression curve and effective stress condition of remolded clay

4.4 Standard Compression Curve from an initial void ratio SCC(e_0)

In most of clay samples, consolidation pressure p has to be increased to as much as 2-4 times larger than p_c in order that the e-log p curves converge into the USC, as shown in Figures 35. When the p is not large enough, the void ratio e is larger or smaller than USC due either to the effect of structure or to the effects of disturbance and low initial void ratio. Especially, in the range of small consolidation pressure, the effect of low initial void ratio is important.

Figure 48 is a typical e-log p curve of remolded clays of different initial void ratios, e_0. As the initial void ratio is getting larger, the void ratio of the same consolidation pressure is shifting upper direction, and all the curves finally converge into a unique line of USC with increase of pressure. The standard compression curve from an initial voids ratio, e_0, is called SCC(e_0) in this study. SCC(e_0) is the e-log p relationship, where the consolidation starts from an initial void ratio e_0 and finally converges into the USC, with the increase of consolidation pressure.

Here, the author induces the SCC(e_0), based on the USC concept and the study on sample disturbance by Okumura (1974).

It is considered that the remolding is an ultimate condition of sample disturbance, and the small amount of effective stress can remain after remolding (Okumura,1974). Figure 49 shows effective stress change when clay of an initial specific volume f_0 (void ratio e_0) is remolded thoroughly (point B → point A) and is consolidated one-dimensionally. The effective stress condition at Point A is obtained as follows:

Figure 50(a) is the relationship between remolded

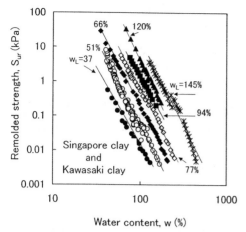

Figure 50(a) Relationship between shear strength of
remolded clay and water content

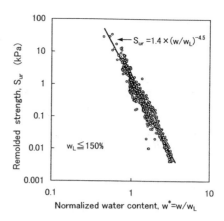

Figure 50(b) Relationship between shear strength of
remolded clay and normalized water content

shear strength c_{ur} and the water content w of clays of different liquid limits. For a give liquid limit, log c_{ur} – log w plot can be expressed by a straight line in a wide range of cur from 10^{-3} kN/m² to 10kN/m² (Tsuchida et al., 1999). Figure 50(b) shows the relationship between s_{ur} and normalized water content w/w_L of the same data. It can be seen that s_{ur} linearly decreases with increase of w/w_L with a relatively narrow band in the bilogarithmic plot. The line of best-fit line is obtained as follows;

$$s_{ur} = 1.4(w/w_L)^{-4.5}$$

Using the above equation, the strength of Point A in Figure 49, s_{uA}, is given as follows;

$$s_{uA} = 1.4(e_0/e_L)^{-4.5} \qquad (19)$$

When the strength increment ratio in thoroughly remolded condition is given as $(s_u/p)_{REM}$, the effective

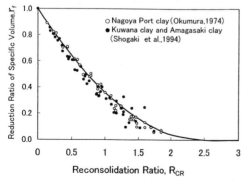

Figure 51 Reduction Ratio of Specific Volume, r_f, and Reconsolidation Ratio, R_{CR}

stress at point A, p_A , is determined as follows:

$$p_A = s_{uA} / (s_u/p)_{REM}$$
$$= 1.4(e_0/e_L)^{-4.5} / (s_u/p)_{REM} \qquad (20)$$

Measured values of $(s_u/p)_{REM}$ for some marine clays ranges from 0.8 to 1.2 (Mikasa ,1988)

Point B in Figure 49 is the effective stress condition of clay of f_0 at the normally consolidated condition on USC. From Eq.(17), the effective stress at point B, p_B, is given as follows;

$$\log p_B = (1.47 - I_{sv0})/0.27 = 5.44 - 3.7 I_{sv0} \qquad (21)$$

$$p_B = 10^{(5.44-3.7I_{sv0})} \qquad (22)$$

where, $I_{sv0} = \ln f_0 / \ln f_L$

According to the comprehensive study on sample disturbance by Okumura (1974), the effect of the sample disturbance on e-log p relationship can be evaluated by disturbance ratio R, which is defined as the ratio of the intact effective stress to the residual effective stress after disturbance or remolding. In this case, R is the ratio p_B/p_A and given as:

$$R = \frac{p_B}{p_A} = \frac{10^{(5.44-3.7I_{sv})}(s_u/p)_{REM}}{1.4(w/w_L)^{-4.5}} \qquad (23)$$

When the disturbed sample is consolidated one-dimensionally, the void ratio at each consolidation pressure becomes smaller than that of sample without the disturbance. The void ratio difference due to the disturbance is determined by the reconsolidation ratio R_{CR}, which is defined by the following equation;

$$R_{CR} = \frac{\ln (p/p_A)}{\ln R} \quad (24)$$

where, p is the consolidation pressure and p_A is the initial effective stress when the consolidation starts.

As shown in Figure 49, the difference of specific volume f from the USC at the same consolidation stress is gradually reduced with increase of consolidation pressure p and R_{CR}. Here the reduction ratio of specific volume r_f, is newly defined as follows;

$$r_f = \Delta(\ln f) / \Delta(\ln f)_0 \quad (25)$$

where, $\Delta(\ln f)$ is the difference of specific volume from USC at the same consolidation pressure, and $\Delta(\ln f)_0$ is the initial value when the consolidation starts (Point A in Figure 49). At the initial condition

of Point A, $r_f = 1$, and $r_f = 0$ when the e-log p curve finally reaches USC by increase of consolidation pressure.

Okumura (1974) and Shogaki and Kaneko(1994) reported the change of e-log p curves of marine clays due to the degree of disturbance. Using the experimental data presented by Okumura and Shogaki et al, the relation between r_f and the reconsolidation ratio R_{CR} is plotted in Figure 51, where the relationship can be successfully expressed by the following equation;

$$\left. \begin{array}{ll} r_f = 0.16(R_{CR} - 2.5)^2 & (R_{CR} \leqq 2.5) \\ r_f = 0 & (R_{CR} > 2.5) \end{array} \right\} \quad (26)$$

Eq.(26) means that, when the sample is remolded at an initial void ratio and then is reconsolidated, the e-log p curve returns to the USC when R_{CR} is larger

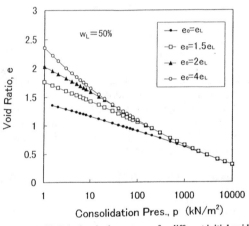

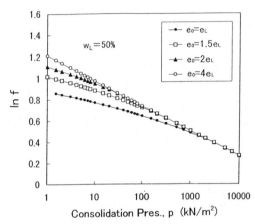

Figure 52(a) Calculated e-log p curves for different initial void ratios (e_L= 1.35)

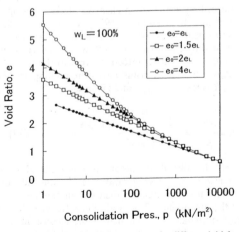

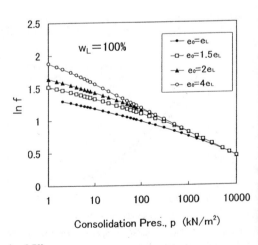

Figure 52(b) Calculated e-log p curves for different initial void ratios (e_L=2.70)

than 2.5, i.e. the consolidation pressure is larger than $R^{2.5} \cdot p_B$. Using Eqs.(17),(23),(24),(25) and (26), the Standard Compression Curve from an initial void ratio SCC(e_0), can be written as follows:

when $p \leqq R^{2.5} p*$

$$
\begin{aligned}
I_{sv} &= 1.47 - 0.27 \log p \\
&\quad - 0.0186(\ln R)\{\ln(p/p*)/(\ln R) - 2.5\}^2
\end{aligned}
$$

when $p > R^{2.5} p*$

$$
I_{sv} = 1.47 - 0.27 \log p
$$

$$\left. \phantom{\begin{aligned}&\\&\\&\\&\\&\\&\end{aligned}} \right\} \quad (27)$$

where, $I_{sv0} = \ln f_0 / \ln f_L$,

$$
R = 10^{(5.44 - 3.7 I_{sv0})} \cdot (s_u/p)_{REM} / 1.4 (e_0/e_L)^{-4.5}
$$

$$
p* = 1.4 (e_0/e_L)^{-4.5} / (s_u/p)_{REM}
$$

Comparing Eq.(27) with Eq.(17), the effect of the initial void ratio, e_0 is given as $-0.0186 \ln R\{\ln (p/p*)/\ln R - 2.5\}^2$, which is the difference from USC.

Figure 52(a) and 52(b) show the calculated stan-

dard compression curves of different initial void ratios as $e_0 = 1.0 e_L$, $1.5 e_L$, $2.0 e_L$ and $4.0 e_L$ (e_L is void ratio at liquid limit) for the conditions of typical Japanese marine clays as $(s_u/p)_{REM} = 1.0$, $w_L = 50\%$ ($e_L = 1.35$) or $w_L = 100\%$ ($e_L = 2.70$). As shown in the figures, the f-log p curves calculated by Eq.(27) agree fairy well with the experiments for the various initial void ratios.

Imai(1978) examined the sedimentation and consolidation behaviors of fluid mud under gravitational force by a series of laboratory tests with sedimentation tubes. In his study, e-log p curves of remolded marine clays, whose initial water contents were from 4 to 20 times larger than the liquid limits, were shown. In Figure 53(a),(b),(c) and (d), Imai's experiments are re-plotted on I_{sv} – log p relationship and the USC of Eq.(17) is shown for the comparison. As shown in Figures 53, the I_{sv}-log p relationship of 4 clays are not depending on the initial water content, and are very close to the USC.

Umehara and Zen (1982) carried out both the

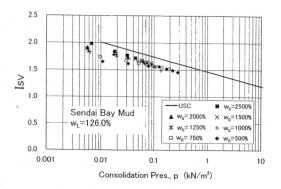

Figure 53(a) I_{sv}-log p curve of Sendai mud
(data from Imai,1978)

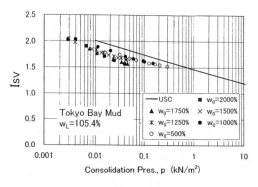

Figure 53(b) I_{sv} log p curve of Tokyo bay mud
(data from Imai,1978)

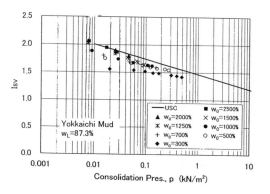

Figure 53(c) I_{sv}-log p curve of Yokkaichi Port mud
(data from Imai,1978)

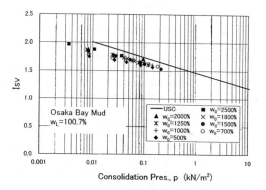

Figure 53(d) I_{sv}-log p curve of Osaka Port mud
(data from Imai,1978)

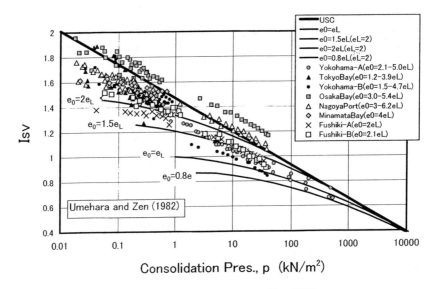

Figure 54 I_{sv}-log p relationship of reconstituted clays (data from Umehara and Zen, 1982)

sedimentation tube test and the constant rate of strain (CRS) test to determine e-log p relationship of remolded marine clays for the wide range of consolidation pressure. In Figure 54, the e-log p curves obtained by Umehara and Zen were plotted on I_{sv}-log p relation and compared with the calculated SCC($e_0=e_L$), SCC($e_0=1.5e_L$) and SCC($e_0=2e_L$). Although some scatters are seen, I_{sv}-log p relation is close to USC when e_0 is large, and I_{sv}-log p relations shift to SCC(e_0) when e_0 becomes smaller.. It can be concluded that the USC and SCC(e_0) obtained Eq.(27) is explaining various e-log p curves of remolded marine clays fairly well.

4.5 Initial void ratio of marine clay and normalized SCC for marine deposit

Holocene marine deposits are normally consolidated fine-grained soils that have been transported by river and tidal current and sedimented for recent thousands of years. Accordingly, by using the concept of standard compression curves, we can obtain the void ratio − effective overburden stress relationship, if an initial void ratio at the start of consolidation is known. It is considered that, even after the sedimentation, the soils with very high water content will behave like fluid, and will be moved easily by the currents and waves. With decreasing the water content, they settle down at some location and the consolidation starts at this moment.

Gomyo et al. (1986) studied the interaction between wave and bottom mud layer by the waterway

experiment. The wave actions made the erosion and the movements of mud layer; at the same time, the wave height was reduced by the loss of the energy due to the interaction. The relationship between the damping ratio of wave height and the water content of mud layer is shown in Figure 55, where H and h is wave height and water depth, respectively, and bentonite ($w_L=160\%$) and kaolin ($w_L=48\%$) were used as the bottom mud. As shown in the figure, the interaction between wave and mud showed a peak at a specific water content, which were about 250% for bentonite and about 90% for Kaolin, regardless of the relative wave height H/h. This phenomenon can be explained as follows; when the water content is

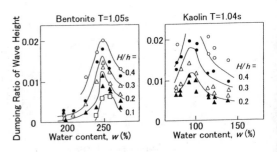

Figure 55 Damping of wave height and water content of bottom mud (Gomyo et al., 1985)

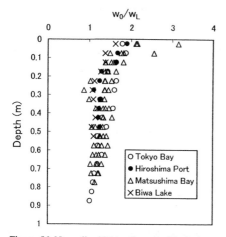

Figure 56 Normalized Water Content of Seabed
(Depth = 0－1m)

tsushima Bay, and Biwa Lake are shown with the depth. Most of marine clays near the sea floor have the water contents of 1.5 to 2.0 times the liquid limits. It can be assumed that the initial void ratios e_0 at the very beginning of consolidation of seabed is considered to be $1.5\text{-}2.0e_L$.

Based on these findings, the void ratio e_0 at the start of consolidation of seabed is considered to be 1.5-2.0 times of the water content. Figure 57(a) and (b) showed the normalized water content, w_n/w_L, with the depth calculated by Eq.(27) under the conditions that $w_0=1.5w_L$, $w_0=2.0\,w_L$, $w_L=50-150\%$, and $(s_u/p)_{REM}=1$. Figure 57(a) indicates that the calculated w_0/w_L at the sea floor agrees well with the field data in Figure 56. However, in Figure 57(b), most of w_n/w_L values in the field are larger than the calculated, when the depth is more than 10m. The reason of this difference seems to be that the structure is formed by the aging effect in the field.

Assuming that the initial void ratios e_0 at the beginning of consolidation of seabed is 1.5 e_L or $2.0e_L$, standard compression curves SCC($e_0=1.5e_L$) and SCC($e_0=2e_L$) are calculated for the cases of $e_L=1.0$, $e_L=2.0$ and $e_L=3.0$. Figure 58 shows the $I_{sv}-\log p$ relationships of SCC($e_0=1.5e_L$) and SCC($e_0=2e_L$).

As shown in Figure 58, I_{sv}-log p relationships of SCC ($e_0=1.5e_0$) and SCC($e_0=2e_L$) are different depending on the liquid limit of soil. However, when accepting ±0.1 difference of I_{sv}, the difference is not so large, the average I_{sv}-log p curve can be determined for SCC($e_0=1.5e_0$) and SCC($e_0=2e_L$) , respectively.

The vales of I_{sv} of SCC($e_0=1.5e_0$) and SCC($e_0=2e_L$) are listed for consolidation pressures in Table 3, where the mean of I_{sv} is also calculated. Here the author calls the relation between the mean I_{sv} value and the consolidation pressure "standard

small, the mud layer has enough strength to be considered as a rigid bottom, making no significant interaction, while, when the water content is large, the mud behaves like a fluid and moves with the same phase of water, making minor interactions. Accordingly, the largest interaction will take place, when the water content condition becomes a certain boundary value. As shown in Figure 55, the water content when the interaction showed the maximum were 1.5 - 2.0 times the liquid limit of mud, which seems to be the boundary for mud between solid and fluid.

Inoue et al. (1990) also pointed out that the shear strength of remolded slurry increased remarkably when the water content becomes less than $2w_L$. This seems to agree with the results of Figure 55.

In Figure 56, the normalized natural water content w_0/w_L of sea floor at Tokyo Bay, Hiroshima Bay, Ma-

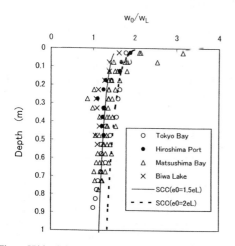

Figure 57(a) Calculated and field water content
(Depth is 0－1m)

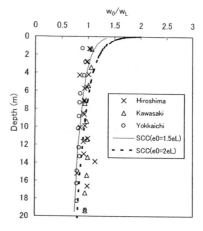

Figure 57(b) Calculated and field water content
(Depth is 0 －20m)

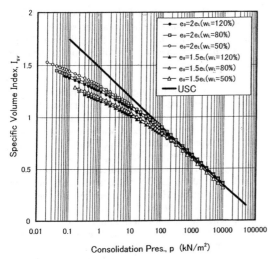

Figure 58 I_{sv} – log p relation of marine clay

Table 3 I_{sv} on Standard Compression

Overburden stress (kN/m²)	I_{sv} on Standard Compression Curve		
	SCC(e_0=1.5e_L)	SCC(e_0=2e_L)	SCC-marine
1	1.17	1.27	1.22
2	1.13	1.22	1.17
5	1.07	1.15	1.11
10	1.02	1.10	1.06
20	0.97	1.04	1.01
50	0.90	0.95	0.93
100	0.87	0.92	0.90
200	0.78	0.81	0.79
500	0.69	0.70	0.69
1000	0.61	0.61	0.61
2000	0.53	0.53	0.53

compression curve for marine clay (SCC-marine)". SCC-marine is the standard e-log p relationship, when the marine deposits have been normally consolidated from the initial void ratio of 1.5-2.0e_L and not having any structure due to the aging. Accordingly, when the initial void ratio of marine clay is larger than that of SCC-marine at a consolidation pressure, the clay has a structure due to the cementation, while the void ratio is smaller than that of SCC-marine, the clay experiences a secondary compression or may be disturbed during sampling.

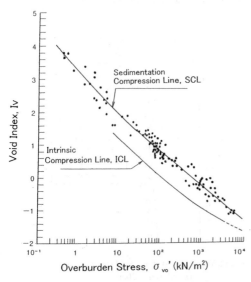

Figure 59 ICL and SCL proposed by Burland

4.6 Intrinsic compression line and Sedimentation compression line by Burland(1990)

As a method to evaluate the structure of clay, Burland (1990) proposed a concept of intrinsic properties of clay. Burland showed that, most of the e-log p relationships of remolded clays, whose initial water contents are ranging from 1.0w_L to 1.5w_L, could be normalized into a unique I_v-log p curve by using the void index, I_v, defined as follows;

$$I_v = (e-e_{100}^*) / (e_{100}^* - e_{1000}^*)$$
$$= (e-e_{100}^*) / C_c^* \qquad (28)$$

where e_{100}^* and e_{1000}^* are void ratios when consolidation pressure are 100kPa and 1000 kPa, respectively. Further, the following equations to determine 2 parameters, C_c^* (= $e_{100}^* - e_{1000}^*$) and e_{100}^*, from the void ratio at liquid limit e_L , are given as follows;

$$C_c^* = 0.256 \, e_L - 0.04 \qquad (29)$$

$$e_{100}^* = 0.109 + 0.679e_L - 0.089e_L^2 + 0.016e_L^3 \qquad (30)$$

Burland called this unique I_v - log p relationship as the intrinsic compression line (ICL), and considered that ICL is inherent to the clay and independent of its natural state, that is, ICL shows the *standard* compression curve when clay have no structure due to aging.

When the void index I_v of normally-consolidated natural deposits is calculate with Eqs.(28)-(30) and plotted with the in-situ effective overburden stress σ_0' , there exists another unique $I_v - \sigma_0'$ relationship, which is called sedimentation consolidation line (SCL). Figure 59 shows the ICL and SCL and the range of variation presented by Burland (1990). According to Burland, natural deposits on SCL have structures due to the aging effects of cementation or bonding, which constitute larger void ratios against the same effective overburden pressure.

In this study, the concept of ultimate standard

70

compression curve, USC is proposed. All e-log p curves finally converge into USC, which is independent of structure due to the aging, initial void ratio and sample disturbance. When the initial void ratio e_0 is given, the standard compression curve, SCC(e_0) can be calculated by Eq.(27). In this sense, the USC and SCC(e_0) are *intrinsic* properties of clay.

A question on ICL concept of Burland is why the initial void ratio have to be $1.0e_L-1.5e_L$ for ICL. Various e-log p curves of remolded clay can be easily produced in the laboratory when the initial void ratios are different. As discussed above, the initial void ratios of natural seabed are considered to be 1.5 –2.0 times of e_L, while the initial void ratio of 1.0–1.5 times of e_L in ICL seems not to have any specific meanings. In the author's opinion, ICL is one of the SCC(e_0), whose initial void ratio is $1.0\,w_L-1.5\,w_L$. Although ICL concept may be useful for the reference of consolidation test of reconstituted clay, it does not have an *intrinsic* meaning on clay property.

Considering that the initial water content of seabed is approximately $1.5w_L-2.0w_L$, the standard compression curves were calculated by Eq.(27) for w_L =70%–110% with assuming that w_0 =1.75w_L . The average I_v - log p relationship of the calculated curves standard compression curve for marine deposits SCC-marine and plotted in Figure 60. As shown in Figure 60, void index I_v of *SCC-marine* is much larger than that of ICL and slightly smaller than that of SCL.

This means that most of the difference between ICL and SCL can be explained by the difference of

The average I_{sv}−log p line of SCC(e_0=1.5e_L) and SCC(e_0=2.0e_L) is named SCC-marine. When in-situ void ratio e_0 is larger than the value given by USC, it is considered that the clay has a structure due to aging. Using the SCC-marine with the in-situ value of I_{sv}, the effect of structure can be evaluated quantitatively. When the difference ΔI_{sv} is obtained as shown in Figure 61, ΔI_{sv} is written as follows;

$$\Delta I_{sv0} = \frac{\Delta(\ln f)}{\ln f_L} = \frac{\ln f_0 - \ln f_{SCC}}{\ln f_L} = \frac{\ln(V_0/V_{SCC})}{\ln f_L} \quad (31)$$

where, f_0 is in-situ specific volume, and f_{SCC} and V_{SCC} are specific volume and volume of soil when the soil is on SCC-marine at the overburden stress, respectively. Eq.(31) can be written as;

$$\varepsilon_v = \ln (V_0/V_{SCC}) = (\Delta I_{sv0})(\ln f_L) \quad (32)$$

where ε_v is expansive volumetric strain to the volume of soil on SCC-marine.

Using Eq.(32), the degree of structure can be indicated by volumetric strain. In the case that ΔI_{sv0}= 0.1 and f_L=3 (w_L=74%), $\varepsilon_v = 0.1 \times \ln 3 = 0.11$, which means that the volume of clay is 11% larger than that without the structure. This clay layer potentially has a large compressibility, because the structure may be destroyed accompanied with consolidation due to the loading.

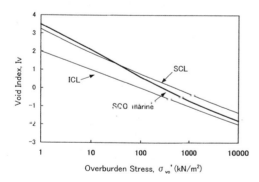

Figure 60   Comparison of void index I_v by ICL , SCL and SCC-marine

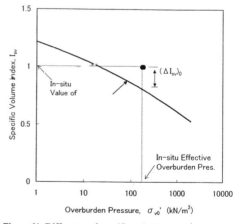

Figure 61 Difference of specific volume ratio, ΔI_{sv}

the initial void ratio at the start of consolidation. The effect of the structure will be shown as the difference in I_v between SCL and SC-Marine, not between SCL and ICL.

a) Holocene seabed
Figure 62 is the relationship between the specific volume ratio I_{sv} and the in-situ effective overburden stress σ_{v0}' at Holocene seabed in Tokyo Bay, Hiroshima Bay, Biwa Lake and Matsushima Bay, the

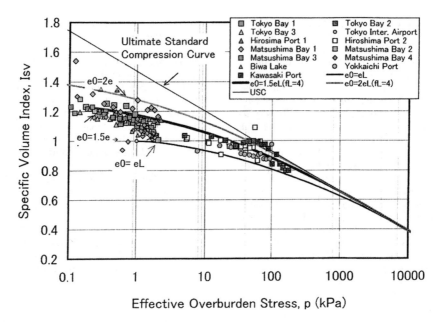

Figure 62 $I_{sv0} - \sigma_{v0}'$ relationship (Holocene clay)

largest depth of which is 20m. The average lines of standard compression curves SCC(e_0=1.5e_L) and SCC(e_0=2.0e_L) in Figure 58 are indicated for the comparison. As shown in Figure 62, the I_{sv}-log σ_{v0}' relationship of natural seabed are plotted around the two SCC(e_0) lines, when p<40kPa. When p>40kPa, some soils show larger void ratios than those determined by SCC(e_0). This means that the void ratios of

Holocene seabed in the shallow depth can be explained mainly by the compression characteristics of clay and the self-weight consolidation, and that no significant effect on the void ratio were seen by the aging.

b) Data collected by Skempton (1970)
The first study on the in-situ void ratio - overbur-

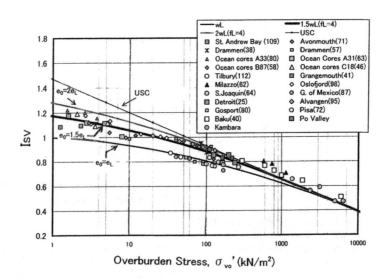

Figure 63 $I_{sv0} - \sigma_{v0}'$ relationship of natural deposits (data from Skempton,1970)

den stress of normally consolidated deposits was presented by Skempton (1970). Figure 63 shows reproduced I_{sv}-log σ_{vo}' relationship of 21 sites collected by Skempton. As shown in this figure, most of data are plotted around SCC($e_0=1.5e_L$), and some are plotted along SCC($e_0=e_L$). In the region of p> 400kPa, the values of in-situ specific volume index are larger than SCC and USC, suggesting that the cementation in the sedimentation process had formed structure and kept the void ratios larger. According to Skempton, when the sedimentation of clay starts at the tidal land, the initial water content become smaller than the case of sedimentation at the sea floor. Although the further study is necessary, this may be a reason of small in-situ void ratio along SCC($e_0=e_L$).

c) Osaka Bay Pleistocene Clay

As mentioned Chapter 3, undisturbed samplings and soil tests of clay up to 400m depths were carried out at the site in Osaka Bay, where the Holocene and Pleistocene deposits continue more than 400m, and the over-consolidation ratios of clay layers are ranging from 1.0 to 1.5 in all the depth (Figure 4).

Figure 64 is the relationship between the in-situ void ratio, e_0, and the overburden stress, σ_{vo}', at the site. As shown in the figure, the void ratio decrease with the overburden stress increase, depending on the liquid limit. The void index by Burland, I_v, is calculated by Eqs.(28)-(30), and plotted in Figure 65, where ICL, SCL and SCC-marine are also shown for comparison. As shown in Figure 64, the values of I_v in Osaka Bay Clay are much larger than SCL and SCC-marine.

The in-situ specific volume index I_{sv} and the effective overburden stress are plotted in Figure 66. The values of I_{sv} of Osaka Bay Pleistocene Clay are much larger than those determined by SCC($e_0=1.5e_L$) or SCC($e_0=2.0e_L$), which means that the clays in these area have larger void ratios or structures due to the cementation effects in the sedimentation process. Taking the mean of ΔI_{sv} in Figure 66 as 0.15, and the

mean water content of liquid limit w_L as 80%, expansive volumetric strain, ε_v, is calculated by Eq.(32);

$$\varepsilon_v = 0.15 \ln(2.70 \times 0.8 + 1) = 0.17$$

This means that, due to the cementation effect during

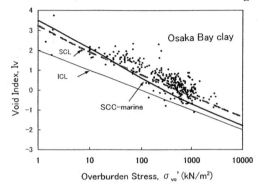

Figure 65 I_v - σ_{vo}' relationship of Osaka Bay Pleistocene clay

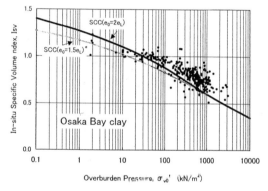

Figure 66 I_{svo}- σ_{vo}' relationship of Osaka Bay Pleistocene clay

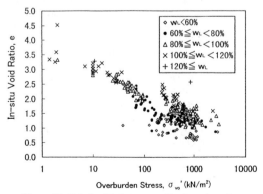

Figure 64   Relationship between void ratio and σ_{vo}' (Osaka Bay Pleistocene clay)

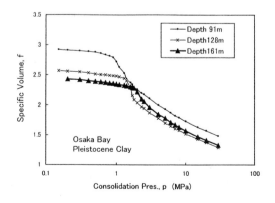

Figure 67   Typical e-log p curve of Osaka Bay Pleistocene clay

73

the sedimentation process, the volume of Osaka Pleistocene Clay layer are getting 17 % larger than those determined only by gravitational consolidation. Figure 67 is typical e-log p curve of Osaka Bay Pleistocene Clay obtained by constant rate strain consolidation test. The extremely large compressibility is observed when the consolidation pressure gets larger than the consolidation yield stress p_c, which seems to be due to the destruction of the structure.

d) Ariake clay

Ariake clay is widely deposited around Ariake Bay in Kyusyu Island with an area of several hundreds of square km. Ariake clay is known as typical sensitive clay in Japan. Torrance and Otsubo (1995) reported that most of Ariake clays have sensitivity larger than 16, with the maximum over 1000. Ariake Bay Research Group (1965) has revealed that the upper Ariake layer above 11m was deposited under a marine condition, while the lower Ariake layer below 11m was deposited under brackish condition. The effect of salt extraction by leaching has been reported to be an important factor causing large sensitivity of Ariake clay (Torrance and Otsubo, 1995)

Figure 68 shows the in-situ $I_{sv} - \sigma_{v0}'$ relationship of Ariake clays, which were taken from various sites around Ariake Bay, with the liquid limit w_L ranging from 51% to 164%. As shown in Figure 68, the values of I_{sv} of in-situ Ariake clay are extremely larger than those of SCC-marine. It is seen that the difference in I_{sv} from SCC-marine is more remarkable as the liquid limit of clay was smaller. The most probable explanation for this result is that Ariake clays sedimented under marine or brackish conditions with a high liquid limit. The leaching that occurred in the post-depositional process decreased the liquid limit, while the in-situ water content had little change. Accordingly, in the case of Ariake clay, the structure was not developed by cementation, but the SCC-marine was lowered by the decrease of liquid limit due to leaching and consequently the value of I_{sv} increase drastically.

Figure 69 is an example of e-log p curve of Ariake clay sample, which was taken by high quality sampler and tested by constant rate strain consolidation test. Although the reason of high ΔI_{sv} value is different from Osaka Bay Pleistocene clay, the similar characteristics of structured clay, such as large compressibility at $p > p_c$ is clearly observed.

4.8 Summary

The concept of standard compression curve (e-log p curve) of clays was proposed. The standard compression curve consists of ultimate standard compression curve, USC, and the standard compression curves from an initial void ratio, SCC(e_0). USC is the e- log p relationship, where e- log p curves converge ultimately with increase of consolidation pressure. When the initial void ratio is small, the standard compression curve is different from USC. SCC(e_0) is the e-log p relationship that starts the consolidation at an initial void ratio e_0, and with the increase of the consolidation pressure, it finally converges into the USC.

Both curves are determined mainly by the liquid limit of clay and the void ratio at the beginning of consolidation. The uniqueness of USC was examined by the consolidation data of 18 marine clays. Using the specific volume index, I_{sv}, which is defined as ln $(1+e)$ / ln $(1+e_L)$, USC was shown as a unique I_{sv} – log p relationship not dependent on the plasticity of clays. I_{sv} – log p relationship of SCC(e_0) was also presented based on the USC concept and the study on sample disturbance by Okumura (1974).

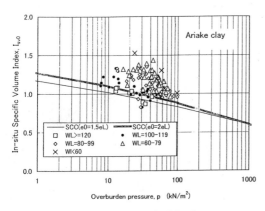

Figure 68 $I_{sv0} - \sigma_{v0}'$ relationship of Ariake clay

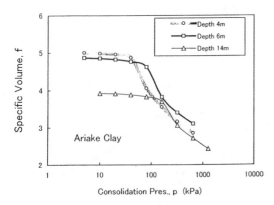

Figure 69 Typical e-log p curve of Ariake clay sample

It is considered that, under the conditions of marine waves and currents, the void ratio of marine clay at the beginning of consolidation is 1.5-2.0 times of liquid limit e_L. By calculating SCC(e_0) with the conditions of $e_0=1.5e_L$ and $e_0=2e_L$, the standard relationship between I_{sv} and the effective overburden pressure, σ_{v0}', was obtained for normally consolidated marine deposits, which is called SCC-marine. By comparing the in-situ specific volume of clay with the value on of SCC-marine, the degree of structure of marine deposits can be evaluated and the followings findings were obtained:

1) The void ratios of Holocene seabed in the shallow depth can be explained mainly by SCC-marine and that no significant effect by aging on the void ratio.

2) As the in-situ values of I_{sv} of Osaka Bay Pleistocene clay at the site of Kansai International Airport are much larger than those of SCC-marine, the clays seems to have well-developed structure due to aging.

3) The values of I_{sv} of in-situ Ariake clay prefecture were extremely larger than those of SCC-marine. The cause of high I_{sv} value is considered to be the decrease of liquid limit by leaching that occurred in the post-depositional process.

5. CONSOLIDATION BEHAVIOR OF OSAKA BAY PLEISTOCENE CLAY

5.1 Separate-type consolidometer

Osaka Pleistocene clay shows a remarkable increase in compressibility when the overburden pressure exceeding the consolidation yield stress is loaded. It is usually revealed with a high consolidation yield stress which is believed to be originated not only by the high overburden pressure, but also by its well-developed structure resulting from long-term secondary compression and aging effects.

It is therefore necessary to adopt more elaborated and sophisticated method for the investigation on this type of clay retaining peculiar compressibility characteristics in the purpose of proper understanding on the development of deformation and pore water pressure.

A separated-type consolidometer test, in which the soil layer is divided into five inter-connected sub-specimens is carried out for Osaka Pleistocene clay. Several researchers have conducted the separated-type consolidation test mostly in the purpose of verification of various consolidation theories(Berre and Iversen,1972, Mesri and Choi,1985, Aboshi et al.,1985, and Imai and Tang,1992). However most of the soils used in their studies were artificial reconstituted soils or undisturbed Holocene clays with less developed soil structure. For the typical condition of Osaka Pleistocene clay with high consolidation yield

stress and with its well-developed structure, a separated-type consolidometer of high capacity was specially designed. With this device, the some particular consolidation characteristics of undisturbed Osaka Pleistocene clay are to be investigated.

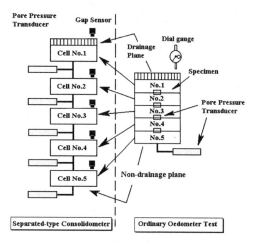

Figure 70   Concept of separated-type consolidation

5.2 Method and procedure

The fundamental concept of separated-type consolidometer test is illustrated in Figure 70. The soil specimen is separated into several inter-connected samples, so that the consolidation characteristics of the soil by thickness variation together with the internal variation of pore water pressure can be examined in connected test cells. Those are usually impossible to be obtained in the conventional type of oedometer test. The bottom drainage plane of one sub-specimen is connected to the top drainage of the next one in the series.

Figure 71 shows the systematic diagram for the separated-type consolidometer test. The cell is made of metallic body to endure the high pressure and drainage between the sub-specimens is also carried by copper pipes to support high pore water pressures during the tests. Pore water pressure is measured at the bottom of each sub-specimen by pressure transducers. To increase saturation of the specimen, backpressure is applied to the drainage surface of the top specimen.

A gap-sensor of non-contact type measuring the gap distance by magnetic pulse between the sensor and the target plate on the loading cap is employed. A soil sample trimmed into five sub-specimens, 60mm in diameter and 10mm in thickness each, was used for this test. Each sub-specimen was placed into the consolidometer cells, labeled No.1, No.2, ⋯, and No.5. This gives a total maximum length of 50mm for the whole drainage path. The drainage was allowed only at the upper boundary of sub-specimen No.1 cell and the drainage volume was measured for the total volume change by a burette in which back-pressure is applied.

Soil properties of Osaka Pleistocene clay used in this test are shown in Table 4. Osaka Pleistocene clay tested in this study was collected from the soil layer of 150~170m depth below the sea water level. The specific volume index I_{sv} is 0.682 and the difference of I_{sv} from SCC-marine, $\Delta I_{sv,}$ is 0.072, which means this layer is highly structured with the expansive volumetric strain of 9.9%. Recently, it has been known that a large amount of Diatom microfossils are contained in Osaka Pleistocene clay and they may a major influential factor for the large void ratio and the high compressibility (Tanaka and Locat,1999).

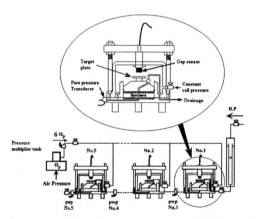

Figure 71 Systematic diagram of the separated-type Consolidometer

To compare the effect of soil structure, reconstituted soils from the same Osaka Pleistocene clay were prepared for the comparison of test results. For the preparation of a reconstituted sample, the soil passed with 106µm sieve, was remolded and mixed thoroughly at the water content of above 180%, which is almost twice the liquid limit of soil to eliminate the possible effect of soil structure. Then it was pressurized in a cell up to 196kPa by step loading until the 95% of consolidation by √t method was obtained.

Table 4. Soil properties of Osaka Pleistocene clay

Natural water Content, $w_n(\%)$	Liquid limit, LL(%)	Plastic limit, PL(%)	Specific gravity, G_s
53.8~57.8	73.6~108	31.1~36.6	2.69~2.72

Figure 72 shows the void ratio – effective stress curve obtained from the consolidation test by constant rate of strain (CRS) comparing the reconstituted and the undisturbed soil used in this test. The consolidation behavior of undisturbed Osaka Pleistocene clay can typically explained by some typical features like low compressibility in the over consolidated range, sudden decrease of void ratio after yielding and non linear relations for the void ratio with logarithm effective stress in the normally consolidated range(also refer to Figure 67).

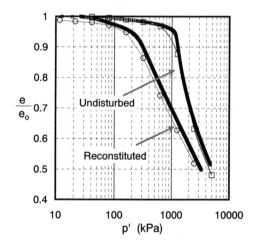

Figure 72 Comparison of e-log p for reconstituted and undisturbed Pleistocene clay

5.3 Test results and discussions

During the separated-type consolidometer test, one-dimensional displacement and excess pore water pressure of each sub-specimen by the increment of loading are measured at a given measuring time. Backpressures were applied by 196kN/m^2 for reconstituted soil and by 392kN/m^2 for undisturbed soil. Five sub-specimens in series were used both for the reconstituted soil and undisturbed soil.

Figures 73 and 74 show the variation of strain and excess pore water pressure as a function of time for each increment of stage loading for the reconstituted and the undisturbed sample respectively. In the case of the reconstituted sample, the delay of compression by the distance from the drainage surface was clearly seen in both over consolidated and normally consolidated range. This delay of compression is due to the fact that sub-specimen nearer to the free drainage boundary is compressed faster by taking place of rapid dissipation of excess pore water pressure. Comparing over-consolidated compression with normally consolidated compression, no distinctive difference of the consolidation behavior was observed except an increase in primary strain and coefficient of secondary consolidation in the normally consolidated range.

However, for the undisturbed sample of Osaka Pleistocene clay, the consolidation behavior showed evident contrast between the normally consolidated and the over-consolidated ranges as it can be seen in Figure 74. In the over-consolidated range, primary consolidation is over almost instantly and all subsequent strains for all sub-specimens continue at nearly constant rate of secondary compression. Excess pore water pressure also showed instant dissipation within a minute, which means almost no excess pore water pressure was developed in the over consolidated region. It is not so difficult to understand the rigidity of the soil structure against overburden pressure.

Meanwhile it is interesting to say that the dissipation characteristics of excess pore water pressure at the loading stage exceeding the consolidation yielding stress of undisturbed soil, displayed very unique aspect. As the stage loading closes to the consolidation yielding stress, the compressibility of soil started to increase dramatically with little development of excess pore water pressure. However once the soil underwent yielding, the soil began to display the delay of compression according to the drainage length to the drainage surface.

Particularly it is distinguishable to see the peculiar development of excess pore water pressure at each subspecimen. Once the variation of excess pore water pressure showed the decreasing tendency at first,

then it increased again to a notable degree after the soil experienced yielding. For the main reason of this, it is considered that excess pore water pressure once barely developed in the over-consolidated range was affected by the rapid increase of compressibility progressing in the loading stage shifting over to the normally-consolidated range. This means that the rate of increase in compressibility became much higher than the rate of excess pore water pressure dissipation when skeletons of Osaka Pleistocene clay with well-developed structure were yielded by a high overburden pressure. However it is necessary to point out that this behavior is highly affected by the size of loading increment and consolidation yielding stress.

In the normally consolidated range, the consolidation behavior of soil showed delay of compression with distinct delay of dissipation of excess pore water pressure from the start of drainage. Those behaviors in normally consolidated range revealed almost similar variation characteristics to the results for reconstituted sample.

In Figure 75, the distribution of degree of consolidation, U_z to the thickness of soil layer according to the length of drainage path is compared with theoretical solution by Terzaghi's one-dimensional consolidation theory. Both soil conditions are for the loading spanning the preconsolidation pressure.

For the reconstituted sample, the distribution of U_z to the depth of soil from measured values is incidentally well consistent with theoretical solutions. Of

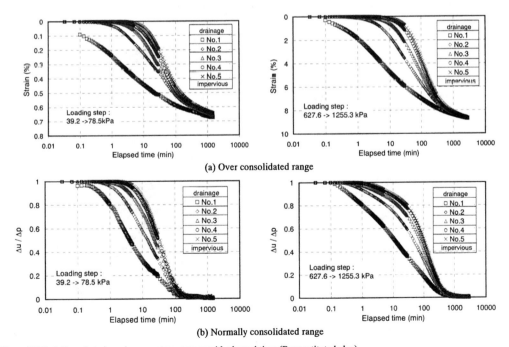

(a) Over consolidated range

(b) Normally consolidated range

Figure 73 Variation of strain and pore water pressure with elapsed time (Reconstituted clay)

course, this is always the same with different type of soils or different test conditions. At any rate, from the figures, it can be said that, even though the primary consolidation of undisturbed soil is once ended up at much higher rate than that of reconstituted one

after yielding, the consolidation process became greatly delayed and showed complexity at the consolidation process.

Relations between void ratio and effective stress are shown respectively for reconstituted and undis-

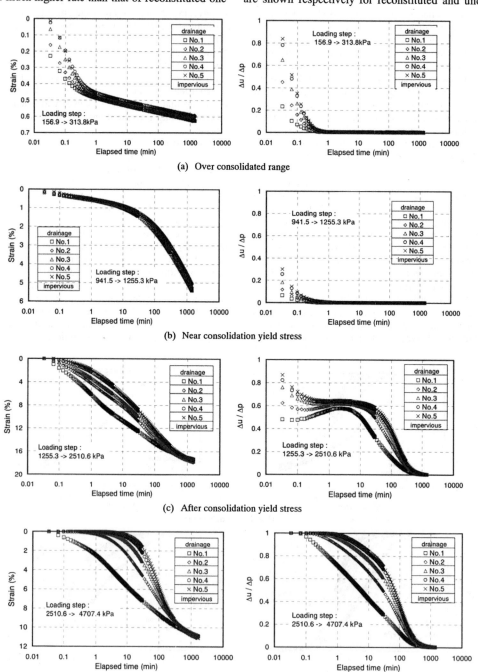

Figure 74 Variation of strain and pore water pressure with elapsed time (Undisturbed clay)

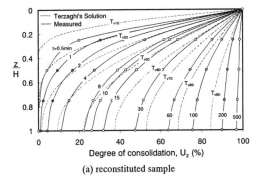

(a) reconstituted sample

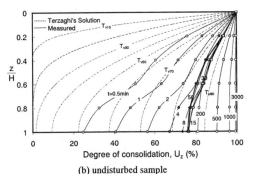

(b) undisturbed sample

Figure 75 Degree of consolidation with depth of soil

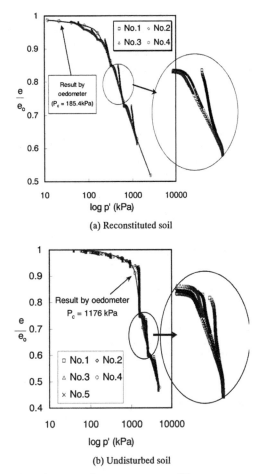

(a) Reconstituted soil

(b) Undisturbed soil

Fig.76 e-log p curve by separate type consolidometer

turbed samples in Figure 76. For the comparison, void ratio is normalized with the initial void ratio, e_o. It is clear that the e-log p' relation from the results by separated-type consolidometer test is not linear either with any load increment or with the length of drainage path for both sample conditions. Due to the fact that, at the soil layer which is near to the drainage boundary, effective stress instantaneously increased, the curve for No.1 layer reveals the e-log p' relation almost close to the ideal no water condition with the largest curvature. However it is notable that varied relation of e-log p' by the length of drainage path for undisturbed sample can be observed at the next load increment after exceeding the preconsolidation pressure.

And also for the undisturbed soil, before the substantial increase of compressibility after preconsolidation pressure, large span of over-consolidated range with respect to the increase of stress implies high resistance of soil structure. It is notable that sudden increase of compressibility can be observed near the consolidation yielding stress at almost constant effective stress for all sub-specimens following the same curvatures. After yielding, tremendous decrease of void ratio took place at subsequent loading pressures after yielding, revealed with almost two third of total decrease in void ratio.

5.4 Summary

Using separated-type consolidometer of high capacity designed in this study, the consolidation characteristics of Osaka Pleistocene clay was examined with the purpose of investigating soil structural effect. Through the tests as a fundamental stage, undisturbed soil in the over-consolidated range showed extremely faster expulsion of excess pore water pressure than the normally consolidated range and that the most part of consolidation settlement is occupied by secondary compression at a given pressure increment.

At the pressure increment spanning the preconsolidation pressure of Osaka Pleistocene clay, effective stress increasing with elapsed time showed sudden decreasing tendency after the effective stress exceeded the preconsolidation pressure. It is estimated that the cause of this unique behavior is to be attributed to the fact that radical increase of compressibility by the yielding of soil structure surpasses the dissipation rate of excess pore water pressure. This behavior is particularly relevant to clays with a well-developed structure such as Osaka Pleistocene clay.

6 CONCLUSIONS

In recent years, as the scale of structures has been enlarged and the construction works have been carried out in deeper sea areas, the loads due to constructions has become close or larger than the consolidation yield stress, p_c, of the Pleistocene clay layer at large depth. In this report, the compression and shear characteristics of aged clay, were reviewed with related to the case study on Kansai International Airport project. The method for evaluating the degree of structure was newly proposed based on the concept of standard compression curve. A unique consolidation behavior of structured clay was also shown. The main conclusions are summarized as follows:

1) The strength gain with time can be divided into 2 components, one of which is associated with volumetric compression, while the other is attributed to the cementation effect. The compression index ratio, r_c is proposed as an index of cementation effect. The value of r_c of Holocene clays and Pleistocene clays in Japan are 1.1-3.0 and 1.1-6.0,respectively, increasing with depth, while r_c of reconstituted young clays are almost 1.0.
2) The high temperature consolidation method was proposed for reproducing the structure of aged clay in laboratory, by consolidating remolded clay slurry under 75° C. This is a useful technique to duplicate aged clay in laboratory.
3) Tan and Tsuchida (1999) investigated the shear strength gain by cementation effect experimentally, and showed that the strength gain by cementation is proportional to elapsed time, and the increment is related to the overburden stress. The strength increment Δs_u can be expressed by the following simple equation;

$$\Delta s_u = 0.3\sqrt{p} \; \Delta(\log t)$$

where, p is an overburden pressure (unit:kN/m^2), and t is an elapsed time after end of primary consolidation. Equation means that the contribution of cementation dominates strength gain with time when the overburden pressure is smaller than 100kN/m^2, while strength gain due to secondary compression becomes the main component when the overburden pressure is larger than 100kN/m^2.
4) The undrained shear strengths and deformation characteristics of Pleistocene clay in Osaka Bay were studied by the recompression method with triaxial tests. Residual effective stresses in Pleistocene clays collected from large depth were about one fifth of effective overburden stresses. The quality of undisturbed samples of Osaka Bay Pleistocene clay was better than anticipated.
5) The Pleistocene clay at large depths is brittle and shifts very rapidly from elastic deformation to plastic failure. The clay obtained from the depth lower than 80m has the peak extension strength at the axial strains of 3 ~ 4 %.
6) Although the safety factor of slip circle which passes through the Pleistocene clay layer was larger than 1.20, the margin of the safety were not so much. The accurate estimation of the shear strength of Pleistocene clays and the field observations and construction control will be necessary to complete the projects safely.
7) The concept of standard compression curve (e-log p curve) of clays was proposed. The standard compression curve consists of ultimate standard compression curve, USC, and the standard compression curves from an initial void ratio, SCC(e_0). USC is the e- log p relationship, where e- log p curves converge ultimately with increase of consolidation pressure. SCC(e_0) is the e-log p relationship that starts the consolidation at an initial void ratio e_0. Both curves are determined mainly by the liquid limit of clay and the void ratio at the beginning of consolidation.
8) Using the specific volume index, I_{sv}, which is defined as ln $(1+e)$ / ln $(1+e_L)$, USC was shown as a unique I_{sv} – log p relationship not dependent on the plasticity of clays. I_{sv}–log p relationship of SCC(e_0) was also presented
9) Considering the initial condition of marine sediments, the standard relationship between I_{sv} and the effective overburden pressure, σ_{v0}', was calculated for normally consolidated marine deposits, which is called SCC-marine. By comparing in-situ specific volume of clay with value on of SCC-marine, the degree of structure of marine deposits can be evaluated. It is shown the void ratios of Holocene seabed in the shallow depth can be explained mainly by SCC-marine with no significant effect by aging on the void ratio, and that Osaka Bay Pleistocene clay seems to have well-developed structure due to aging, because the in-situ values of I_{sv} are much larger than those of SCC-marine.
10) Using separated-type consolidometer of high capacity designed in this study, the consolidation characteristics of Osaka Pleistocene clay was examined with the purpose of investigating soil structural effect. Through the tests as a fundamental stage, undisturbed soil in the over-consolidated range showed extremely faster expulsion of excess pore water pressure than the normally consolidated range and that the most part of consolidation

80

settlement is occupied by secondary compression at a given pressure increment.

11) At the pressure increment spanning the consolidation yield stress of Osaka Pleistocene clay, effective stress increasing with elapsed time showed sudden decreasing tendency after the effective stress exceeded the consolidation yield stress. It is estimated that the cause of this unique behavior is to be attributed to the fact that radical increase of compressibility by the yielding of soil structure surpasses the dissipation rate of excess pore water pressure. This behavior is particularly relevant to clays with a well-developed structure such as Osaka Pleistocene clay.

ACKNOWLEDGEMENT

The study of mechanical properties of Osaka Bay Pleistocene clay has been carried out in Soil Mechanics Laboratory of Port and Harbour Research Institute for about 20 years. A number of the Author's colleagues have contributed to this study and their assistance is gratefully acknowledged. Dr. Y. Watabe performed a lot of triaxial tests with recompression method, and the separate-type consolidation test was carried out by Dr. Kang, M-S. Dr. Hong, Z. contributed the works on standard compression curve. He would also like to express thanks to Mr. Mizuno, K. and Hikiyashiki, H. in preparing the manuscripts.

REFERENCES

Abdrabbo, M.(1990):"Correlation between index tests and compressibility of Egyptian Clays", *Soils and Foundations*, Vol.30, No.2, pp.128-132.

Aboshi, H, Matsuda, M. and Okuda, M. (1981): "Preconsolidation by separate type consolidometer", Proc. of 10th ICSMFE, Vol.3. pp. 572~579.

Arai, Y.(1991):"Construction of an artificial offshore island for the Kansai International Airport", Proceedings of International Conference on Geotechnical Engineering for Coastal Development, Geo-Coast'91,Port and Research Institute. Vol.2, pp.927-943.

Arai, Y., Oikawa, K. and Yamagata, N. (1991) : "Large scale sand drain works for the Kansai International Airport", Proceedings of International Conference on Geotechnical Engineering for Coastal Development, Geo-Coast'91, Port and Research Institute. Vol.1, pp.281-286.

Burland, J., B. (1990):"On the Compressibility and Shear strength of Natural Clays", *Geotechnique*, Vol.40, No.3, pp.329-387.

Butterfield, R.(1979):"A natural compression law for soils (an advance on e-log p)",*Geotechnique*, Vol.29, No.4, pp.469-480.

Berre, T. and Iversen, K.(1972): "Oedometer tests with different specimen heights on a clay exhibiting large secondary compression", *Geotechnique* 22. No.1. pp.53-70.

Berre, T. and Bjerrum, L. (1973): "Shear strength of normally consolidated clays", Proc. of the 8th ICSMFE, pp.39-49.

Endo, H., Oikawa, K., Komatsu, A. and Kobayashi, M. (1991):"Settlement of diluvial clay layers caused by a large scale man-made island", Proceedings of International Conference on Geotechnical Engineering for Coastal Development, Geo-Coast'91,Port and Research Institute. Vol.1, pp.177-182.

Gomyo, M., Yauchi, E., Sakai, K., Ohtsuki, T. and Itosu (1985): "On the effect of the mud properties on the interaction between wave and the mud", 33rd Meeting of Coastal Engineering, JSCE, pp.322-326 (in Japanese).

Hanzawa, H. (1982): "Undrained strength characteristics of alluvial marine clays and their application to short term stability problems." Thesis of Dr. Eng., University of Tokyo.

Hanzawa, H. and Kishida, T. (1982): "Determination of in-situ undrained strength of soft clay deposits", Soils and Foundations, vol.22, No.2, pp.1-14.

Horie, H., Zen, K., Ishii, I. and Matsumoto, K. (1984): "Engineering properties of marine clay in Osaka bay, (Part-1) Boring and sampling", *Technical Note of The Port and Harbour Research Institute*, No.498, pp.5-45. (in Japanese)

Hong, Z. and Tsuchida, T.: "On compression characteristics of Ariake clays", *Canadian Geotechnical Journal*, No.36, pp.807-814, 1999.

Imai, G. and Tang, Y. X.(1992): "A constitutive equation of one-dimensional consolidation derived from interconnected tests", *Soil and Foundations,* JSSMFE. Vol. 32. No. 2. pp.83-96.

Imai, G. (1978): "Fundamental Studies on one-dimensional consolidation characteristics of fluid mud", Thesis of Dr. Eng. University of Tokyo.

Ishii, I., Ogawa, F. and Zen, K. (1984) : "Engineering properties of marine clays in Osaka bay, (Part-2) Physical Properties, consolidation characteristics and Permeability", *Technical Note of the Port and Harbour Research Institute*, No.498, pp.47-86. (in Japanese)

Inoue, T., Tan, T. S. and Lee S.L. (1990):"An investigation of shear strength of slurry clay", *Soils and Foundations*, Vol.30, No.4, pp.1-10.

Jamiolkowski, M., Ladd, C.C., Germaine, T.T. and Lancellota, R. (1985): "New developments in field and laboratory testing of soils", State of the Art Report, Proc. of the 11th ICSMFE, pp.57-153.

Japanese Society on Soil Mechanics and Foundation Engineering (1966): Investigation, design and construction in soft ground, Soil Engineering Library 1, pp.92-94. (in Japanese)

Japanese Society on Soil Mechanics and Foundation Engineering (1992): "Fall cone method, Report of Research Committee III", Proceedings of symposium on new method of physical soil testing", pp.69-90. (in Japanese)

Japanese Geotechnical Society (1995): "Method for obtaining undisturbed soil sample using thin-walled tube sampler with fixed piston." JGS 1221-1995.

Kanda, K., Suzuki, S., and Yamagata, N. (1991): "Offshore soil investigation at the Kansai International Airport.", Proc. of GEO-COAST '91, Vol.1, pp.33-38.

Kang, M.-S. and Tsuchida, T. (2000): Experimental study on the consolidation properties of Osaka Pleistocene clay by separated-type consolidometer test, IS-Yokohama2000, 2000.9.(submitted)

Kawakami, H., Abe, H. and Iwasaki, K. (1978): "Influence of suction on unconfined compression strength." Proc. of 13th annual meeting of JSSMFE, pp.313-316. (in Japanese)

Kitazume, M. and Terashi, M. (1994):"Effect of stress-strain characteristics on slope failure", Proceedings of international conference on centrifuge test", Centrifuge 94, Balkema, pp. 611-616.

Kobayashi, M. (1994):"Geotechnical engineering aspects of Kansai International Airport Project", Proceedings of 39th Geo-engineering Symposium, JSSMFE, pp.1-6. (in Japanese)

Ladd, C.C. and Lambe, W. (1963): "The strength of undisturbed clay determined from undrained tests", ASTM, STP-361, Laboratory Shear Testing of Soils, pp.342-371.

Ladd, C. C. and Foott, R. (1974): "New Design Procedure for Stability of Soft Clay", Proc. ASCE, GT7, pp.763-786.

Locat, J. and Lefebvre, G. (1985): "The Compressibility and Sensitivity of an Artificially sedimented Clay Soil: The Grande-Baleine Marine Clay, Quebec, Canada", Marine Geotechnology, Vol.6, No.1.

Mesri, G. and Choi, Y. K. (1985): " The uniqueness of the end-of-primary (EOP) void ratio-effective stress relationship", Proc. of the 11th ICSMFE, Vol. 2, pp587-590.

Mesri, G. and Godlewski, P. M.(1977): "Time- and stress-compressibility inter-relationship", Journal of the Geotechnical Engineering Division. ASCE. Vol.103. No.5. pp.417-430.

Mesri,G.: "New design procedure for stability of soft clays", Discussion, ASCE, GT4, Vol.101, pp.409-412, 1975.

Mikasa, M.(1988):"Liquid limit test by cyclic consolidation", 23rd Annual Conference of J.S.M.F.E.,pp.267-268. (in Japanese)

Mitachi, T. and Fujiwara, Y. (1987): "Undrained Shear Behavior of Clays Undergoing Long-term Anisotropic Consolidation", Soils and Foundation, Vol.27, No.4, pp.45-61.

Mitachi, T. and Kudoh, Y. (1996): "Method for predicting in-situ undrained strength of clays based on the suction value and unconfined compressive strength.", Journal of Geotechnical Engineering, JSCE, No.541 / III-35, pp.147-157. (in Japanese)

Nakashima, Y., Kozawa, K., Kiyama, M. and Maegawa, H. (1995): Back analysis of settlement in Osaka Port undersea tunnel, Proc. of International symposium on compression and consolidation of clayey soils, IS-Hiroshima 95', Vol.1, pp.725-732.

Ogawa, F. and Matsumoto, K. (1978): Correlations of geotechnical parameters in port and harbor area, Report of the Port and Harbour Research Institute, Vol.17, No.3, pp.7-20. (in Japanese)

Okumura, T.(1974): "Studies on the disturbance of clay soils and the improvement of their sampling techniques", Technical Note of Port and Harbour Research Institute. No. 193. (in Japanese).

Shimizu, M. and Tabuchi, T. (1993): "Effective stress behavior of clays in unconfined compression tests", Soils and Foundations, Vol.33, No.3, pp.28-39.

Shogaki, T., Kaneko, M., Moro, H. and Mihara. M. (1995): "Method for predicting in-situ undrained strength of clays by unconfined compression test with suction measurements.", Proc. of Sampling Symposium, JSSMFE. (in Japanese)

Shogaki, T. and Kaneko, M.: "Effects of sample disturbance on strength and consolidation parameters of soft clay", Soils and Foundations, Vol.34, No.3, pp.1-10,1994.

Skempton, A.W.(1944) : "Note on the Compressibility of Clays", Q. J. Geological Society. Vol.100, pp.119-135.

Skempton A. W.(1970): "The Consolidation of Clays by Gravitational Compaction", Quarterly Journal of Geological Society of London, Vol.125, pp.373-412, 1970.

Tan, Y.X. and Tsuchida, T.: The development of shear strength for sedimentary soft clay with respect to aging effect, Soils and Foundations, Vol.39, No.6, pp.13-24, 1999.

Tanaka,H. and Tanaka,M.(1994) : Determination of Strength of Clay by means of Vane Shear Test, Report of the Port and Harbour Research Institute, Vol.33, No.4, pp.1-17. (in Japanese)

Tanaka, H., Tanaka, M. and Hamouche, K.K.(1997a): "Applicability of unconfined compression test to overseas countries' clays." Proc. of 41th Symposium on Geotechnical Engineering, pp.61-66. (in Japanese)

Tanaka, H. and Tanaka, M. (1997b): "Applicability of unconfined compression test to European clays", Proc. of the 14th ICSMFE, pp.209-212.

Tanaka, H. and Locat, J. (1999): "A microstructural investigation of Osaka Bay clay: the impact of microfossils on its mechanical behavior." Canadian Geotechnical Journal, Vol.36, pp.493-508.

Tanaka, M., Tanaka, H., Mitsukuri, K. and Suzuki, K. (1995): "Relation of unconfined compression strength and suction on undisturbed clay samples." , Proc. of 30th annual meeting of JGS, pp.621-622. (in Japanese)

Tavenas, F. and Leroueil, S. (1977): "Effects of stresses and time on yielding of clays", Proc. of the 9th ISFMFE, Vol.1, pp.319-326.

Terzhagi, K. and Peck, R. B. (1967): Soil Mechanics in Engineering Practice, 2nd ed. John Wiely and Sons, New York.

Tsuchida, T. (1990): "Study on determination of undrained strength of clayey ground by mean of triaxial tests." Technical Note of The Port and Harbour Research Institute, Ministry of Transport, Japan, No.688. (in Japanese)

Tsuchida, T.(1991) :"A New Concept of e-log p Relationship for Clays", Proc. of 9th ARCSMFE., Bangkok, Vol.1, pp.87-90.

Tsuchida, T. (1994):"A unified concept of e-log p relationship", Proc. of 13th Conference of ISSMFE, New Delhi, Vol.1, pp.71-74.

Tsuchida, T. (1995): "Unified model of e-log p relationship with the consideration of the effect on initial void ratio", Proc. of International Symposium on Compression and Consolidation of Clayey Soils, Is-Hiroshima'95, Vol.1, pp. 379-384.

Tsuchida, T. (1999):"Unified model of e-log p relationship of clay and the interpretation of natural water content of marine deposits", Characterization of soft marine clays, pp.185-202. Balkema.

Tsuchida, T. (2000): "Evaluation of undrained shear strength of soft clay with consideration of sample quality", *Soils and Foundations*, Vol.40, No.3, pp.29-42.

Tsuchida, T., Kikuchi, Y., Nakashima, K. and Kobayashi, M. (1984): "Engineering properties of marine clays in Osaka bay, (Part 3) Static characteristics of shear.", *Technical Note of the Port and Harbour Research Institute*, No.498, pp.87-114. (in Japanese)

Tsuchida, T., Mizukami, J., Oikawa, K. and Mori, Y. (1989): "New method for determining undrained strength of clayey ground by means of unconfined compression test and triaxial test." *Report of the Port and Harbour Research Institute*, Ministry of Transport, Japan, Vol.28, No.3, pp.81-145. (in Japanese)

Tsuchida, T. and Kikuchi, Y. (1991): "K_0-consolidation of undisturbed clays by means of triaxial cell." *Soils and Foundations*, Vol.31, No.3, pp.127-137.

Tsuchida, T., Kobayashi, M. and Mizukami, J. (1991):"Effect of aging of marine clays and its duplication by high temperature consolidation":*Soils and Foundations*, Vol.31, No.4, pp.133-147.

Tsuchida, T., Hong, Z., Watabe, Y. and Ogawa, F. (1999): "Relationship of undrained shear strength versus normalized water content for remolded clayey soils", 34th Annual conference on geotechnical engineering, JGS, Vol.1, pp. 543-544. (in Japanese)

Torrance J.K. and Otsubo M. (1995) : Ariake Bay Quick Clay: A Comparison with the General Model, Soils and Foundations,Vol.35,No.1,pp11-19.

Umehara Y. and Zen, K. (1982):"Consolidation characteristics of dredged marine bottom sediments with high water content", *Soils and Foundations*, Vol.22, No.2, pp.40-54.

Watabe, Y. (1999): "Mechanical properties of K_0-consolidation and shearing behavior observed in triaxial tests for five worldwide clays –Drammen, Louiseville, Singapore, Kansai and Ariake clay", Characterization of Soft Marine Clays, Balkema.

Watabe, Y., Tsuchida, T. and Adachi, K. (2000a): "Undrained shear strength of Pleistocene clay in Osaka Bay and its effect on the stability of a large scale seawall structure." Proc. of IS-Yokohama (submitted)

Yashima, S. (1986) :Pulverulence and properties of pulverulence, Chemical Engineering Series, No.10, Baihukan, pp.54-100. (in Japanese)

Considerations on stability of embankments on clay

S.Leroueil
Université Laval, Ste-Foy, Québec, Canada

D.Demers
Ministère des Transports du Québec, Québec, Canada

F.Saihi
Golder & Ass., Montréal, Québec, Canada

ABSTRACT: The paper examines the different approaches that have been proposed for the evaluation of stability of embankments on soft clays. Stability of embankments constructed in one stage is well controlled. However, experience shows that engineers have to be careful when they want to apply to a given geological context an empirical approach developed in a different one. For stage-constructed embankments, there are indications that the methods proposed up to now have a tendency to underestimate stability.

1 INTRODUCTION

Stability of embankments on soft clays is evaluated by about 10 different methods. It is generally well controlled and failures are seldom observed. However, progresses have been recently made on the comparison of the different approaches used in different countries and by different engineers. This paper examines these approaches and how they compare, and presents some practical conclusions.

2 STABILITY OF EMBANKMENTS ON SOFT CLAYS: GENERALITIES

Stability of embankments on soft clays is generally evaluated by limit equilibrium stability analyses. These methods do not consider strains and make the implicit assumption that the stress-strain behaviour of the soil is ductile. Ladd (1991) defined three types of stability analysis: (a) effective stress analysis (ESA); (b) total stress analysis (TSA); and (c) undrained strength analysis (USA). A total stress analysis, as often used in single-stage construction analysis, is based on a fixed strength profile. In undrained strength analysis, the *in situ* undrained shear strength is computed as a fraction of the pre-shear vertical effective stress. This latter type of analysis is mostly used for evaluating the stability of stage-constructed embankments.

Tavenas et al. (1980) and Ladd (1991) examined effective stress analysis. They showed that the factor of safety (F_{ESA}) calculated on the basis of pre-shear effective stresses overestimates the margin of safety available for further loading. In effective stress analyses, the factor of safety F_{ESA} is defined as the ratio of the shear stress at failure τ_f to the applied shear stress τ, for the same normal effective stress ((0) – (1)) in Fig. 1. However, due to pore pressures generated during loading, the stress path in the τ-σ'_n diagram follows a curve such as ((0)-(2)) in Fig. 1. The corresponding factor of safety is then τ_{fu}/τ. As indicated on Fig. 1, F_{ESA} is larger than the undrained factors of safety, F_{USA} or F_{TSA}. It is only at failure that points (1) and (2) coincide and that the two approaches give the same factor of safety equal to 1.0. This approach is thus not recommended for stability analysis during undrained loading.

An alternative to limit equilibrium stability analyses is the use of deformation analyses, generally based on finite element methods. A key aspect however is the representativity of the considered constitutive equations. In particular, natural soft clays are anisotropic due to their deposition under $K_{o.nc}$ conditions, and this is reflected by the shape of their limit state curves (Diaz-Rodriguez et al., 1992) and by the strengths that can be mobilized in compression and in extension, characteristics which are very different from those given by Cam-clay models.

Hight (1998) considered the anisotropy of natural clays in the analysis of the Saint-Alban test embankment (La Rochelle et al., 1974). The soft clay and its crust were modelled using the MIT E3 soil model (Whittle, 1993). Figure 2 shows the limit state curve of the Saint-Alban clay as well as the stress paths predicted by the MIT E3 model for $K_{o.nc}$ normally consolidated soil subject to undrained triaxial compression, triaxial extension and simple shear. The influence of anisotropy on strength is obvious, the triaxial compression strength being about 3 times the extension strength.

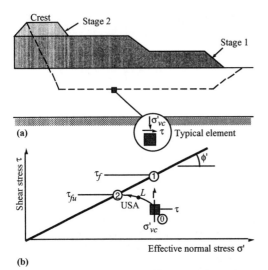

(a)

(b)

Fig. 1 - Comparison of effective stress and undrained strength analyses for evaluating stability during staged construction (after Ladd, 1991)

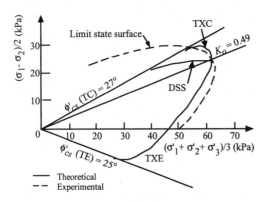

Fig. 2 - Undrained stress paths for normally consolidated Saint-Alban clay (From Hight, 1998)

Figure 3 shows strength profiles obtained experimentally and deduced numerically for triaxial and plane strain conditions for the Saint-Alban clay deposit. Below the crust that extends to a depth of 2.0 m, the soft clay has an overconsolidation ratio of 2.2 and shows strengths increasing linearly with depth.

The model predicted a failure height of 3.9 m, in fact the height at which the trial embankment failed (La Rochelle et al., 1974). Figure 4 shows the strength mobilized around the predicted failure surface compared to other available strengths along the same surface. The horizontal axis shows the distance around the failure surface, identified in the inset.

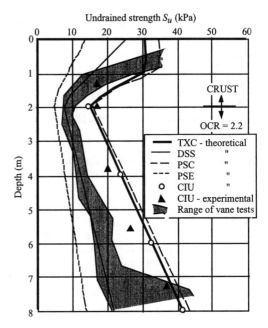

Fig. 3 - Undrained strength profiles in the soft clay deposit at Saint-Alban (From Hight, 1998)

The mobilized strength varies enormously, reducing from plane strain compression under the centre of the embankment to plane strain extension beyond the toe. The average strength mobilized at failure was found to be 1.25 times the direct simple shear strength.

Several general remarks can be made from this study:

1 The influence of anisotropy on strength mobilized along a shear surface may be important.

2 This influence, however, can be less important than indicated in Fig. 4 due to the strain softening behaviour of natural clays, particularly in compression tests, and progressive failure. It is worth noting that for the Saint-Alban test embankment and others, Trak et al. (1980) came to the conclusion that the average mobilized strength was close to the undrained strength measured at large strains.

3 It is important to remember that consolidation occurs during the early stages of construction, when the soil is overconsolidated and has a high coefficient of consolidation. This has a considerable effect on the stress paths followed, the soil becoming normally consolidated in most cases (Tavenas & Leroueil, 1980; Leroueil & Tavenas, 1986). It does not seem that this aspect has been considered by Hight (1998) but deformation methods generally can include consolidation.

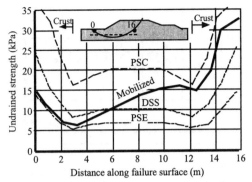

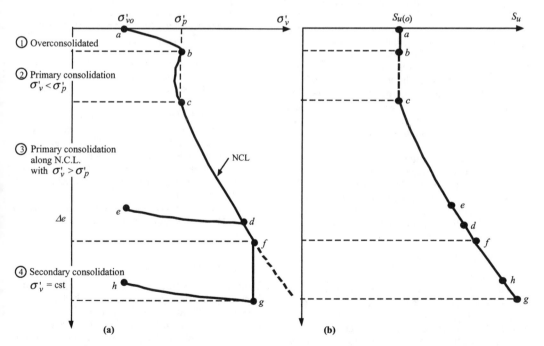

Fig. 4 - Distribution of mobilized undrained strength along failure surface for the Saint-Alban test embankment (From Hight, 1998)

Deformation methods could also be used to evaluate the evolution of deformations and stability with time, considering possible changes in loading conditions and consolidation of the clay foundation. Deformation methods certainly represent the future for analysing the behaviour of clay foundations.

However, at the present time, they have not been well calibrated against actual failures.

Considering total stress analysis (TSA) and undrained strength analysis (USA), it is convenient to refer to the vertical compression of the clay deposit to discuss strength parameters. As indicated in Fig. 5, four different domains can be defined. Domain 1, which is the overconsolidated domain in which the soil is during its first stage of construction. In some cases, an increase in pore pressure is observed after the end on construction, which results in a decrease in vertical effective stress after the passage of the preconsolidation pressure. Such a behaviour is not unusual and has been observed, in particular, by Crooks et al. (1984), Kabbaj et al. (1988) and St-Arnaud et al. (1992). This stage has been defined as Domain 2. Domain 3 is associated with vertical stresses in excess of the preconsolidation pressure, on the normal compression line (NCL). Finally, if soil foundation is in secondary consolidation, i.e. with deformations under constant vertical effective stress, it is in Domain 4. In the following sections, the undrained shear strength of clay under embankments will be examined for these 4 domains.

Fig. 5 - a) Variation of void ratio with vertical effective stress under an embankment; b) Variation of undrained shear strength with void ratio

3 DOMAIN 1: SINGLE-STAGE CONSTRUCTION

In Domain 1, while the clay foundation is in its overconsolidated range, the stability of embankments is usually estimated on the basis of one or several shear strength profiles (TSA). These profiles are undrained shear strength profiles, even if is known that there is significant consolidation during the early stages of construction. There are other factors such as strain rate (typically 10% variation per logarithm cycle of strain rate; Kulhawy & Mayne, 1990), stress axis rotation, stress path, etc. that are also different under embankments and during tests. It results that the methods used for stability analyses are to some extent empirical.

The stability analysis methods used for single-stage constructions have been described elsewhere (e.g. Leroueil & Jamiolkowski, 1991). They will be briefly described here with emphasis only on new elements.

3.1 Preconsolidation pressure

On the basis of correlations proposed by Bjerrum (1972), Mesri (1975) concluded that the average strength at failure under an embankment is equal to 0.22 times the preconsolidation pressure of the clay ($S_{u(m)} = 0.22\ \sigma'_p$). Larsson (1980) and Jardine & Hight (1987) calculated ratios between the average mobilized strength and the preconsolidation pressure for 35 embankment and foundation failures. The results are shown in Fig. 6a. It can be seen that $S_{u(m)}/\sigma'_p$ slightly increases from about 0.19 at a plasticity index of 10 to about 0.28 at a plasticity index of 80. As noted by Leroueil et al. (1985) and Ladd (1991), the upper values often correspond to organic clays. These results, which refer to strength mobilized in clay foundation at failure, can be used as a reference.

Data presented by Tanaka (1994) indicate that the mobilized strength in Japanese clays would be in the order of, or larger than $0.3\sigma'_p$, independently of the plasticity index. It can be seen from Fig. 6a that such values are larger than those gathered by Larsson (1980) and Jardine & Hight (1987).

3.2 Recompression technique

Having in mind what has been recently demonstrated by Hight (1998) and reported in Figs. 2, 3 and 4, Ladd (1969) proposed a methodology in which the undrained shear strength is obtained from a laboratory program of CK_oU shear tests having different modes of failure: compression under the central part of the embankment, direct simple shear

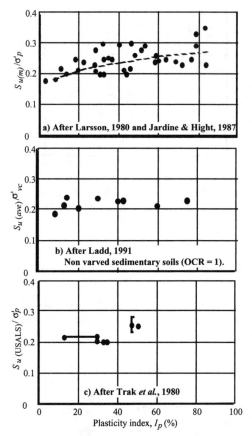

Fig. 6 - Strength ratios deduced from different approaches: a) directly deduced from failures; b) deduced from SHANSEP approach; and c) deduced from USALS

(DSS) where the failure surface is close to horizontal and extension near the toe of the embankment. Along the same lines, Bjerrum (1973) suggested the Recompression technique in which the specimens are reconsolidated to *in situ* stresses prior to shearing. While attracting since it apparently takes anisotropy into account by considering different strength profiles, this approach still suffers from not taking strain rate effects and initial consolidation into account.

3.3 SHANSEP technique

In order to minimize the effects of sampling disturbance on the strengths measured with the Recompression technique, Ladd & Foott (1974) suggested the SHANSEP technique in which the specimens are first consolidated well beyond the *in situ* preconsolidation pressure and maintained under these conditions or rebounded to varying OCRs before being subjected to undrained compression, extension and direct simple shear tests. Ladd et al. (1977) showed that the results can generally be written as follows:

$$\left(\frac{S_u}{\sigma'_{vc}}\right)_{oc} = \left(\frac{S_u}{\sigma'_{vc}}\right)_{nc} OCR^m \qquad (1)$$

where m is a parameter approximately equal to 0.8. Because of destructuration of the soil generated by loading to stresses in excess of the preconsolidation pressure, Ladd (1991) specified that "SHANSEP" is strictly applicable only to mechanically overconsolidated and truly normally consolidated soils exhibiting normalized behavior". As it is more and more accepted that most natural clays are microstructured (Mesri ,1975; Tavenas & Leroueil, 1985; Burland, 1990; Leroueil & Vaughan, 1990) the applicability of SHANSEP approach seems to be limited. It can, however, provide minimum strength values to the recompression approach when soil samples are disturbed.

An analysis of the results presented by Ladd (1991) indicates that the S_u/σ'_p ratio deduced from the SHANSEP approach is in the order of or smaller than 0.22. Indeed, Ladd (1991) obtained for normally consolidated clays the strength ratios shown in Fig. 6b, where $S_{u(ave)}$ is the average of the strengths measured in compression, extension and direct simple shear tests. This ratio is in the order of 0.22 and, according to Equation 1 and typical m values smaller than 1.0, would give even smaller S_u/σ'_p ratios for overconsolidated soils. In comparison with results presented in Fig. 6a, it seems that this approach could be conservative, which was recently confirmed on 2 sites by Hanzawa (2000).

3.4 Field vane test

With the exception of Japan and a few other countries, the field vane shear strength $S_{u(v)}$ is the most often used strength for the evaluation of the stability of embankments on clays. However, back analyses of numerous failures have indicated that this strength often overestimates the average strength mobilized in situ. For this reason, Bjerrum (1972) suggested the use of a correction factor μ. This correction factor, μ_1 in Fig.7, decreases from 1.0 to 0.6 when the plasticity index increases from 20 to 100.

3.5 Unconfined compression (UC) test

The unconfined compression test is the most usual test for defining the strength mobilized at failure under embankments in Japan (Nakase, 1967; Ohta et al., 1989). It gives reasonable results and is normally not corrected. However, analysing 24 failures of embankment foundations, Ohta et al. (1989) indicated that a correction factor (μ_2 in Fig. 7) should be applied ($S_{u(m)} = \mu_2 q_u/2$).

Tanaka (1994) compared the strengths deduced from unconfined compression tests, $q_u/2$ and field vane tests, for 7 different Japanese soft clay sites. This comparison shows that the vane shear strength is generally higher than $q_u/2$, as shown in Fig. 8a. If the correction factor suggested by Ohta et al. (1989), μ_2 in Fig. 7, is considered, the ratio of the vane shear strength over the mobilized strength in situ at failure would be approximately equal to 1.2, independently of the plasticity index (Fig. 8b). This would correspond to a correction factor for the vane shear strength of 0.83 for Japanese clays, irrespective of I_p. This result is quite different from what has been observed elsewhere (see Bjerrum's correction factor μ_1 in fig. 7).

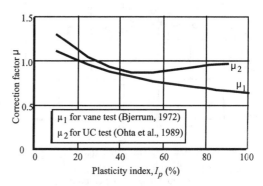

Fig. 7 - Correction factors proposed for vane tests and unconfined compression tests

3.6 Direct simple shear (DSS)

"For preliminary design and for final design of less important projects involving 'ordinary' soils with low to moderate anisotropy", Ladd (1991) recommends the use of CKoU direct simple shear (CK₀UDSS) tests results for stability analyses. For non-varved normally consolidated clays, he observed strength ratios S_u/σ'_{vc} equal to 0.225 for 16 clays plotting above the A-line in the plasticity chart and 0.26 for 9 silts and organic soils.

3.7 Direct shear tests (DST)

Hanzawa (1991, 2000) suggests the use of undrained direct shear test strength obtained at a rate of 0.25 mm/min and then multiplied by a correction factor for strain rate effect equal to 0.85. Stability analyses of 7 Asian embankments among which 5 failed confirmed the validity of this approach (Hanzawa, 2000).

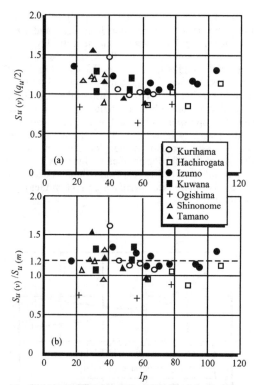

Fig. 8 - Strength ratios versus the plasticity index :
a) $S_{u\ (v)}/(q_u/2)$ from Tanaka (1994); b) $S_{u\ (v)}/S_{u\ (m)}$
where $S_{u\ (m)}$ is equal to $q_u/2$ corrected by μ_2 (Fig. 7)

3.8 USALS

Trak et al. (1980) suggested the use of the undrained strength measured at large strains in triaxial compression tests (USALS) for design purposes. On a limited number of cases, they obtained the ratios $S_{u(USALS)}/\sigma'_p$ shown in Fig. 6c. These ratios increase with the plasticity index in a manner very similar to the $S_{u(m)}/\sigma'_p$ ratios shown in Fig. 6a. More recent studies have shown however that $S_{u(USALS)}/\sigma'_p$ can also be smaller than 0.20, indicating that this approach could underestimate *in situ* strength.

3.9 CPT or CPTU tip resistance

The tip resistance measured in cone penetration test is related to the undrained shear strength of the clay as follows:

$$q_T = \sigma_{vo} + N_{kT}\,S_u \qquad (2)$$

where q_T is the tip resistance corrected for the pore pressure acting behind the tip, σ_{vo} is the overburden pressure and N_{kT} is the cone factor. Because of the advantages of CPTU test, in particular the fact that it

gives a continuous soil profile, it is thought that it could be directly used for the evaluation of a representative field strength profile. However, the piezocone has not been well calibrated against failures yet and is generally used in combination with other methods (e.g. Hanzawa, 2000).

3.10 Conclusion

Several conclusions can be deduced from previous considerations on stability of embankments constructed in one stage:
- There are several methods, all based on undrained shear strength profiles, that are available. As clay behaviour during the early stages of construction is not perfectly undrained and each type of test is associated with a stress path and strain rate that are different from those existing under embankments, all these design methods have to be considered as semi-empirical.
- All the proposed methods seem to give reliable results, at least regionally. There are indications, however, that the SHANSEP and USALS methods could be conservative.
- The Japanese experience (Tanaka, 1994), which seems at variance with experience gained elsewhere, indicates that empirical correlations may not be applicable in geological contexts different from the one where they have been developed. It also indicates that the plasticity index may be not a key parameter when referring to the mechanical behaviour of clays.
- Considering that several methods are available, it is recommended to consider at least 2 different approaches and to take into account local experience for stability analyses. If they give similar results, one can be confident; if it is not the case, one has to be more careful.

4 DOMAIN 2

Domain 2 corresponds to development of creep and possible microstructure collapse after passage of the preconsolidation pressure (Leroueil, 1996). This may result in a temporary reduction in effective stress, as shown in Fig. 5a. However, there is no evidence of a decrease in strength. This can, at least partly, be explained on the basis of laboratory observations schematically presented in Fig. 9. When a clay specimen is consolidated in its normally consolidated range to stress conditions such as A in Fig. 9 and then subjected to undrained shear, excess pore pressures develop and failure is reached at point F. If a second specimen is also consolidated under stress conditions at point A and then, drainage closed, let under the same total stresses during several days or weeks, due to creep, there is an increase in pore pressure, and a decrease in effective stress to a point such as B. When this specimen, at the same void ratio as the previous one, is sheared in undrained conditions, the observed effective stress path is initially similar to that obtained with an overconsolidated soil. Then, it reaches and follows the stress path observed with the first specimen up to failure at point F. This means that, even with different pre-shear vertical effective stresses and thus factors of safety defined in term of effective stress (τ_{efA}/τ_m or τ_{efB}/τ_m in Fig. 9), two specimens having the same void ratio have the same undrained shear strength. As a consequence, the undrained shear strength for Domain 2, from b to c in Fig. 5b, is considered constant and equal to the undrained shear strength of the intact soil, $S_{u(o)}$ in Domain 1.

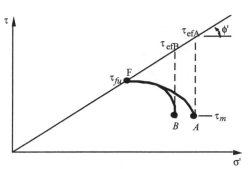

Fig. 9 - Effective and undrained strengths in a normally consolidated clay

5 DOMAINS 3 AND 4

Domain 3 represents the usual situation of clay foundations during staged construction. The soil under the embankment is then normally consolidated and practically undrained during further loading. One difficulty comes from the fact that, contrary to embankments constructed in one stage for which the proposed methods have been calibrated against failures, there is practically no well-documented failure of stage-constructed embankments. As a consequence, the proposed methods generally are extensions of methods used for embankments constructed in one stage.

Two types of stability analysis methods are used for staged construction: Undrained strength analyses and Total stress analysis.

5.1 *Undrained strength analyses (USA)*

In undrained strength analyses, the undrained shear strength S_u is defined as a fraction of the pre-shear vertical effective stress, as follows:

$$S_u = \alpha \sigma'_{vc} \tag{3}$$

where σ'_{vc} is the vertical effective stress under the embankment and α is a factor that depends on the soil and local mode of failure that can be compression, extension or simple shear in SHANSEP approach. Leroueil et al. (1985) suggested the use of a strength profile established with $\alpha = 0.25$ in Eq. 3.

One difficulty with USA is the uncertainty associated with the determination of the vertical effective stress profiles under an embankment that has settled significantly and where arching phenomena may develop. Parry (1954, 1972) showed that the vertical stress applied under the centre of a triangular fill with a width L, a maximum height H and a unit weight γ decreases from 0.84 γH when there is no settlement to 0. 55 γH and 0.26 γH when the settlement increases to 0.0043L and 0.0225L respectively (Fig. 10). This phenomenon can also be illustrated by observations made by Tavenas et al. (1978). During staged construction of the Rang Saint-Georges embankment in Québec, observed pore pressure increases were as shown in Fig. 11. While during the first stages of construction, r_u (= $\Delta u/\Delta \gamma H$) was at its maximum under the centre of the embankment and very close to 1.0, after about 3 years and settlements in the order of 1.5 m, r_u became much smaller under the centre and higher under the slopes and the berms of the embankment, confirming the effect of arching.

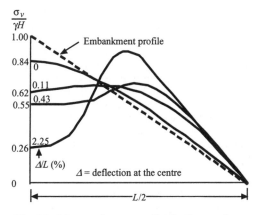

Fig. 10 - Measured pressure distributions under a model embankment (after Parry, 1954, 1972)

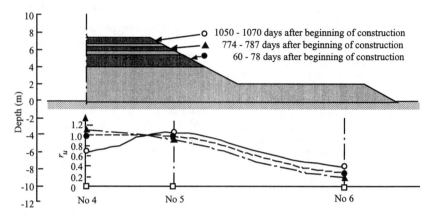

Fig. 11- Distribution of the r_u coefficient at various stages of construction of the Rang Saint- Georges fill (from Tavenas *et al.*, 1978)

There are also aspects of stress axis rotation that have been neglected, even in SHANSEP and Re-compression approaches. Ladd & Edgers (1972), Ladd (1991) and De Groot & Finocchio (1995) compared the undrained shear strength measured in conventional direct simple shear (CK$_o$UDSS) test and the undrained shear strength measured in direct simple shear test in which a shear stress $\tau_{hc(CAU)}$ is applied to the specimen during consolidation, prior to undrained shear (CAUDSS). Figure 12 shows the ratio of the undrained shear strength measured in both tests per-formed on normally consolidated clays, $S_{u(CAU)}/S_{u(CKoU)}$, against the shear stress ap-plied during consolidation, $\tau_{hc(CAU)}$, normalized with respect to $S_{u(CKoU)}$. It can be seen that the measured strength typically increases by 20% when the con-solidation shear stress ratio, $\tau_{hc(CAU)}/S_{u(CKoU)}$ in-creases to 0.85. Jardine et al. (1997) confirmed the effect of shear stress applied during consolidation on shear strength on the basis of tests performed with hollow cylinder apparatus. It thus appears that, for a soil element such as the one shown in Fig. 1, the undrained shear strength is larger than the one ob-tained from conventional DSS tests in which speci-mens are consolidated without any shear stress ap-plied. An aspect also evidenced by Ladd (1991) is that application of τ_{hc} during consolidation induces an increase of the strain softening behaviour of the soil.

5.2 Total stress analyses (TSA)

TSA approaches are generally based on strength pro-files obtained by in situ testing. The tests considered here are the vane test and the piezocone.

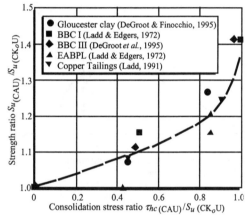

Fig. 12 - CAUDSS strength ratio vs consolidation stress ratio for different soils (after DeGroot & Finocchio, 1995)

Vane test. From observations made by Tavenas et al. (1978) and by Law (1985), it appears that the ra-tio S_u/σ'_v obtained under embankments is smaller than the ratio S_u/σ'_p obtained in the same clay, in-tact. Tavenas et al. (1978) associated this fact to changes in microstructure when the soil becomes normally consolidated and suggested not to correct the vane shear strength measured under embank-ments. Observations made by Demers (2000), (see next paragraphs), tend to confirm this approach.

Piezocone. As $q_T = \sigma_v + N_{kT} S_u$, ($q_T - \sigma_v$) can be used for the evaluation of the undrained shear strength. In this paper, N_{kT} is used for the intact soil and N_{kT}^* is used for the same soil under embank-ments and both are defined in reference with the vane shear strength, uncorrected.

Demers (2000) performed vane tests and piezocone tests outside and under embankments situated on several sites from the Province of Quebec. Figure 13 indicates the strength increase as evaluated with the field vane and the piezocone at Saint-Hyacinthe. Both tests show a clear strength increase. Figure 14 presents data obtained for the embankment C at Saint-Alban. Figure 14a shows the initial preconsolidation pressure profile and the final vertical effective stress profile. It can be seen that the preconsolidation pressure has been exceeded only up to a

depth of about 7 m. Accordingly, the piezocone shows an increase in ($q_T - \sigma_v$) at depths where the soil has become normally consolidated (in Domain 3) and no increase at depths where the soil remained in its overconsolidated range (Domain 1), (Fig. 14b). Demers (2000) determined N_{kT} and N_{kT}^* for 7 embankments situated on 4 different sites. He systematically observed N_{kT}^* values larger than the corresponding N_{kT} values. The ratio of these cone factors is shown against I_p in Fig. 15. It can be seen that this ratio is smaller than 1.0 and very close to the line showing Bjerrum's correction factor (μ_1 in fig. 7) as a function of the plasticity index. This is in agreement with the idea of correcting the vane shear strength of the intact clay for single-stage constructions and not correcting the vane shear strength for staged constructions, while the clay is normally consolidated and destructured. This also indicates that strength increases under embankments can be directly estimated from the change in net tip resistance ($q_T - \sigma_v$).

Figure 16 shows the strength increase evaluated from the net tip resistance [($q_T - \sigma_v$)/($q_{To} - \sigma_{vo}$) − 1, %] against the increase in effective vertical stress in excess of the preconsolidation pressure [$\sigma'_v/\sigma'_p - 1$, %]. There is a good agreement between the two strength increases for soil conditions in Domain 3 (as indicated by the numbers). On the other hand, it can be seen that for soil conditions indicated in Domain 4, the strength increase deduced from piezocone test is larger than the one deduced from the increase in effective stress. This can be explained as follows: During secondary consolidation, the verti-

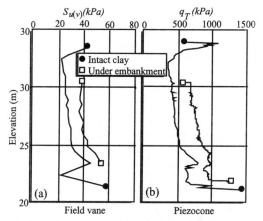

Fig. 13 - Vane strength (a) and net tip (b) resistance increase under an embankment at Saint-Hyacinthe (from Demers, 2000)

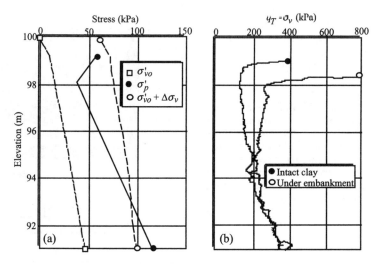

Fig. 14 - Vertical effective stress (a) and net tip resistance (b) under an embankment at Saint-Alban (from Demers, 2000)

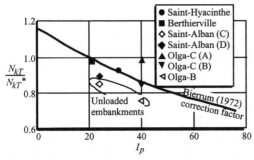

Fig. 15 - Ratio of N_{kT} determined on intact clay and $N_{kT}*$ determined under embankment versus I_p (from Demers, 2000)

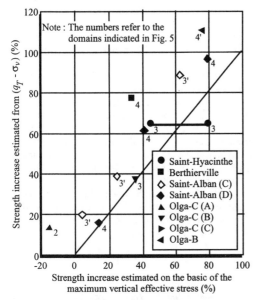

Fig. 16 - Strength increase estimated from the piezocone net tip resistance vs strength increase estimated on the basis of the maximum vertical effective stress (from Demers, 2000)

cal effective stress is constant (from f to g in Fig. 5a) but, as void ratio decreases, the undrained shear strength increases as indicated in Fig. 5b. Part of the strength increase could also be due to restructuration of the clay as the vertical strain rate becomes small (Leroueil & Marques, 1996). In Domain 4, the strength increase deduced from piezocone tests thus seems more representative than the one estimated on the basis of vertical effective stress.

5.3 Olga test embankments.

Several of the methods previously described have been considered on the site Olga, about 1000 km

north of Montreal. The Olga-B test embankment was built in 1971 and dismantled in 1990. The fill material was then used to built the Olga-C test embankment. This latter embankment has several sections, Section A without vertical drains and Sections B, C and D with vertical drains with different spacing.

These 2 embankments are shown in Fig. 17. Investigations were performed in 1977 and 1991. In the layers of interest, the vertical strains had the following values: 18% and 20% under the Olga-B embankment; 6, 13 and 16.5% under the Olga-C embankment.

Strength increases were evaluated with different approaches, and the results are summarized in Fig. 18:

- Figure 18a shows the strength increase defined as the increase in vertical effective stress in excess of the preconsolidation pressure.
- With the piezocone, the strength increase was directly defined on the basis of $(q_T - \sigma_v)$. Results are shown in Fig. 18b.
- With the vane shear test, the reference strength was the vane shear strength measured on the intact clay and corrected with Bjerrum's correction factor (μ_1 in Fig. 7). The vane shear strength measured under embankments was not corrected, in agreement with Canadian practice previously described. The results are shown in Fig. 18c.

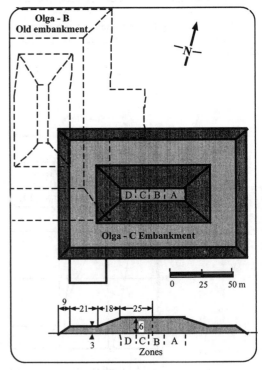

Fig. 17 - Olga test embankments (after St-Arnaud et al., 1992)

94

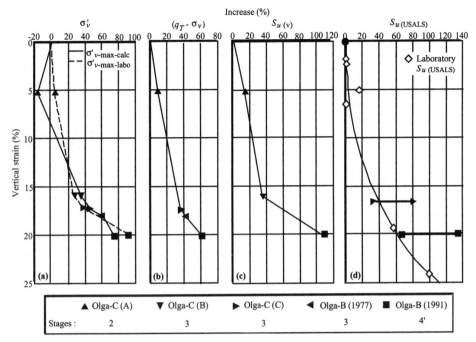

Fig 18 - Strength increase under Olga test embankments

- Strength increase was also evaluated on the basis of the large strain strength measured in undrained compression tests, $S_{u(USALS)}$. The results are shown in Fig. 18d. The open diamonds correspond to results obtained on intact clay specimens compressed to different vertical strains and void ratios before being subjected to triaxial undrained compression test. The black triangles and squares were obtained on samples taken under the Olga-B and Olga-C embankments and reconsolidated to stresses close to in situ values in triaxial cells before being subjected to an undrained compression test.

Several remarks can be made on the basis of the results shown in Fig. 18:
- The first one is that the different methods give similar results.
- The second one is that Olga-B in 1991, which had been in secondary consolidation during about 10 years, has a tendency to show strength increases slightly larger than those indicated by the general trend. This could be associated with some progressive restructuration of the clay with time (aging).
- The third one is that the strength increase follows the compression curve. For example, on Olga site, we need a vertical strain of about 20% to get 60% strength increase. This also indicates that strength increase could be directly estimated from

the compression curve at the design stage or from settlements during the construction and consolidation stages.

5.4 Conclusion

Several approaches exist to evaluate strength increase under embankments constructed in several stages, and they apparently give safe results. Several remarks have, however, to be made:
- With the exception of the SHANSEP or Recompression method, strength increases are evaluated with reference to "average shear strength". They thus underestimate the strength increase available under the central part of an embankment, which corresponds to a compression mode, and should thus be corrected accordingly.
- As previously indicated, consolidation with shear stress applied during consolidation induces an increase in strength that is neglected in the proposed methods.

It seems to result from these 2 remarks that present stability analyses tend to underestimate the strength that can be mobilized under embankments built in several stages. This could explain why failures of embankments constructed in several stages are seldom observed.

6 STRENGTH OF THE WEATHERED CRUST

Because it is generally weathered and fissured, the mobilized strength of the surface crust of clay deposits is smaller than that measured in vane tests. It is also generally considered that this strength is larger than the minimum strength in the underlying intact clay. Tavenas & Leroueil (1980) and Lefebvre et al. (1987) suggested that the operational strength in the crust is given as shown in Fig. 19.

7 LONG-TERM STABILITY OF EMBANKMENTS ON SOFT CLAY

7.1 Long term stability

On land, there is no special difficulty to estimate the factor of safety, except that, for different reasons such as a rise of the water table into the embankment, there could be some excess pore pressures in comparison with an assumed water table at the base of the embankment. This aspect has to be considered in stability analyses. These analyses are generally performed in effective stress.

7.2 Partly submerged dike subjected to tide

A particular case of interest is for dikes partly submerged and subjected to tide. If the soil foundation has high hydraulic conductivity and coefficient of consolidation (e.g. sand with a c_v value of 10^{-1} m^2/s), pore pressures in the soil deposit are hydrostatic and follow sea level.

If the foundation has low hydraulic conductivity and coefficient of consolidation (e.g. clay with a c_v value of 10^{-8} m^2/s), the soil cannot change its void ratio during a tidal cycle, typically 12 hours and 24 hours in the Atlantic and Pacific Oceans respectively. As a consequence, the vertical effective stress remains essentially constant and the change in pore pressure is equal to the change in total stress.

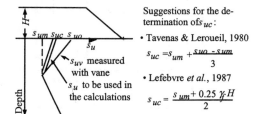

Suggestions for the determination of s_{uc}:

• Tavenas & Leroueil, 1980

$$s_{uc} = s_{um} + \frac{s_{uo} - s_{um}}{3}$$

• Lefebvre et al., 1987

$$s_{uc} = \frac{s_{um} + 0.25\, \gamma_c H}{2}$$

Fig. 19 - Correction to obtain operational strength of a crust beneath an embankment.

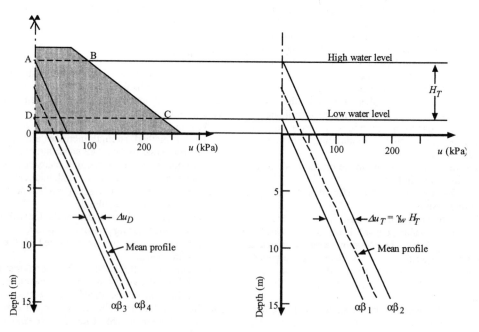

Fig. 20 - Pore pressure variations under a dike on a low c_v clay deposit, subjected to cyclic variation of the water level

Outside the zone influenced by the dike, the change in total stress is equal to the change in water level times the unit weight of water, γ_w, and water pressures in the soil deposit are hydrostatic and follow the see level. As indicated on the right side of Fig. 20, if see level changes by H_T, the change in pore pressure in the entire deposit is $\Delta u_T = \gamma_w H_T$.

Under the dike, the change in total stress associated to a change in water level is not equal to Δu_T but equal to Δu_D defined as follows:

$$\Delta u_D = \Delta u_T \left[1 - 2 \left(\frac{\gamma_w + \gamma_f - \gamma_{fsat}}{\gamma_w} \right) I_{ABCD} \right] \qquad [4]$$

where γ_f and γ_{fsat} are respectively the unit weights of the fill in drained and saturated conditions, and I_{ABCD} is the influence factor at the considered depth of the volume of fill (ABCD in Fig. 20) in between the high and low water levels. In fact, for symmetry reason, the influence factor for both sides of the fill is $2 I_{ABCD}$.

If it is assumed that the area is subjected to regular cyclic variations of the see level, the mean pore pressure profile corresponds to the mean water level, as shown by the dashed lines in Fig. 20.

In term of stability, worse conditions are reached at low water level. At that time, the water profile below the dike is $\alpha\beta_3$ with pore pressures higher than those corresponding to the profile $\alpha\beta_1$ outside the dike.

8 CONCLUSION

This paper reviews the different methods which are used for the evaluation of stability of embankments on soft clays.

For single-stage constructions, there have been many embankment failures that have been back-analysed with reference to different strength profiles. Depending on local experience and people involved, the strength profiles can be determined in different ways. However, stability of embankments is now well controlled and failures are seldom observed. It is worth mentioning that experience recently gained in Japan (Tanaka, 1994) evidences that: first, we have to be careful when we want to use empirical correlations that have been developed in a given geological context in a different one; second, plasticity index may be not a reliable indicator of the mechanical characteristics of clays.

For stage-constructed embankments, practically no well documented failures have been reported. As a result, the methods used, which generally are extensions of methods developed for embankments constructed in one stage, are not well calibrated but seem to give similar results. There are however indications that there is a general trend for these methods to underestimate the stability of embankments in these conditions.

REFERENCES

Bjerrum, L. 1972. Embankments on soft ground. Proc. ASCE Specialty Conf. on Earth and Earth-Supported Structures, Purdue University, Vol. 2: 1-54.

Bjerrum, L. 1973. Problems of soil mechanics and construction on soft clays and structurally unstable soils (collapsible, expansive and others). Proc. 8th Int. Conf. on Soil Mechanics and Foundation Engineering, Moscow, Vol. 3:111-159.

Burland, J.B. 1990. On the compressibility and shear strength of natural clays. Géotechnique, 40(3): 329-378.

Crooks, J.H.A., Becker, D.E., Jefferies, M.G. & McKenzie, K. 1984. Yield behaviour and consolidation – 1: pore pressure response. Proc. ASCE Symp. on Sedimentation Consolidation models: Prediction and validation, ASCE, New York, N.Y., 356-381.

DeGroot, D.J. & Finocchio, D.W. 1995. Anisotropically consolidated direct simple shear behavior of a destructured sensitive clay. 48th Canadian Geotechnical Conf., Vancouver, Vol. 2: 599-606.

Demers, D. 2000. Le piézocône: Outil d'investigation et de dimensionnement dans les sosl argileux. Ph.D. Thesis in preparation, Université Laval, Québec.

Diaz-Rodriguez, J.A., Leroueil, S. & Aleman, J.D. 1992. Yielding of Mexico City clay and other natural clays. J. Geotechnical Engineering Division, ASCE, 118 (GT7): 981-995.

Hanzawa, H. 1991. A new approach to determine the shear strength of soft clay. Proc. Int. Conf. on Geotechnical Engineering for coastal development – theory and practice of soft ground – Geo-Coast'91, Yokohama, Vol.1: 23-28.

Hanzawa, H. 2000. Use of direct shear and cone penetration tests in soft ground engineering. Proc. Int. Symp. on Coastal Geotechnical Engineering in Practice (IS-Yokohama), Yokohama, Vol.2, in press.

Hight, D.W. 1998. Anisotropy in soils – Its measurement and practical implications. 2nd GRC Lecture, Singapore. Also in the 38th Rankine Lecture presented in London on 18 March 1998.

Jardine, R.J. & Hight, D.W. 1987. The behaviour and analysis of embankments on soft clays. Embankments on soft clays. Bulletin of the Public Work Research Center, Athens, pp. 159-244.

Jardine, R.J., Zdravkovic, L. & Porovic, E. 1997. Panel contribution: Anisotropic consolidation including principal stress axis rotation: Experiments, results and practical implications. Proc. 14th Int. Conf. On soil Mechanics and Foundation Engineering, Hamburg, Vol. 4, pp. 2165-2168.

Kabbaj, M., Tavenas, F. & Leroueil, S. 1988. In situ and laboratory stress-strain relations. Géotechnique, 38(1), 83-100.

Kulhawy, F.H. & Mayne, P.W. 1990. Manual of estimating soil properties for foundation design. Geotechnical Engineering Group, Cornell University, Ithaca.

Ladd, C.C. 1969. The prediction of in situ strees-strain behaviour of soft saturated clay during undrained shear. Proc. NGI Bolkesjö Symp., Norway, pp.14-20.

Ladd, CC. 1991. Stability evaluation during staged construction. J. Geotechnical Engineering Div., ASCE, 117(4): 540-615.

Ladd, C.C. & Edgers, L. 1972. Consolidated-undrained direct simple shear tests on saturated clays. Research Report R72-82, No. 284, MIT.

Ladd, C.C. & Foott, R. 1974. A new design procedure for stability of soft clays. J. Geotechnical Engineering Div., ASCE, 100(7): 763-786.

Ladd, C.C., Foott, R., Ishihara, K., Schlosser, F. & Poulos, H.G. 1977. Stress-deformation and strength characteristics: S.o.A. Report. Proc. 9th Int. Conf. on Soil Mechanics and Foundation Engineering, Tokyo, Vol. 2: 421-494.

La Rochelle, P., Trak, B., Tavenas, F. & Roy, M. 1974. Failure of a test embankment on a sensitive Champlain clay deposit. Canadian Geotechnical J., 11(1): 142-164.

Larsson, R. 1980. Undrained shear strength in stability calculation of embankments and foundations on soft clays. Canadian Geotechnical J., 17(4): 526-544.

Law, K.T. 1985. Use of field vane tests under earth structures. 11th Int. Conf. on Soil Mechanics and Foundation Engineering, San Francisco, Vol.2: 893-898.

Lefebvre, G., Paré, J.J. & Dascal, O. 1987. Undrained shear strength in the surficial weathered crust. Canadian Geotechnical J., 24(1): 23-34.

Leroueil, S. 1996. Compressibility of clays: Fundamental and practical aspects. J. Geotechnical Engineering, ASCE, 122(7): 534-543

Leroueil, S. & Jamiolkowski, M. 1991. General Report, Session 1: Exploration of soft soil and determination of design parameters. Int. Conf. on Geotechnical Engineering for Coastal Development, Geo-Coast-91, Yokohama, Vol. 2: 969-998.

Leroueil, S. & Marques, M.E.S. 1996. State of the Art on the importance of strain rate and temperature in geotechnical engineering. ASCE Convention, Washington, Geotechnical Special Publication, No. 61: 1-60.

Leroueil, S. & Tavenas, F. 1986. Discussion on "Effective stress paths and yielding in soft clays below embankments" by Folkes and Crooks. Canadian Geotechnical J., 23(3): 410-413.

Leroueil, S. & Vaughan, P.R. 1990. The general and congruent effects of structure in natural soils and weak rocks. Géotechnique, 40(3): 467-488.

Leroueil, S., Tavenas, F., Trak, B., La Rochelle, P. & Roy, M. 1978. Construction pore pressures in clay foundations under embankments. Part 1: The Saint-Alban test fills. Canadian Geotechnical J., 15(1):54-65.

Leroueil, S., Magnan, J.-P., & Tavenas, F. 1985. Remblais sur argiles molles. Lavoisier, Technique et Documentation, Paris. Also in English, Embankments on soft clays (1990), Ellis Horwood, Chichester, England.

Mesri, G. 1975. Discussion on "New design procedure for stability of soft clays. J. Geotechnical Engineering Div., 101(4): 409-412.

Nakase, A. 1967. The $\Phi = 0$ analysis of stability and unconfined compression strength. Soils and Foundations, 11(4): 433-443.

Ohta, H., Nishihara, A., Iisuka, A., Morita, Y., Fukagawa, R. & Arai, K. 1989. Unconfined compression strength of soft aged clays. Proc. 12th Int. Conf. on Soil Mechanics and Foundation Engineering, Rio de Janeiro, Vol. 1: 71-74.

Parry, R.H.G. 1954. Measurement of pore pressure distribution across the base of triangular section of granular masses. M. Sc. Thesis, Melbourne

Parry, R.H.G. 1972. Stability analysis for low embankments on soft clays. Proc. Roscoe Memorial Symposium, pp. 643-668.

St-Arnaud, G., Morel, R. & Lavallée, J.G. 1992. Comportement de la fondation argileuse traitée avec des drains synthétiques sous le remblai d'essai Olga-C. Internal Report, Hydro-Québec, Service géologie et structures, Montréal.

Tanaka, H. 1994. Vane shear strength of a Japanese clay and applicability of Bjerrum's correction factor. Soils and Foundations, Vol. 34(3): 39-48.

Tavenas, F. & Leroueil, S. 1980. The behaviour of embankments on clay foundations. Canadian Geotechnical J., 17(2): 236-260.

Tavenas, F. & Leroueil, S. 1985. Discussion to Session 2-B on Laboratory Testing. Proc. 11th Int. Conf. on Soil Mechanics and Foundation Engineering, San Francisco, Vol.5: 2693-2694.

Tavenas, F., Blanchet, R., Garneau, R. & Leroueil, S. 1978. The stability of staged-constructed embankments on soft clays. Canadian Geotechnical J., 15(2): 283-305.

Tavenas, F. Trak, B. & Leroueil, S. 1980. Remarks on the validity of stability analyses. Canadian Geotechnical J., 17(1): 61-73.

Trak, B., La Rochelle, P., Tavenas, F., Leroueil, S. & Roy, M. 1980. A new approach to the analysis of embankments on sensitive clays. Canadian Geotechnical Journal, 17(4): 526-544.

Whittle, A.J. 1993. Evaluation of a constitutive model for overconsolidated clays. Géotechnique, 43(2): 289-313.

Stability and instability: Soft clay embankment foundations and offshore continental slopes

R.J.Jardine
Department of Civil & Environmental Engineering, Imperial College of Science Technology and Medicine, London, UK

ABSTRACT: The first part of the paper reviews the stress paths imposed on soft clays by embankments and considers their influence on the undrained shear strengths mobilised in the field. The shortcomings of conventional design procedures are discussed, making a key distinction between single and multi-stage construction. New research with Hollow Cylinder Apparatus is used to highlight the foundations' anisotropic shear strength characteristics, and simplified parametric stability analyses are presented for the multi-stage case. Comments on stability monitoring by ground movement observation are offered before introducing the far less familiar, and less well understood, topic of large offshore landslides.

INTRODUCTION

Background

The Yokohama International Symposium (IS) on Coastal Geotechnical Engineering in Practice shares many themes with the successful 'Geocoast' Conference (PHRI 1991). A key topic at Geocoast '91 was the prediction of foundation stability for embankments and other structures built on clays. The proceedings contain several keynote presentations, and many papers, that discuss sampling, soil testing, analysis and full scale behaviour. This paper presents a personal view of how some of these areas have developed over the last nine years. The main focus is on assessing the stability of embankments built on soft clays. Wherever possible, the research described will be used to help explain phenomena or trends that have been observed in the field. The intention is to show how practice can benefit from the research undertaken over the past decade. A better understanding the basic mechanics should allow engineering predictions to be made far more reliably.

Scope of the paper

The paper has two main subjects. The issues considered first are:

1. Reviewing the stress paths imposed on the foundation soils and their influence on the operational shear strengths mobilised in the field.

2. The shortcomings of conventional design procedures

3. The key distinction between single and multi-stage construction

4. New research with Hollow Cylinder Apparatus' (HCA) to identify the anisotropic shear strength characteristics associated with both single stage and multi-stage cases

5. Parametric stability analyses considering the multi-stage case

6. Comments on stability monitoring by ground movement instrumentation

Following from the above, the final sections of the paper consider briefly the far less familiar, and less well understood topic of large offshore landslides. Marine geophysical surveys have identified a series of very large deepwater landslides in several continental margin areas. It is now known that such slope failures may generate tsunami waves that can have a major impact on nearby coastlines (Driscoll et al 2000). A short introduction is given to some topical cases, emphasising their scale and posing hypothetical explanations as to their causes. It is shown that these events are hard to explain in terms of conventional soil mechanics; investigating their mechanics will be one of the major challenges for coastal geotechnical engineering.

1. STABILITY OF EMBANKMENTS ON SOFT CLAY

1.1 *Single stage and multi-stage embankments*

In this paper it is assumed that large scale failures in soft clays take place under largely undrained conditions. Embankments whose construction rates, and clay consolidation characteristics, leave the undrained shear strengths (S_u values) remaining practically unchanged during construction are termed *single stage*, even though they may involve multiple filling stages. The term *multi-stage* is reserved for cases where significant strength gains take place through consolidation, either by natural means, or with the assistance of, for example, vertical drains.

1.2 *Effective stress analysis and ways of estimating operational undrained shear strengths*

An important series of papers was published on *single stage* embankments in the 1970's including those by Bjerrum (1973), Ladd and Foot (1974), Mesri (1974), Tavenas and Leroueil (1980), Hanzawa (1979) and others. As discussed later, relatively little was written about multi-stage structures, but two key points to emerge concerning *single stage* stability were:

I. Conventional effective stress approaches are non-conservative when considering undrained failures in contractant soils

II. Establishing the operational undrained shear resistances developed in the field is the key to reliable stability analysis

III. However, the operational S_u values are rate dependent and anisotropic.

IV. Natural clays can be sensitive, or brittle: their large strain S_u values can fall far below those applying at peak, so creating scope for progressive and runaway failures.

Figures 1 and 2 illustrate the first point, showing the Mohr circles of effective stress for elements of clay that have been consolidated to an initial (anisotropic) K_0 state of stress prior to undrained shearing. We focus here on the normal effective stress σ'_{no} and shear stress developed on the plane A-A' - that of maximum obliquity, on which we would expect failure to eventually occur. In conventional effective stress stability calculations we assume implicitly that the ultimate shearing resistance available on the failure surface is $c' + \sigma'_{no} \tan\phi'$, which is equivalent to

assuming that σ'_n remains unchanged on A-A' as loading proceeds to failure. The sequence of Mohr circles that matches this supposed condition is shown (assuming $c' = 0$), along with the corresponding effective stress path (plotted in t, s' coordinates). However, it is immediately clear that the projected path is quite different to that expected for a low OCR soft clay experiencing undrained loading, and leads to far higher ultimate shear resistances than would be measured: the approach is both unrealistic and unreasonable.

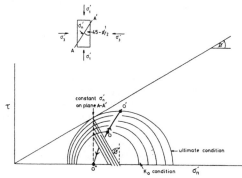

Figure 1. Effective stress conditions within a triaxial element corresponding to constant σ'_n on plane A-A'

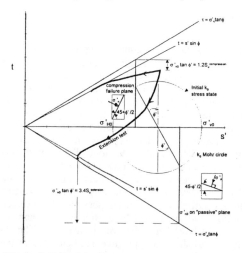

Figure 2. Effective stress paths for normally consolidated clay in compression and extension, showing also limits expected if σ'_n remained constant on planes of peak obliquity

Figure 2 shows data from undrained triaxial compression and extension tests on a typical, uncemented, relatively insensitive soft clay. The undrained shear strengths available in these two modes are significantly different, and brittleness is seen in compression. Note also that the effective stress paths pass well to the left of the curved constant σ'_n trajectory indicated in Figure 1. The value of σ'_{no} tanϕ' applying in compression over-estimates the compression test S_u by 20%, while the error in extension would be 245% (noting that the less steeply inclined failure plane identified in Figure 2 is critical for this case).

Following from the above, undrained (total stress) approaches have been retained to deal with single stage stability. Jardine and Hight (1987), Hight, Jardine and Gens (1987), Lerouiel and Jamiolkowski (1992) and others summarise the attempts made to account for the dependency of S_u on sampling disturbance, consolidation history, loading stress path, strain rates and progressive failure by means of empirically developed procedures and correction factors. Popular methods of deriving operational S_u profiles include:

I. Applying empirical correction factors (correlated to plasticity index I_p) to S_u values from field vane or CPT tests, these factors being based on back analyses of field collapses

II. Similar plots relating operational undrained shear strength to the vertical effective yield stresses, σ'_{vy} seen in oedometer tests on good quality samples

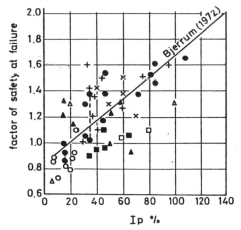

Figure 3. Data base of collapses, showing factors of safety calculated from uncorrected vane test S_u values, after Tavenas and Leroueil (1980)

Figures 3 and 4 indicate the scatter associated with both types of procedure. The vane test plot shows the factors of safety computed with uncorrected strengths against I_p, drawing from Tavenas and Leroueil's data base of collapses, while Figure 4 plots the spread of operational S_u/σ'_{vy} values interpreted from another data base by Jardine and Hight

(1987). The scatter around Bjerrum's (1973) vane correction trend line is typically $\mp$35%, while that for the oedometer method is about $\mp$15%.

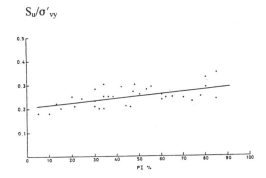

Figure 4. Reinterpretation of field collapses to show relationship between operational S_u values and oedometer yield stresses σ'_{vy}, Jardine and Hight (1987)

The central weaknesses of such approaches are: (i) the correlations with I_p are poor, (although there is scope for adopting more reliable local correlations) and (ii) they offer no scope for taking account of site specific features such as the anisotropy and sensitivity of particular clays (which vary greatly with material type and geological setting), or the geometry of the embankment and soil layering.

The soft clays' shearing properties may be better characterised by carrying out undrained compression, extension and simple shear laboratory tests on samples that have been re-consolidated anisotropically to overcome sampling disturbance. Figure 5 presents a series of trend lines established by Jardine and Hight (1987) from a data base of uncemented, relatively 'young' clays. The trend identified for embankment collapses (in Figure 4) is also plotted. It appears that the 'embankment collapse' curve passes, roughly, mid way between the active and passive plain strain trend lines. It also passes reasonably close to the mean of triaxial compression, extension and direct simple shear. But these are only approximate trends; the laboratory tests do not model the stress conditions imposed by embankments perfectly, and it is not exactly clear how the test results should be applied to obtain better reliability in practical design. However, significant developments have taken place in the last ten years that allow improvements to be made. Findings that have emerged from research by the Author and his coworkers are considered below.

S_u/σ'_{vy}

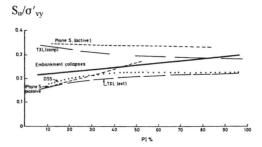

Figure 5. Trend lines for undrained strength ratios for low OCR clays considering laboratory tests and field collapses: Jardine and Hight (1987).

1.3 Sampling disturbance

Recent years have seen a new emphasis on sampling quality, particularly in the UK, Europe, Scandinavia and the USA. Prior to work by the NGI (Lacasse et al 1985), and the Bothkennar characterisation study (Hight et al 1992), there had been insufficient appreciation that many standard samplers cause far greater disturbance than the more advanced Canadian soft clay tools, or the best Japanese piston sampling techniques. The sampling issue will not be pursued here, except to note that important improvements have taken place in UK practice, and that the contributions made by the Imperial College group are included in the summaries made by Hight and Jardine (1993) and Hight (1995), (2001).

1.4 Natural structure and sensitivity

The second aspect to consider is the greater appreciation of the differences between peak and critical states; the importance of natural structure in comparison with the intrinsic behaviour of reconstituted materials (Burland 1990); and the basic weaknesses of correlations based on Ip. Alternative frameworks have been derived to account for clay sensitivity in these terms, as illustrated in Figures 6 and 7 drawing on the work of Smith (1992), Smith Jardine and Hight (1992) and Jardine and Chow (1996). The first figure contrasts the idealised oedometer behaviour of a natural sample (the Intrinsic Compression Curve, or ICC) with that of the same soil when reconstituted and compressed from slurry. The yield point of the intact sample lies in the region defined by Vaughan, Maccarini and Mohktar (1988) as 'structure permitted space' and we can define void index measures as shown in the figure. Most significantly, the ratio of the intact sample's vertical yield stress (σ'_{vy}) to that to the value of $\sigma'_v{}^*$, the vertical effective stress projected at the same void ratio on the ICC, is a direct measure of the

clay's sensitivity. The latter reflects the geological origin (including minerals and energy at time of deposition), any cementing and other post depositional experiences (including chemical weathering or leaching).

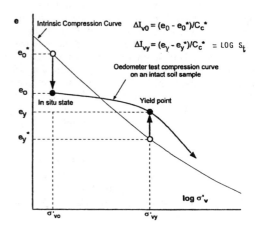

Figure 6. Intact and intrinsic oedometer curves and new sensitivity definitions, after Jardine and Chow (1996).

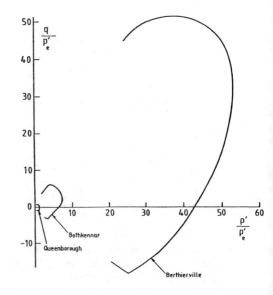

Figure 7. Normalised State Boundary Surfaces for three clays of different geological origins, after Smith (1992)

The same ideas can be applied to the more general shearing behaviour of soft clays. Figure 7 presents the State Boundary Surfaces (SBS) found by Smith (1992) from probing tests on three clays, all

102

defined in relation to their respective ICC's. This involves normalising each test point (q, p' with current void ratio e) by p'_e, the mean effective stress applying on the clay's ICC curves at the current e value. This procedure immediately draws attention to the different origins of the three clays. The outermost SBS corresponds to Berthierville clay, a sensitive glacio-lacustrine clay from eastern Canada that contains ample crystalline rock-flour and was laid down under very low energy cold-water conditions. The least extensive SBS is provided by Queenborough clay, a peaty clay from the Thames Estuary near London, which was derived from the erosion of highly weathered mud-rocks and was laid down in relatively energetic temperate tidal conditions and subjected to aerial exposure and plant action. In between is the Bothkennar clay which also contains ample rock-flour and was deposited in shallow marine/estuarine conditions, where the energy levels fell between those described for Berthierville and Queenborough. Bothkennar clay is known to be lightly bonded by organic cements (Hight et al 1992).

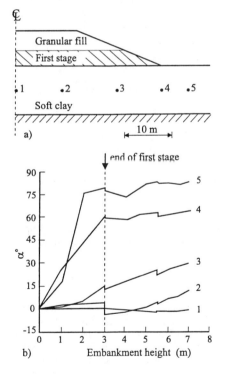

a)

b)

Figure 8. Stress system developed under embankment loading

The general pattern of behaviour observed in Smith's tests is that, when the effective stress paths (all normalised in terms of p'_e) derived from drained

radial probing tests reach the SBS's shown in Figure 7, they either (in the case of sensitive soils) turn around and describe new SBSs' that shrink inwards towards the plot's origin as strains accumulate, or remain static - as in the case of the insensitive Queenborough clay. Such tests provide the basic building blocks for new models that capture more realistically the basic behaviour of sensitive natural clays.

1.5 Understanding the stress conditions imposed by embankments

One of the most important developments over the past decade has been the widespread use of numerical methods to overcome the severe limitations of conventional limit equilibrium analysis. The latter implicity assume rigid plastic (usually isotropic) shearing behaviour, provide no information prior to collapse and do not, in any rigorous sense, correspond to predictive stress analyses. Non-linear elastic plastic Finite Element analysis provides more powerful insights, particularly when combined with realistic soil models and a coupled consolidation formulation. Figure 8 shows one example given by Jardine and Smith (1991) in which an idealised version of the Queenborough embankment (Nicholson and Jardine 1981) was considered using the programme ICFEP written by Professor David Potts at Imperial College. The soil foundation clays were idealised using the Modified Cam clay model. The key points to note are:

I. The stress conditions developed at points 1 to 5 in the foundation show a systematic tendency for the axis of the major principal stress σ'_1 to rotate from the vertical by an angle α during single stage loading. Points on the centre-line showed α close to zero, while those near the toe developed rotations of up to 90°.

II. Plane strain conditions applied. Analysis of the variations of σ'_2 during loading showed that the intermediate principal effective stress parameter b varied from 0 to 1.0, depending on the initial OCR and loading conditions, tending to values of around 0.5 when failure was approached.

In this case b is defined as:

$$b = (\sigma'_2 - \sigma'_3)/(\sigma'_1 - \sigma'_3)$$

Soil models may now be adopted for use in codes such as ICFEP that can address, to some extent, the potential sensitivity and anisotropy of natural soft clays. The MIT-E3 model described by Whittle (1993) is one example. New insights can be obtained into the behaviour of embankment foundations though numerical models that incorporate these features. For example, Hight (2001) reports a numerical analysis of the St Alban embankment failure in which the soft clays were represented by MIT-E3, with the anisotropic input parameters being developed from site-specific advanced laboratory tests. He found that the FEM approach led to a satisfyingly accurate prediction for the field collapse. However, he also showed that the S_u values mobilised in the failure of the trapezoidally shaped embankment would not apply as well to other embankment shapes (specifically one involving a loading berm). The ratio between the operational strengths and those developed in, for example, direct simple shear laboratory tests, would inevitably vary with soil characteristics and embankment geometry.

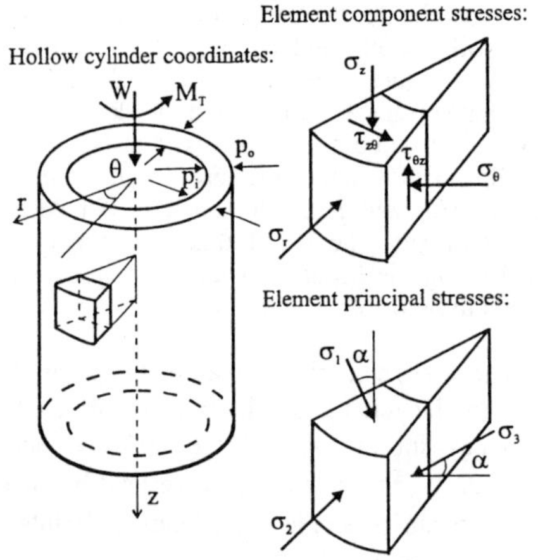

may all be changed independently and the axis of σ'_1 may be rotated as required in a vertical plane. The resulting strains may be monitored by suites of local strain sensors; Hight et al (1983). Figures 10 and 11 show the arrangements adopted in an HCA developed recently by the Author for use with samples of outside diameters of up to 200mm, with space for additional elements, such as local strain sensors, control components, geophysical equipment or resonant column oscillators.

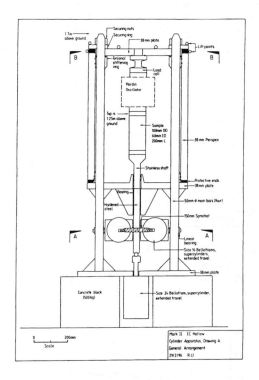

Figure 9. Simulation of embankment stress system in Hollow Cylinder tests

Figure 10. General arrangements of Mark II Imperial College Hollow Cylinder Apparatus, after Jardine 1996.

Element component stresses:

Hollow cylinder coordinates:

Element principal stresses:

1.6 Experimental characterisation of S_u anisotropy

Technological advances have also taken place in soil testing. One that has most concerned the Author is the use of Hollow Cylinder Apparatus to establish how soils respond to loading along general stress paths, such as those illustrated in Figure 8. The scheme given in Figure 9 illustrates the basic principles: a hollow specimen is subjected to combinations of axial load, inner and outer hydraulic pressures, and torque. In a well designed system these actions allow four dimensions of control, over the stress conditions within the sample's wall: σ'_1, σ'_2, and σ'_3

Equipment of this type has been used for several research projects at Imperial College. Figure 12 shows data gathered by Menkiti (1995) in HCA tests on a clay-sand (H-K), plotting data from HCA tests on K_0 consolidated samples that had been compressed to a vertical effective stress $\sigma'_v = \sigma'_c$. The samples were sheared at OCR = 1, under nominally plane strain conditions (with b kept at 0.5) and with the σ'_1 axis rotated to fixed values of α, providing a close match to the conditions developed during single stage embankment loading (see Figure 8). A trace is shown in Figure 12 indicating how the peak values of S_u/σ'_{vc}

vary with α. Also shown are the peak values from more conventional triaxial compression and extension tests, and simple shear experiments. None of these familiar tests plots precisely onto the plane strain line. We can see that the intermediate principal stress parameter b influences behaviour, although α - the orientation of the σ'_1 axis - clearly has a stronger effect on the peak strengths.

formed by Menkiti (1995) on normally consolidated H-K clay-sand. It is clear that the path involves a continuous reduction in $t = (\sigma'_1 - \sigma'_3)/2$ as the sample is sheared, and that the peak horizontal shear stress τ_{zx} is reached after a relatively small rotation of the σ'_1 axis (α = 26°).

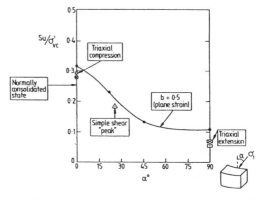

Figure 12. Variations of S_u/σ'_{vc} with α from Hollow Cylinder and other tests on a normally consolidated clay-sand; after Menkiti (1995)

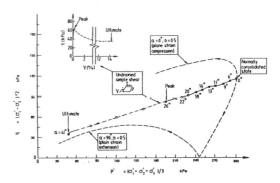

Figure 13. Effective stress path and variations in α from a Hollow Cylinder simple shear test on a normally consolidated clay-sand; after Menkiti (1995)

Figure 11. Photograph of Mark II Imperial College Hollow Cylinder Apparatus (without control systems and internal transducers etc)

Further insights may be gained into the undrained simple shear test, which is often considered appropriate for embankment stability assessments, even though the stress conditions inside simple shear samples cannot be defined fully from the limited measurements made in routine tests. The normal practice of measuring the boundary shear stress τ_{zx} and inferring the vertical effective stress σ_z' does not provide enough information to plot the Mohr circles of stresses, or the undrained effective stress paths. However simple shear tests carried out in the HCA allow the full set of effective stresses to be monitored. Figure 13 shows the effective stress path followed during an undrained test of this type per-

The pattern of anisotropy illustrated in Figures 12 and 13 is common to many normally consolidated soils; Jardine et al (1997) and Jardine and Menkiti (1999). Figure 14 presents the variations of S_u/p'_0 with α that were seen in similar undrained HCA tests on four soils (clays and sands) performed by Shibuya (1985), Menkiti (1995) and Zdravkovic (1997). HCA tests provide the crucial means by which anisotropic soil models can be tested and improved.

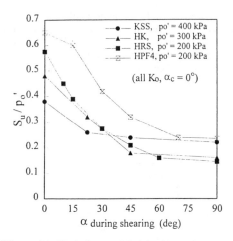

Figure 14. Variations of S_u/p' with α from `plane-strain' Hollow Cylinder tests on four normally consolidated soils; after Jardine et al (1997)

2. MULTI-STAGE EMBANKMENTS

2.1 *Introduction*

While many papers have been written about the stability of single stage embankments, whose foundations do not improve through consolidation during construction, relatively little attention has been given to multi-stage embankment stability.

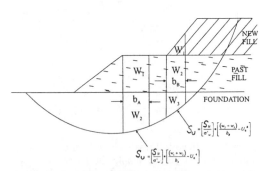

Figure 15. Consolidated undrained approach for multi-stage stability analysis

Figure 15 outlines one scheme that has been adopted for multi-stage limit equilibrium stability calculations. It is convenient to assume that the operational undrained shear strength S_u increases linearly with the vertical consolidation stress σ'_{vc}. Method of slices procedures may then to be modified to (i) distinguish between the consolidated stress conditions that apply just before or after new fill is placed (ii) take account of any excess pore water pressures that are acting prior to adding the next stage of loading and (iii) compute the value of σ'_{vc} acting at the base of each slice, and hence the appropriate S_u value. S_u may then vary from slice to slice, and between loading stages. Care is also needed to model those parts of the foundation that might remain overconsolidated throughout construction, for example the region beyond the embankment's toe, or possibly within a well-developed drying crust. The mechanics are fairly straightforward: the crucial problem is choosing operational shear strength ratios S_u/σ'_{vc} to represent the strengths associated with consolidation under the complex effective stress paths developed under embankment loading.

2.2 *Parametric multi-stage stability calculations*

Simplified calculations can illustrate the importance of S_u/σ'_{vc} in multi-stage stability analysis. In this case we consider the idealised soft clay profile shown in Figure 16 in which the initial profile of σ'_{vo} is based on ground water at ground level, with $\gamma' = 6$ kN/m^3, and a 'typical' profile of initial yield stress σ'_{vy} is specified.

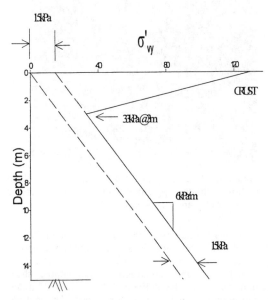

Figure 16. Vertical yield stress profile assumed for multi-stage parametric study

The profile is combined with the simple wedge limit equilibrium approach outlined in Figure 17 to consider a trapezoidal embankment made from sand.

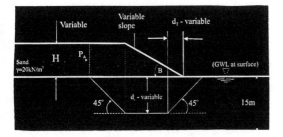

Figure 17. Simplified limit equilibrium mechanism for illustrative study

The lateral thrust produced by the fill is calculated from Rankine earth pressure theory, with the fill's unit weight and K_A value taken as $20kN/m^3$ and 0.33 respectively. The approximate 'consolidated-undrained' stability is assessed by updating the undrained shear strengths during construction according to a series of fixed S_u/σ'_{vc} ratios, starting with the initial profile given in Figure 16. The non-uniform excess pore water pressure regime is modelled by adopting a water table set at ground level, plus a variable degree of excess pressure specified by an average parameter R_u^*:

$$R_u^* = u^*/\gamma_{fill}\, h$$

Where h is the height of fill above any particular point and the excess pore pressures u^* is that anticipated at the point. For convenience u^* is defined here in terms of the full excess pressures acting just after a small increment of fill ΔH has been placed. The value of R_u^* is assumed to be constant across whole loaded area; no excess pore pressure is considered beyond the toe. The minimum factor of safety is assessed by considering a range of depths d_c for the basal surface and locations d_T for the toe wedge boundary. Around 100 combinations need to be assessed for each cross-section.

The above approach can be used to study the inter-relationships between the Factor of Safety (F_T), the assumed S_u/σ'_{vc} ratio, the fill slope angle, β, and the centre-line height, H. A total of 200 cases was considered to generate the plots shown in Figures 18 a) to d). In each instance the depth of the base wedge was varied and the position of the toe wedge changed incrementally to locate the critical failure mechanism. The charts show curves of H against cotan β for fixed values of R_u^*. Separate plots are given for each value of $S_u/(\sigma'_{vc}F_T)$.

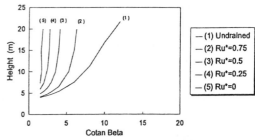

(a) $S_u/[\sigma'_{vc}/F_T] = 0.30$

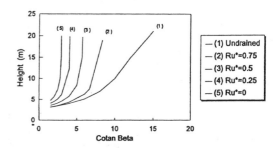

(b) $S_u/[\sigma'_{vc}/F_T] = 0.24$

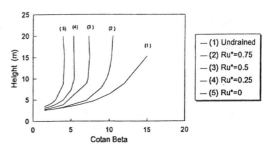

(c) $S_u/[\sigma'_{vc}/F_T] = 0.20$

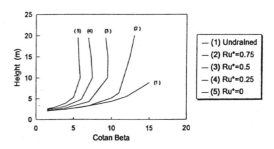

(d) $S_u/[\sigma'_{vc}/F_T] = 0.17$

Figure 18 a), b), c) and d). Results from simplified parametric multi-stage analysis.

Some key points are:

1. During the first few stages of loading stability is relatively insensitive to the side slope angle and to $S_u/(\sigma'_{vc} F_T)$.

2. At greater heights the results become highly sensitive to $S_u/(\sigma'_{vc} F_T)$.

3. When the operational $S_u/(\sigma'_{vc} /F_T)$ drops below 0.2, very flat side slopes are required to construct high embankments using a consolidated undrained multi-stage procedure.

4. If the allowable $S_u/(\sigma'_{vc} F_T)$ exceeds ≈ 0.25 then multi-stage structures can be raised to almost any height (on flat ground) using side slopes between 1:3 and 1:2.5, provided the pore pressure ratio R_u^* can be kept well below 0.5.

5. The results are also very sensitive to R_u^* once the embankment height starts to grow above the safe single-stage construction height.

However, the central difficulty remains: what value should we take for S_u/σ'_{vc} under multi-stage conditions?

2.3 Hollow Cylinder Tests to simulate consolidation involving principal stress rotation

The finite element results illustrated in Figure 8 came from an analysis in which single and multi-stage conditions were modelled using the isotropic Modified Cam Clay (MCC) model. It was found that points within the foundation consolidated under a range of conditions. The most important area for stability analysis is that under the side slopes (point 3 in Figure 9) where the σ'_1 axis gradually rotated from $\alpha = 0$ before loading, to 10 to 20° at the end of the first undrained stage, and then on to 20 to 40° at the end of multi-stage construction.

The isotropic MCC calculations indicated that it would be safe to design on the basis of the S_u/σ'_{vc} ratios that apply to single stage construction, taking the nominal effective overburden pressure as being equal to σ'_{vc}, as in Figure 16. However, the MCC model does not capture any aspects of the strong anisotropy that is often seen in undrained shearing, and HCA tests are required to explore how clays respond after being consolidated in a region where the principal stress axes are rotating.

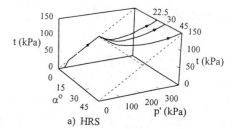

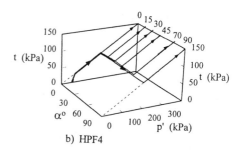

Figure 19. HCA tests to assess effects on S_u/σ'_{vc} ratios of consolidation involving principal stress axis rotation. After Jardine et al (1997)

Porovic (1985) and Zdravkovic (1997) have performed such HCA tests on samples of sand and silt. The samples were first consolidated to $p' = 200$ kPa under K_0 conditions and then, as shown in Figure 19, compressed further to $p' = 400$ kPa while keeping the same t/p' ratio, but also rotating α (in Porovic's tests on Ham River sand), or after carrying out a drained α rotation stage (in Zdravkovic's tests on HPF4 silt). After reaching their final consolidation stress points, each sample was sheared to failure under undrained conditions, with $b = 0.5$, while maintaining α constant at the value applied during the earlier fully drained consolidation process. These tests are termed $\alpha = \alpha_c$ tests.

The results of the HCA tests are summarised in Figures 20 and 21. Figure 20 contrasts the ratios of S_u/σ'_{vc} developed at various α values after either $\alpha = \alpha_c$, or K_0, consolidation. The strengths developed under $\alpha = \alpha_c$ conditions are far higher than those available after K_0 consolidation, a feature that would greatly aid the stability of a multi-stage embankment. Considering a mid-slope location such as point 3 in Figure 8, where α climbs to around 30°, the peak undrained strength ratios rise by more than 50%.

However, two further factors that need to be considered are:

- The gains S_u/σ'_{vc} depend on the number of filling stages and the degree of drainage. The HCA tests involved full drainage and infinitely small loading steps. Such perfection is impossible to achieve in practical construction.

- The gains in strength are also associated with a switch towards more brittle behaviour, as shown in Figure 21 by data from Porovic's tests. Although multi-stage structures develop relatively high S_u/σ'_{vc} ratios, they may be more prone to runaway failure modes than single stage structures.

Jardine et al (1997) and Zdravkovic and Jardine (2001) present more information on the Hollow Cylinder tests, describing the way that anisotropy is modified by consolidation involving α rotation. The general conclusion is that reducing the size of the loading steps and ensuring a high a degree of consolidation between stages offers scope for surprisingly high S_u/σ'_{vc} ratios to develop.

2.4 Multi-stage failure at Fair Haven, Vermont USA

Reports of failures among multi-stage embankments are rare, perhaps as a consequence of their relatively high S_u/σ'_{vc} ratios, or their more closely observed performance. However, Haupt and Olson

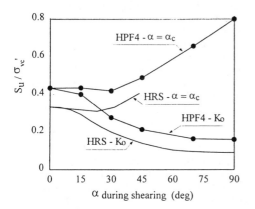

Figure 20. Differences between S_u/σ'_{vc} ratios developed in undrained HCA tests on sand and silt after K_0 consolidation and α rotation paths shown in Figure 19; after Jardine et al (1997)

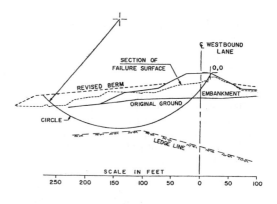

Figure 22. Multi-stage failure at New Haven; after Haupt and Olson (1972)

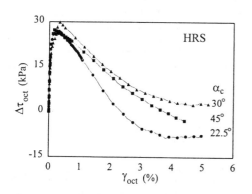

Figure 21. Brittle response seen in Porovic's HCA tests on Ham River sand in undrained HCA tests following α rotation paths shown in Figure 19; after Jardine et al (1997)

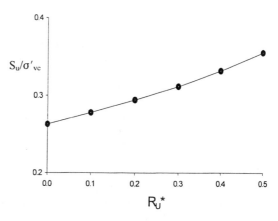

Figure 23. Relationship between pore pressure ratio and S_u/σ'_{vc} from back analysis of New Hampton shown in Figure 22.

(1972) report a collapse that involved a 15m high embankment built on varved clay-silt at Fair Haven, Vermont, USA. The Fair Haven case history, which is illustrated in Figure 22, has been re-analysed by the Author. Making use of all the available data, and noting that the piezometer records indicate an average R_u^* value greater than 0.35 shortly before failure, back-analysis indicates that the operational S_u/σ'_{vc} ratio must have exceeded 0.32, as shown on Figure 23. Single stage failures on similar `lean' and laminated strata are normally expected to develop ratios around ~ 0.2 (see Figure 4 and Jardine and Hight 1987), and it is re-assuring that the difference between the multi-stage and single-stage S_u/σ'_{vc} ratios is within the range anticipated from the HCA tests.

3. ASSESSING STABILITY FROM GROUND MOVEMENTS

3.1 Introduction

It is clear from the earlier sections that routine stability calculations may be subject to considerable error. One way of avoiding failures without adopting overly conservative designs is to monitor construction for signs of instability. The key points to note at the outset are:

1. Ground movements due to volume changes can be large even when stability is adequate.

2. Instability is primarily linked to lateral spreading.
3. It is useful to develop non-dimensional monitoring criteria that can set performance limits to movements, strains, and their rates.

4. The criteria applying to multi-stage and single stage cases can be quite different.

5. An observational approach is only valid when a viable control loop is in place, with reliable instruments that have been protected to ensure a high survival rate and a short replacement time cycle.

6. The Record and Respond Time (RRT) required to obtain, process, assess and respond to the monitoring information is the critical control parameter

It is suggested that a potential failure can only be controlled if effective corrective action can be taken before any specified strain limits are exceeded by, say, 10%. This is equivalent to placing a limit on the shear strain rate so that $d\gamma_{max}/dt$ does not exceed 10% of the [critical γ_{max} /RRT].

3.2 Developing performance criteria for single stage embankments

It is generally recognized that single stage stability is best gauged by monitoring the horizontal shear strains developed near to the embankments' toes by means of inclinometers.

Two trial embankments built in the Thames estuary are used to illustrate typical patterns of toe movements, before presenting a summary of several other cases and proposing some tentative general guidelines. Both embankments were raised in the 1970's as part of the Thames tidal flood defences programme. The Mucking Flats trial (described by Pugh 1976) involved a trapezoidal embankment under which part of the crust had been removed, giving the section shown in Figure 24, while the Littlebrook trial described by Marsland and Powell (1977) included a stepped loading berm.

Figures 24 and 25 include plots of how the maximum horizontal shear strains γ_{vh} calculated from the indicated (near toe) inclinometer positions varied with the Factor of Safety, F_T. F_T is approximated here as the final fill height at collapse divided by the current maximum fill height. Also shown is the maximum rate of horizontal shear strain, $d\gamma_{vh}/dt$. Figure 26 presents a typical data plot from the Mucking Flat trial, showing the characteristic acceleration of strain with time as failure is approached. Figures 24 and 25 show that the acceleration of strains and strain rates becomes most marked once F_T falls below 1.2, and this condition will be taken here as the limit to acceptable performance. In both of the Thames trials we see that the safety limit is reached when the maximum γ_{vh} grows to around 2.5%, and $d\gamma_{vh}/dt$ becomes larger than 0.25% per day.

It is interesting that the embankment shape did not seem influence the shear strain criteria. As illustrated in Figure 27 the loading berms placed at Littlebrook extended the failure mechanism into a non-circular type that passed underneath the berms on a sub-horizontal plane. We also note that the strain rate criteria cited above imply that the Record and Respond Time (RRT) would have to be less than 24 hours for F_T to be kept above 1.2 through ground movement observations.

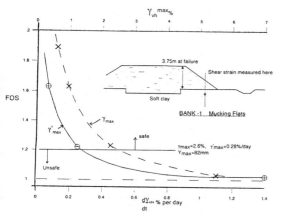

Figure 24. Mucking Flats Thames trial flood embankment – 1. Scheme and plots of maximum shear strain and strain rate against F_T

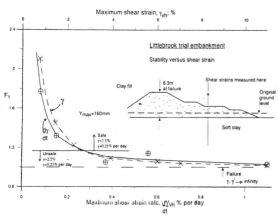

Figure 25. Littlebrook Thames trial flood embankment. Scheme and plots of maximum shear strain and strain rate against F_T

Further information is compiled in Table 1 from eight case histories that involved instrumented trial embankments taken to failure. The parameters listed are the limits interpreted for a Factor of Safety equal to 1.2. The data set suggests that stability may be-coming marginal in a single stage structure when any of the following limits is exceeded in 'ordinary' soft clays:

I.   A maximum lateral movement of 80mm with standard trapezoidal embankments up to 7m high, or 120mm when loading berms or very flat slopes are specified

II.   A maximum γ_{vh} of 2.0%

III. A maximum $d\gamma_{vh}/dt$, of 0.2% per day

Different criteria may apply to cases involving peats, or very lean clays. The peat layers loaded at Bouldin Island, for example, did not respond according to the proposed criteria; nor did the lean Portsmouth NH clays.

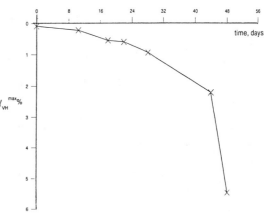

Figure 26. Mucking Flats Thames trial flood embankment – 1. Shear strain against time (near failure).

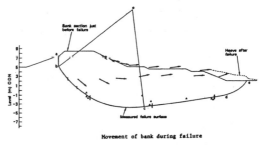

Figure 27. Final failure pattern for Littlebrook trial flood bank; after Marsland and Powell (1977)

3.3 *Performance criteria for multi-stage embankments*

As mentioned earlier, little information exists regarding the failure of multi-stage embankments. It is therefore difficult to set performance criteria that have been checked against field behaviour. Also, multi-stage construction varies from (i) cases involving involve just two or three stages of filling,

Table 1. Lateral movement parameters from eight single stage embankments all at Factors F_T = 1.2. Note: Bouldin Island case involved peats; while failure was in lean sensitive clays at Portsmouth NH

Case	Max H, m	Max Y, mm	Max γ_{vh}, %	Max $d\gamma_{vh}/dt$, %/day
Kings Lynn, Norfolk, UK	6.8	70	2.5	2.25
Portsmouth, NH, USA	6.6	40	0.3	0.10
Lanester, Brittany, France	3.9	130	3.1	1.90
Littlebrook, Kent, UK	6.3	150	2.5	0.25
Mucking Flats, Essex, UK	4.0	82	2.6	0.28
Muar, Malaysia	5.4	140	2.8	0.11
Cuzbac-les-Ponts, France	4.5	120	2.5	1.75
Bouldin Island, CA, USA	3.5	247	11	0.03

Table 2. Movement parameters from four multi-stage embankments. Note that the centrifuge H has been scaled up the equivalent 1g dimension

Case, notes	H, m	γ_{fill} kN/m³	Slope $\tan^{-1} \beta$	Design $S_u / \sigma'_{vc} F_T$	$\Delta Y/\Delta S$
Queenborough: safe	7.0	22 & 13	2.3	0.20	0.20
Sandwich, south: safe	7.5	22	3.0	0.23	0.22
Sandwich, north: Very safe	9.5	22	3.0	N/A	0.07
Centrifuge model sand on Kaolin: failed	11	16.9	1.9	N/A	0.8 to 1.0

each one of which could be considered as being similar to a single stage loading event, through to (ii) embankments that are built in many stages, with each thin layer causing only small strains that may be hard to separate from the background pattern of consolidation and creep movements.

It appears that the key is to compare maximum lateral movements ΔY developed at the toe with maximum settlements under the fill, ΔS, as proposed by Tavenas and Leroueil (1980). Table 2 provides a summary of the ground movement patterns seen in three UK case histories where stability was completely satisfactory. Also shown is data from a multi-stage test performed to failure in a centrifuge by Almeida et al (1985). The Queenborough and Sandwich south case histories involved low permeability soft clays that needed closely spaced vertical drains and an extended construction period (during which pore pressures were generally only partially dissipated, giving significant R_u^* values) to be completed. The Sandwich north case, however, involved intercalated sands and clays; pore pressures dissipated very quickly giving R_u^* close to zero throughout construction.

The data in Table 2 suggest that stability should be sufficient when:

I.   The single stage criteria are met comfortably for each individual lift

II.   The global trends for $\Delta y/\Delta S$ remain below 0.3

These criteria should be considered tentative; different limits might apply to cases involving peats, very lean clays, or other atypical soils.

4.   LARGE OFFSHORE LANDSLIDES

4.1 Introduction

The final part of this paper touches upon a new challenge for coastal geotechnical engineering: large deepwater submarine landslides. A series of such slides have been identified in continental margin areas; see for example Ashi (1999). As mentioned at the start of the paper, such failures can generate tsunami waves that can have severe consequences. For example, a relatively small slide that took place offshore Papua New Guinea in July 1998 led to possibly 2,000 lives being lost; Driscoll et al (2000). The same authors suggest that the outer continental shelf off southern Virginia and North Carolina in the USA might be in the initial stages of a large scale slope failure. A system of en echelon cracks has been discovered by swath bathymetric geophysical imaging that indicates down slope movements of the order of 50m near to the site of a much earlier (Pleistocene) large scale slide (Albemarle-Currituck), where the continental slope angles are of the order of 8°.

Figure 28. Vicinity map for Storegga slide; after Bugge et al (1987).

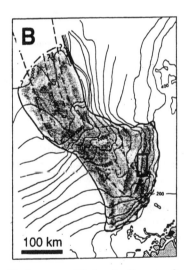

100 km

Figure 29. Storegga slide headwall area; after Bugge et al (1987).

The potential scale of continental slope landslides is indicated in Figures 28 and 29, showing the extent of the Storegga slides which took place around 8,000 years ago. The slides ran out for 800 km, their volume amounted to 5,600 km^3 and their total area (112,200 km^2) is similar to that of some European countries. The headwalls are found at water depths of around 300m, while the toe of the slide ran down to a water depth of around 3km. Dawson et al (1988)

and others have found and mapped geological evidence that the Storegga slide generated tsunamis which deposited debris at levels many metres above sea level in Scotland, Scandinavia and the Faroe Islands. Apart from its scale, the most remarkable feature of the Storegga slides is the low slope angles (around 1°) on which they were thought to have formed. Investigating the triggering mechanisms for such slides is one of the major challenges for coastal geotechnical engineering.

4.2 Potential triggering mechanisms

Hypotheses that have been put forward to explain the initiation of the large slides include:

I. Super-sensitive soil layers. It has been suggested that sheets of very loose carbonate sands/oozes, or highly sensitive clays could have been draped over the continental marginal slopes. Their resistance to either static or cyclic undrained loading might be sufficiently low for them to liquefy or flow when disturbed by even a minor earthquake or other sudden loading event.

One difficulty with this explanation is that it supposes a very uniform depositional regime within areas that may be known to have substantial local variations. Another problem is that super-sensitive layers have yet to be identified through borings.

II. Major earthquake loading. Most locations are likely to experience at least some significant seismic events over periods extending for tens of thousands of years. Given the apparent age of some of the slides, it is possible that earthquakes contributed to the slope failures.

However, it is difficult to imagine that even very large earthquakes could deliver sufficient energy to cause, on their own, massive landslides on extraordinarily gentle slopes. It would seem that some other agency must be present that can reduce the operational shear strengths to very low values.

III. Disassociation of CH_4 gas from hydrate compounds. Under conditions of low temperature and high pressure hydrocarbon gases (mostly methane) can combine with water to

113

make ice like compounds termed gas hydrates. If the sea temperature rises (geothermal or interglacial warming), or sea pressure falls (sea level reducing during glaciation), then the gas can be released from the hydrate 'ice', as shown in Figure 30 (as modified by Ashi 1999 from Kayen and Lee 1991). If such gas was able to concentrate into a pandemic coarse grained sand layer, it could cause the pore water to be replaced by gas, and give a potentially low effective stress, low shear strength, layer on which failure might start.

Gibson (1958) analysed a series of idealised cases by means of consolidation theory, dering the solutions shown in Figure 31. A far wider variety of cases, and much more realistic soil properties, may now be considered in quantitative numerical analyses.

V. It is possible to construct scenarios in which failures could develop through a combination of the above factors. For example, a large earthquake would be more likely to trigger a large slide if the soil profile was underconsolidated.

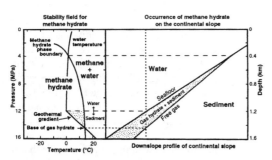

Figure 30. Stability field and conceptual occurrence of methane hydrate, after Ashi (1999).

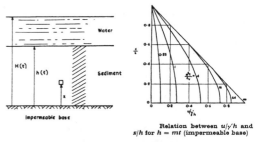

Figure 31. Gibson (1958) solution for consolidation under sedimentation rate m = dh/dt; excess pore water pressure isochrones for fixed $m^2 t/c_v$ ratios.

As with the super-sensitive soil hypothesis (I above) the hydrate mechanism requires that a uniform layer of material exists over a wide area.

IV. Under-consolidation caused by rapid deposition. Continental margins may be the sites of rapid sediment deposition. The Mississippi delta is one well-known example where the rates of accretion are such that excess pore water pressures are unable to dissipate fully leading to under-consolidation at some locations. It is possible that at certain times (for example around the end of a glaciation) the rates of deposition become so large that only minimal dissipation is possible within thick clay rich layers. The undrained shear strengths available within these layers would then be unusually low, and the layers susceptible to landsliding.

It is clear that research is necessary to understand continental margin landslides. Three key topics are: assessing the risks posed to coastal regions by potential landslide triggered tsunamies; investigating the potential link between the landslides and hydrocarbon reserves and understanding the risk such slides might pose to deepwater hydrocarbon extraction.

5. SUMMARY

This paper has summarized some recent advances in characterising soft clay behaviour, and in the approaches available for stability assessment. Key improvements have been made in sampling, soil testing, and in soil modelling. Particular emphasis has been placed on distinguishing between single and multi-stage embankments, showing how design parameters may be developed for both types of structure. Parametric studies have been presented to help illustrate the main aspects of multi-stage stability, and guidance has been given on the practical assessment of stability by field ground movement monitoring. Finally, the new geocoastal challenge posed by large deepwater landslides has been described and reviewed.

6. ACKNOWLEDGEMENTS

The Author wishes to acknowledge the contributions made to the work described in the paper by four former PhD students: Dr P Smith, Dr C Menkiti, Dr E Porovic, Dr L Zdravkovic. In addition, thanks are due to Mr Gerinsa Anak Tan who helped to prepare many of the figures, and to the British Council and Japanese Monobushi who supported a study visit to Japan during which much of the paper was prepared.

7. REFERENCES

Almeida M.S.S., Davies, M.C.R. and Parry, R.H.G. 1985. Centrifuged embankments on strengthened and unstrengthened clay foundation. *Geotechnique, 35, No.4, pp 425-441*

Ashi. J. 1999. Large submarine landslides associated with decomposition of gas hydrate. *Landslide News, June 1999, pp 17-20.*

Bjerrum, L. 1973. Problems of soil mechanics and construction on soft clays and structurally unstable soils (collapsible, expansive and others). *Proc. 8th ICSMFE, Moscow, Vol.3, pp 111-159.*

Bugge, T., Befring, S., Belderson, H., Eidvin, T., Jansen, E., Kenyon, N.H. Holtedahl, H., and Sejrup, H.P. 1987 A giant three-stage submarine landslide off Norway. *Geo-Marine Letters, Vol. 7, pp. 191-198.*

Burland, J. B. 1990. On the compressibility and shear strength of natural clays. *30th Rankine Lecture, Geotechnique, Vol 40, No 3, p329-378.*

Driscoll, N.W., Weissel, J.K. and Goff, J.A. 2000. Potential for large scale submarine slope failure and tsunami generation along the US mid-Atlantic coast. Geology; May 2000, v. 28, No 5; pp 407-410.

Gibson, R.E. 1958. The progress of consolidation in a clay layer increasing in thickness with time. *Geotechnique, Vol 8, pp 171-182.*

Hanzawa H. 1979. Undrained strength characteristics of an alluvial marine clay in the Tokyo Bay. *Soils and Foundations, Vol.19, No.4, pp*

Hight, D.W., Gens A. and Symes, M.J. 1983. The development of a new hollow cylinder appparatus for investigating the effects of principal stress rotation in soils.. *Geotechnique, 33, No.4, pp 355-384.*

Hight, D.W., Jardine, R. J. and Gens, A. 1987. The behaviour of soft clays. Embankment on soft clays, *Public Works research Center, Athens, Ch.2, pp 33-158.*

Hight, D.W, Bond, A.J. and Legge, J.D. 1992. Characterisation of Bothkennar clay: an overview. *Geotechnique, 42, No 2, pp 303-348.*

Hight, D.W. and Jardine, R. J. 1993. Small strain stiffness and strength characteristics of hard London Tertiary clays. *Proc. Int. Symp. on Hard Soils - Soft Rocks, Athens, Greece, Vol 1, Balkema, Rotterdam, pp 533-522.*

Hight, D.W. 1995. Drilling, boring, sampling and description – moderator's report. Advances in site investigation practice. *Proc ICE Conference 1995. Ed Craig. Thomas Telford, London, pp 337-359.*

Hight, D.W. 2001. Soil Characterisation: the importance of structure, anisotropy and natural variability. *38th Rankine Lecture, forthcoming in Geotechnique.*

Jardine R.J. and Hight, D.W. 1987. Laboratory and Field Techniques for obtaining design parameters. *Embankments on soft ground, Public Works Research Center, Athens, Chapter 4, pp 245- 296.*

Jardine, R.J. and Smith, P.R. 1991. Evaluating design parameters for multi-stage construction. *Geo-coast '91 Int. Conf., Yokohama, Port and Harbour Research Institute, Yokosuka, Vol 1, pp 197-202.*

Jardine, R.J. and Chow, F.C. 1996. New design methods for offshore piles. *MTD Publication 96/103, MTD (now CMPT) London.*

Jardine, R. J., Zdravkovic, L., and Porovic, E. 1997. Anisotropic consolidation including principal stress axis rotation: experiments, results and practical implications. *Invited panel contribution, Proc XIVth ICSMFE, Hamburg, Volume 4, Balkema, Rotterdam, pp 2165-2168.*

Jardine, R. J. and Menkiti, C. O. M. 1999. The undrained anisotropy of K_0 consolidated sediments. *Proc. 12th ECSMGE, Amsterdam. Balkema, Vol. 2, pp 1101-1108*

Kayen, R.E. and Lee, H.J. 1991 Pleistocene slope instability of gas hydrate-laden sediment on the Beaufort Sea margin. *Marine Geotechnology, Vol. 10, pp 125-141.*

Lacasse, S., Berre, T. and Lefebvre, G. 1985. Block sampling in sensitive clays. *Proc. 11th ICSMFE, San Francisco, Vol 2.pp 887-892.*

Ladd, C.C. and Foott, R. 1974. A new design procedure for stability of soft clays. *JGED, ASCE, Vol.100, GT7, pp 763-786.*

Leroueil, S. and Jamiolkowski, M. 1992. Exploration of Soft Soil and Determination of Design

Parameters. *Geo-coast '91 Int. Conf., Yokohama, Port and Harbour Research Institute, Yokosuka, Vol. 2*

Menkiti, C.O. 1995. Behaviour of clay and claey-sand, with particular reference to principal stress rotation. *PhD Thesis, University of London.*

Mesri, G. 1975. Discussion of 'New design procedures stability of soft clays' by Ladd & Foott, *JGED, ASCE, 101, No. 4, pp 409-412.*

Nicholson, D.P. and Jardine, R.J 1981. Performance of vertical drains at Queenborough Bypass. *Geotechnique, Vol 31, No 1, pp 67-90..*

PHRI - Port and Harbour Research Institute. 1991. *Geo-coast '91 Int. Conf., Yokohama, Port and Harbour Research Institute, Yokosuka, 2 Vols.*

Porovic, E. 1995 Investigations of soil behaviour using a resonant column torsional shear hollow cylinder apparatus. *PhD Thesis, University of London.*

Shibuya, S. 1985 Undrained behaviour of granular materials under principal stress rotation. *PhD Thesis, Imperial College, University of London.*

Smith, P.R. 1992 Properties of high compressibility clays with reference to construction on soft ground. *PhD Thesis, University of London.*

Smith P.R., Jardine R.J. and Hight D.W. 1992 On the yielding of Bothkennar clay. *Geotechnique, 42, No 2, pp 257-274.*

Tavenas, F. and Leroueil, S. 1980 The behaviour of embankments on clay foundations. *Canadian Geot. Journ., Vol.17, No.2, pp 236-260.*

Vaughan, P.R., Maccarini, M. and Mokhtar, S.M. 1988. Indexing the engineering properties of residual soil. *Quarterly Journal of Engineering Geology, Vol 21, No 1, pp 69-84.*

Whittle, A.J. 1993 Evaluation of a constitutive model for overconsolidated clays. *Geotechnique, Vol 43, No 2, pp 289-314.*

Zdravkovic L. 1996 The stress-strain-strength anisotropy of a granular medium under general stress conditions. *PhD thesis, Imperial College, University of London*

Zdravkovic L. and Jardine, R.J., 2001 The effects on anisotropy of principal stress rotation during consolidation. *Geotechnique (In Press).*

Project reports

Coastal Geotechnical Engineering in Practice, Nakase & Tsuchida (eds)
© 2002 Swets & Zeitlinger, Lisse, ISBN 90 5809 151 1

Coastal reclamation projects in Singapore

T.S.Tan, C.F.Leung & K.Y.Yong
Centre for Soft Ground Engineering, National University of Singapore, Singapore

ABSTRACT: Singapore is a very small country with a constantly increasing population. Land has and will continue to be the major constraint on all aspects of economic and infrastructure development. As a result of this constraint, in all developments in Singapore, two primary concerns prevail, namely, space creation and optimization of space utilization. For coastal geotechnical engineering, the most significant impact is the very large number of land reclamation projects for such a small country. In this paper, some of these projects are discussed. The lack of land also creates a problem for land reclamation, locating suitable sources of fills. Recent projects have all relied mainly on imported fills, but significant research is also being carried out to develop alternate fill materials. The concentration is on the use of seabed clay and excavated clays from land based construction activities. In this paper, some of the progress and the actual implementation in the use of such fill materials are discussed.

1 INTRODUCTION

Singapore is a very small country of 680 sq km today (and still growing) and has a population of 3 million citizens and 1 million permanent residents and foreign citizens (and still growing also). But since its independence in 1965, this small city-state has progressed in all fields. Undoubtedly, this has arisen mainly from the phenomenal economic growth she has witnessed. However, the economic well-being derived from these advancements will translate into greater demands for all kind of resources. People expect better living standard and quality of living such as freer traffic, less pollution, more natural parks and recreation facilities; the list goes on. Thus, a critical future task is to develop sustainable infrastructures that can support economic growth and improves on quality of life. Without this linkage, economic growth on its own is unsustainable and will not benefit the people.

This challenge is particularly daunting for Singapore, a country with scarce natural resources. For Singapore to survive as an independent and sovereign nation she must be able to meet swiftly the national objectives of security, economic development, social development, and creation of a new nation. Right from the start, the approach adopted by the government is based on pragmatism, referred as "Management by Objectives" by the leader of the country during that period, Mr Lee Kuan Yew (Lee, 2000). This applies especially to the development of

infrastructures; it is always objective-driven, and not technology or publicity driven.

1.1 Land is the first critical resource

Since the independence of Singapore in 1965, the country has always faced the challenge of a growing population crowding into a small island. Even the availability of drinking water was a major challenge as the catchment area is too small for the country to be self-sufficient and through the years, she has continued to rely on Malaysia to provide water; by way of a treaty that was part of the terms of separation in

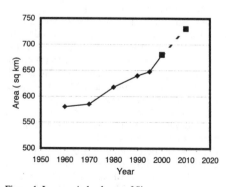

Figure 1. Increase in land area of Singapore

1965. Thus one of the most important considerations in infrastructure development is to build urban systems to meet basic needs, while minimizing space utilization. At the same time, to increase room for manoeuvre in the system, a process of physical space creation was initiated. To a large extent, coastal geotechnical developments have been very much affected by these two considerations.

It is recognized right from the onset of independence in 1965 that lack of land space is a critical issue which impacts on all other issues. To partly alleviate this highly restrictive condition, land reclamation was a top priority national effort. Fig. 1 shows the increase in land space from land reclamation since 1965. The city-state has grown from an original area of 580 square kilometers (sq km) in 1965 to an estimated area of 662 sq km today, an increase of 17%. Ongoing and already approved projects will add at least another 68 sq km in the next ten years while those being planned will add at least another 50 sq km in the following ten years. As could be seen in Fig. 1, short-term perturbations, such as economic crises in 1974, 1985-1987 and 1998-1999 did not slow down the land reclamation programme. The need to plan ahead and invest, even during "bad times" is evident from this figure illustrating the fact that short-term perturbations do not derail long-term objectives.

2 MAJOR RECLAMATION PROJECTS

Figure 2a shows a map of Singapore before the massive reclamation efforts of the last 35 years; and Fig. 2b shows a map of Singapore to day, adapted from Lui and Tan (2001). The physical changes to the coastline of Singapore that have occurred in the last decade are significant. The scale of this change gives a good indication of the intensity of reclamation activities to date.

Many of the coastal reclamation projects are directly linked to other major infrastructure developments such as airport, seaport container port, industrial and residential needs. In Table 1, some of the major projects of recent years are listed.

An example of such a link is the very recent construction of an island known as Jurong Island. The Economic Development Board of Singapore saw the potential to form a chemical hub in Singapore, but due to the severe space constraint on the main island, decided to build the hub around a cluster of offshore islands on the south west coast of Singapore. This project was approved in 1991, and ten years later, the island is almost completed, adding an additional 2000 hectares to Singapore's land.

Initially, most of the fill materials came from hill-cut material locally. But since the early 1980s, the depletion of that source of fill has led to the import of sand from the region. The high cost of land in

Singapore had made it economically feasible to continue reclaiming land from the sea even with imported fill.

In spite of the hectic pace of reclamation in the first 30 years since independence, the pace of reclamation has further accelerated. In a recent announcement of the Concept Plan to guide the future planning of Singapore, it was stated that Singapore would have a shortage of 4000 hectares, even after factoring in all the planned reclamation activities in the future.

Table 1 Major land reclamation activities (non-exhaustive)

	Year	Site	Area (ha.)
1	1964-1976	Kallang Basin	199
2	1966-1985	East Coast	1525
3	1975-1977	West Coast	86
4	1978-1979	Pasir Ris	44
5	1983-1988	Ponggol	276
6	1976-1978	Changi Airport	662
7	1983-1986	Changi Airport	181
8	1985-1989	Tuas	600
10		Pulau Tekong	540
11	1985-1990	Northeast Reclamation	472
12	1990-1994	Marina Bay	43.6
13	1997-2001	Northeast Phase IV	155
14	1995-2001	Jurong Island	1659
15	2001-	Jurong Island Phase IV/Tuas	2000
16	2001-	Pulau Tekong/Ubin	1500

(part of the data adapted from a report on 20 August 2000 by Lianhe Zaobao Newspaper)

Newer reclamation projects need to go into deeper water as most shallow coastal water has been reclaimed. As a result, the volume of fill needed has also increased significantly. In one of the more recent reclamations to be awarded, the reclamation at Jurong Island Phase 4 and Extension to Tuas View,

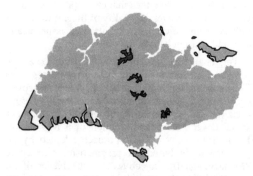

Figure 2a. Map of Singapore before major reclamation activities

Figure 2b. Map of Singapore today

the water depth exceeds 20m in some locations and the average water depth is over 15m. This means the demand for fills will increase. The search for alternate fills, especially using dredged and excavated clays, has become an important research activity.

3 RESEARCH AT NUS ON DEVELOPMENT OF ALTERNATE FILLS

As reclamation volume is always very large, in developing alternate fills, it is always important to consider two factors, namely availability of material and cost of transportation. As a result of these two considerations, at the National University of Singapore, the main focus is on the use of seabed clay. Two different approaches are being evaluated. The first is to dredge the clay as slurry and pump the slurry into a pond and treat the slurry to gain strength and reduce water content in a rapid way. In the second approach, the clay is directly dredged from the seabed as lumps and placed. Both methods have been or being tested on actual field projects and the ensuing discussion will focus on the geotechnical issues relating to these two approaches.

3.1 Tuas Reclmation

Reclaimed work started in Tuas at the western end of Singapore in March 1984 and completed in March 1988. The project was undertaken by the Jurong Town Corporation (JTC) and involved a reclaimed area of 637 ha and altogether about 68.5 million m³ of fill was used. Large quantity of soft marine clay underneath the 14-km perimeter bund had to be dredged and replaced with sand to improve the bund stability.

As it was difficult to find suitable dumping grounds for the dredged marine clay, the dredged clay was dumped within the reclamation area up to a

maximum thickness of 4.5 m. The settlement of the dumped clay is a concern for the JTC engineers. Extensive instruments were installed in the dredged clay dumped area to monitor the settlement of the fill and individual soil layers comprising sand fill, dredged clay, in-situ marine clay and stiff residual soils of sedimentary origin. The settlement-time response for a typical instrument cluster is shown in Figure 3. For this cluster, the in-situ marine clay is relatively thin and the settlement of the dredged clay is found to be significant as compared to that of the insitu marine clay. On the other hand, relatively little settlement was noted for the stiff soil strata.

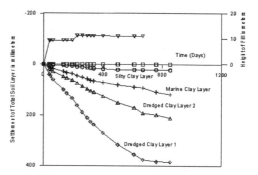

Figure 3. Settlement-time responses for different soil layers (after Hee et al., 1999)

Figure 4 illustrates the variation of fill settlement versus the thickness of dredged marine clay and in-situ marine clay after one year of fill placement. When the thickness of the dredged clay is less than 30% of that of in-situ marine clay, it can be deduced that the fill settlement is mainly governed by the thickness of the in-situ marine clay. Detail results and interpretation of the field data are given in

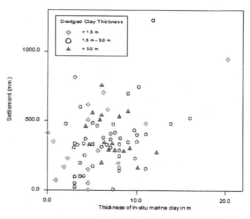

Figure 4. Variation of ground surface settlement with thickness of dredge marin clay and insitu marine clay (after Hee et al., 1999)

Chong et al. (1998) and Hee et al. (1999).

3.2 Changi South Bay Reclamation

In 1988, a reclamation project with an area of 40 ha was set aside to test out research ideas developed at NUS. The site is next to Changi International Airport, which was reclaimed 10 years earlier. In this project, the reclamation was divided into two sections. In one section, the reclamation fill is slurry while in the other section, dredged lumpy fill was used (see Figure 5 from Karunaratne et al., 1991).

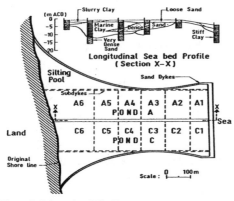

Figure 5. Schematic of Pilot Test

In both cases, sand is used to cap the reclaimed fill. Once the cap is sufficiently strong to support light machinery, vertical drains were installed and surcharging applied. Once this stage is reached, the main geotechnical challenges are resolved.

The main issue in that project is that when slurry is used as a fill material, the high water content of the slurry will mean that the consolidated fill is very small. Thus, the main challenge is to consolidate the slurry while it is being filled up, so as to increase the volume of fill. As slurry has water content many times the liquid limit of the soil, actually a small load will have a significant effect on the consolidation. Using this principle, when the slurry reached a certain height, sand particles were rained in a controlled fashion to form a thin intermediate drainage layer. In this project, the sand was rained by pumping it hydraulically through perforated pipes. By adjusting the speed of pumping, the quantity of sand spread can be controlled. After the first intermediate sand layer is built, slurry is pumped again and another layer is built up in a similar fashion.

Figure 6 shows the typical density profiles at different sections after the clay slurry has been pumped from a point near Section A2. It is clear that particulate sorting has occurred, and at points further away,

the fines are carried there resulting in a lower density compared to points nearer the pumping source. This is a problem that has to be tackled if this method is to be used on a large scale.

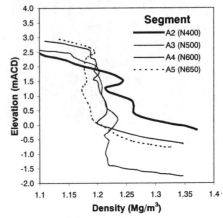

Figure 6. Density after pumping of clay slurry
(from Karunaratne et al., 1991)

After about seven days, when the sedimentation of the slurry is deemed to be over (Tan et al., 1990), sand is been rained. Depending on the thickness of sand spread per pass, and on the combinations of multiple passes of varying thickness, the formation of the sand cap is different as shown in Figures 7a and 7b.

It is clear that through this method, small surcharge load can be placed on the slurry to greatly accelerate the consolidation while filling is continuing. The main challenge is in the way sand will penetrate into a slurry and how the sand will disperse in the slurry if it indeed penetrates into it.

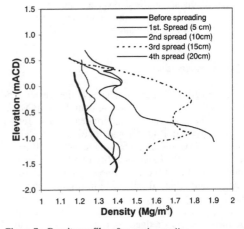

Figure 7a Density profiles after sand spreading

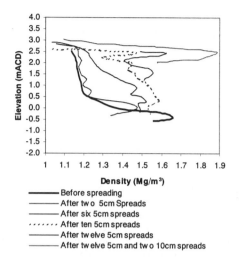

Figure 7b Density profiles after sand spreading

Legend for figure:
- Before spreading
- After two 5cm Spreads
- After six 5cm spreads
- After ten 5cm spreads
- After twelve 5cm spreads
- After twelve 5cm and two 10cm spreads

3.3 New Container Terminal

In the mid-nineties, a new Container Terminal was constructed at Pasir Panjang in the west coast of Singapore. The project was undertaken by the Port of Singapore Authority (PSA) and involved a relatively large reclaimed area and a long wharf line. Gravity caissons were employed as the wharf front retaining structure. The vertical and lateral movements of the caisson under working loads was a major concern for the PSA engineers as the movements could affect the performance of the quay cranes that rest on the caissons.

During the design stage in the early nineties, a joint NUS-PSA research study was initiated to examine the caissons movements using the centrifuge modelling technique. Details of the centrifuge model study are reported in Leung et al. (1997). It is established that the placement of rock sill below the caisson base and preloading the caisson can be effective means in reducing the caisson movements under working load conditions. During the port construction, every caisson was preloaded and reloaded and the performance of the caisson followed a reasonably close pattern to those observed from the centrifuge tests. Details of the field caisson movements under the preloading/reloading process are presented in Tan et al. (1999).

In the course of port construction, some navigation channels needed to be dredged and stiff residual soils of sedimentary origin is the dominant seabed materials. The disposal of dredged soils is always a problem in the land-scarce Singapore. The use of dredged stiff clay for land reclamation can serve the dual purpose of solving the environmental problem of finding dumping grounds for dredged soils and establishing alternate fill materials for land reclama-

tion. Another joint NUS-PSA research study was thus initiated to examine the performance of reclamation fill made up of dredged stiff clay lumps. A schematic layout of such lumpy fill profile is shown in Figure 8. As large voids exist between stiff clay lumps, the closing up of inter-lump voids is a major concern.

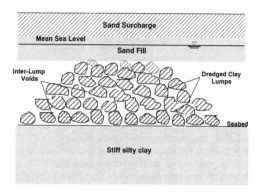

Figure 8. Schematic profile of reclamation fill with dredged stiff clay lumps (after Leung et al., 2001)

Large diameter one-dimensional compression tests and centrifuge model tests were conducted to evaluate the settlement of lumpy fill. Figure 9 shows the variation of void ratio in the lumpy fill with loading pressure as compared to that of a homogeneous clay fill. It is evident that a lumpy fill is a two-porosity soil system consisting of clay lumps having a lower porosity and coefficient of permeability and voids between lumps having a much bigger porosity and coefficient of permeability. As the lumpy fill experi-

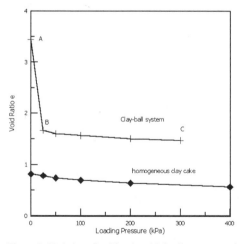

Figure 9. Variation of void ratio with loading pressure for lumpy fill (clay-ball system) and homogeneous clay (after Leung et al., 2001)

ences an initial large settlement due to the closing up of the voids, preloading is necessary to ensure that such large settlement would not occur under working condition. Details of the experimental results are presented in Leung et al. (1997), Manivannan et al. (1998) and Leung et al. (2001).

A small test area of about 50 m by 100 m was adopted where dredged stiff clay lumps were placed. The interpretation of the field test data is currently in progress. Preliminary observations indicate that the observed fill settlements followed a similar trend as that established in the experimental studies.

3.4 Use of Lumpy Fills

Using the technology developed in studies at the National University of Singapore, a new project was initiated and awarded in November 2000. In this new project, the proposed area to reclaim is 1500 ha. and the location is an offshore island in the eastern coast of Singapore, Pulau Tekong, which is opposite to Changi International Airport. In this project, a large section, about 400 ha. will be devoted to receive dredged seabed material from the construction of sand key as part of this project, and also excavated and dredged materials from all other sources in Singapore. This is regarded as a high priority project because the disposal of such poor construction soils is becoming a major problem in land scarce Singapore. Thus by adopting this system approach, two issues handled by two different agencies can be resolved, allowing for a solution to a disposal problem while making the waste material becomes useful. In that sense, technology has been used to provide the leverage to solve two problems with one solution.

The main geotechnical challenge in this project is the ability to characterise the highly non-homogeneous lumpy fill, understand the long-term consolidation behaviour of such a fill and develop appropriate design guide for the use of such fill. As construction activities for this project are just about to commence, currently, there is no field data available.

4 CONCULDING REMARKS

Scarcity of land is recognised to be one of the major constraints facing Singapore since the day of its independence. To overcome this constraint, a two-prong pragmatic approach was adopted, and will continue to be applied. This is to create new physical space on the one hand and to build large-scale infrastructures with an aim to optimise the usage of land.

As a result of this philosophy, land reclamation has been one of the major construction activities in Singapore, adding more than 17% new land area to Singapore, and will add at least another 120% in the future. Even with this intensive effort, the current Concept Plan for the future planning of Singapore is projecting a shortfall of 4000 hectares.

As most shallow coastal areas have been reclaimed, new reclamations are now in deeper water; and this means increasing need for suitable fill material. Intense efforts to develop new alternate material have been ongoing for the last twenty years, with concentration mainly on the use of slurry clay, dredged clay and excavated clay. Two main groups of fills are involved, mainly slurry clay and lumpy clay fills.

This paper gives a brief overview of the land reclamation activities, the research on the development of alternate fill materials and the associate geotechnical problems. In Singapore, both slurry as well as dredged or excavated clays are increasingly used for land reclamation purpose. This actually serves two purpose; it helps in the disposal of soils that are considered as waste and recycle them as useful material of economic value.

ACKNOWLEDGEMENTS

The efforts of many officers in various government agencies and corporations who have supplied relevant information and data for this paper are also deeply appreciated. In particular, engineers from the Housing and Development Board, Jurong Town Corporation, PSA Corporation, the Maritime Port Authority of Singapore and SPECS Consultants have contributed to this work.

REFERENCES

1. Chong, Y.L., Leung, C.F., Tan, S.A. and Lim, T.H. (1998), Performance of dredged clay as reclamation fill, Proc 13th Southeast Asian Geotechnical Conference, Taipei, Vol. 1, pp. 763-768.
2. Hee, A.M., Lim, T.H., Chong, Y.L., Leung, C.F. and Tan, S.A. (1999), Tuas reclamation project: a case study, Proc 5th Int Symp on Field Measurements in Geomechanics, Singapore, pp. 467-474.
3. Karunaratne, G.P., Yong, K.Y., Tan, T.S., Tan, S.A., Lee, S.L. and Vijiaratnam, 1991. "Land reclamation using layered clay-sand scheme," Proceedings of the International Conference on Geotechnical Engineering for Coastal Development, GEO-COAST '91, Yokohama, Coastal Development Institute of Technology, Japan, pp. 335-340.
4. Lee, K.Y. (2000), From Third World to First, The Singapore Story 1965-2000, by Times Media Private Limited, Singapore.

5. Leung, C.F., Lee, F.H. and Khoo, E. (1997), Behavior of gravity caisson on sand, Journal of Geotechnical and Geoenvironmental Engineering, ASCE, Vol. 123, pp. 187-196.
6. Leung C.F., Wong, J.C. and Tan, S.A. (1997), Assessment of lumpy fill profile using miniature cone penetrometer, Proc 14th Int Conf on Soil Mechanics and Foundation Engineering, Hamburg, Vol. 1, pp. 527-530.
7. Leung, C.F., Wong, J.C., Manivannan, R. and Tan, S.A. (2001), Experimental evaluation of consolidation behaviour of stiff clay lumps in reclamation fill, Geotechnical Testing Journal, ASTM, Vol. 24, pp. 145-156.
8. Lianhe Zaobao Newspaper, Report on 20 August 2000, Singapore Press Holding, Singapore (in Chinese).
9. Lui, P.C. and Tan T.S. (2001), Building Integrated Large-Scale Urban Infrastructures: Singapore Experience, Journal of Urban Technology, The Society of Urban Technology. (under review)
10. Tan, S.A., Wong, Y.K., Lay, K.H. and Leung, C.F. (1999), Performance of gravity caisson, Proc 5th Int Symp on Field Measurements in Geomechanics, Singapore, pp. 297-302.
11. Tan, T.S., Yong, K.Y., Leong, E. and Lee, S.L., 1990a. "Sedimentation of Clayey Slurry." Journal of Geotechnical Engineering, ASCE, Vol. 116, No. 6, pp. 885-898.

Coastal Geotechnical Engineering in Practice, Nakase & Tsuchida (eds)
© 2002 Swets & Zeitlinger, Lisse, ISBN 90 5809 151 1

Geotechnical issues on airport projects constructed in Japanese coastal area

M.Kobayashi

Kobayashi Softtech Inc., Tokyo, Japan

ABSTRACT:. This paper introduces some important geotechnical issues evolved and how they were faced in constructing three of such coastal airports constructed on soft land reclaimed, namely, Tokyo International Airport (TIA), Kansai International Airport (KIA) and Central Japan International Airport (CJIA). As the TIA was constructed on the reclaimed land, which consists of dredged slurry and waste soil, the large scale ground improvement had been carried out. Recently, a new ground improvement technique has been developed and used in this project to strengthen the seismic resistance of runway against the anticipated strong earthquake motion. In addition, the light–weight soil method was developed and applied in the new taxi–way, which were constructed on existing railway tunnel. KIA was constructed in an artificial island of 5km offshore in Osaka Bay. The 1st phase construction completed in 1994, when the airport opened with a single runway. The issues related to the 2nd phase construction project, which started July 2000 were introduced. The improvement of Holocene clay layer and the prediction of settlements of deep Pleistocene cay layers are key problems in this project. The construction of CJIA also started in August 2000. CJIA is to be constructed on the artificial island in Ise Bay. A large volume of cement treated clay made by the pneumatic flow mixing method are to be used as the fill in the first time. As the clays used for the cement treatment are taken by dredging of navigation channel in Nagoya Port nearby, the use of pneumatic flow mixing method in CJIA will be a unique case as a large scale recycling of waste soils.

1 INTRODUCTION

As more than 70% Japanese land are mountainous area, the reclamation of shallow seawater has been a traditional way of life for Japanese people to develop the country. It seems to be quite natural that most of airports in Japan are located in the coastal reclaimed lands. Fig.1 shows the airports in coastal areas in Japan and Asian countries, where included are the major airports in this area such as, Tokyo International Airport and Kansai International Airport (Japan), Changi Airport (Singapore), Hon Kong airport, Shanghai Airport (China) and Inchon Airport (Korea). In addition, the expansion of existing airports or the construction of new airports have been vitally carried out in this area in order to meet the rapid globalization and the growing demand for air transportation,

When constructing airport in coastal area, it inevitably need the technology of soft ground engineering. In this project report, the author introduces several

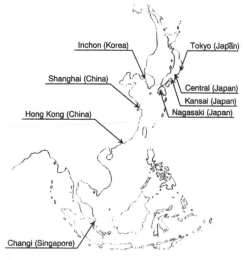

Fig.1 Airport constructed in coastal area in Eastern Asia

geotechnical issues and the new technology related to airport projects in Japanese coastal areas in recent 10 years.

The expansion of Tokyo International Airport (TIA, Haneda Airport) have continued since long. As the airport was constructed on the reclaimed land, which consists of dredged slurry and waste soil, the large scale ground improvement had been carried out. Recently, a new ground improvement technique has been developed and used in this project to strengthen the seismic resistance of runway against the anticipated strong earthquake motion. In addition the light-weight soil method was developed and applied in the new taxi-way, which were constructed on existing railway tunnel.

The 1st phase of Kansai International Airport (KIA) was constructed in an artificial island of 5km offshore in Osaka Bay. The construction started in 1987 and completed in 1994, when the airport opened with a single runway. As the report on the 1st phase project including geotechnical issues were made by Arai (1991), the author discuss the issues related to the 2nd phase construction project, which started July 2000. The improvement of Holocene clay layer and the prediction of settlements of deep Pleistocene cay layers are key problems in this project.

The construction of Central Japan International Airport (CJIA) also started in August 2000. CJIA is to be constructed on the artificial island in Ise Bay. Because there are not so much origins of fill materials for reclamation near the construction site, a large volume (about 9 million cubic meters) of cement treated clay are to be used as the fill. Related to the project, the pneumatic flow mixing method has been newly developed and succeeded to reduce the cost of cement treated soil drastically. As the clays used for the cement treatment are taken by dredging of navigation channel in Nagoya Port nearby, the use of pneumatic flow mixing method in CJIA will be a unique case as a large scale recycling of waste soils.

This paper introduces some important geotechnical issues evolved and how they were faced in constructing three of such coastal airports constructed on soft land reclaimed, namely, TIA, KIA and CJIA.

2 TOKYO INTERNATIONAL AIRPORT PROJECT

Today, Tokyo International Airport (Fig.2) plays a central role in domestic air transport, which handles about 320 flights daily with 45 airports throughout the country. In fiscal 1997, it was Japan's largest airport handling some 50,000,000 people, or nearly 60% of the domestic air passengers, and some 580,000 tons of air cargo. The airport is also frequently used as a place to welcome national and public guests from abroad and serves as a major base for aircraft maintenance and night stay of airplanes. TIA, also called Haneda Airport, which is playing such a vital role, is required to have sufficient aseismic capacity - which has become a strong social demand today. Hence, efforts are being made to improve the seismic resistance capacity of its facilities.

2.1 Overview of Offshore Expansion Project

The grand "Tokyo International Airport Offshore Expansion Project" that aims to secure the functions required of Haneda Airport now and in the future entails construction of new runways, taxiways, and aprons, as well as new passenger terminals, cargo handling facilities, and maintenance facilities, on the waste disposal grounds of Tokyo Metropolis. In addition, roads, railways, and other means of access will be developed. All these works need to be carried out at sites which adjoin the existing airport in service. Therefore, the existing facilities will be moved to the new sites on a step-by-step basis so as not to impair the functions of the existing airport. The Offshore Expansion Project was started in January 1984. The new A-Runway, completed in Phase I, was put into operation in July 1988, and the West Terminal, completed in Phase II, was put into service in September 1993. In the first half of Phase III, the new C-Runway was completed and put into operation in March 1997. So far, the new A-Runway and C-Runway, their taxiways, aprons, a terminal building, and several means of access have been completed. Now project has entered the second half of Phase III, accomplishment of which will complete the entire project.

Fig.2 Tokyo International airport offshore expansion project

The site of the Offshore Expansion Project was previously a waste disposal site. The waste disposal grounds were very soft, since the original ground which consisted of soft clay layers was first filled up with sludge dredged from the Tokyo Bay and then covered with waste soil from other construction sites. Fig.3 shows typical soil profile of TIA.

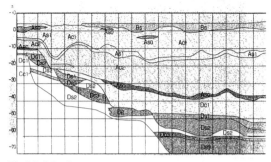

Fig.3 Soil Profiles in Tokyo international airport

The maximum water content of the soil was higher than 200%. Besides, the areas set aside for Phases I to III of the project differ markedly in soil properties. Such grounds not only have large total settlement but also have substantial differential settlement.

In order to secure a stable, level ground for the airport within a limited period of time, large-scale ground improvement work was carried out using efficient methods. In the ground improvement, based on the results of investigations of marked changes in and consolidation characteristics of the soil layers, suitable methods, mainly the vertical drain method, were employed.

The total length of drains placed for the entire project reached nearly 80,000 km. The ground was adequately pre-loaded to promote consolidation of the soil.

The ground improvement works began in 1984, which was completed on the most parts of the project site in 12 years.

The areas for Phases I and II of the project were nonuniform grounds reclaimed from Tokyo Bay, by first removing the existing alluvial sand layer (A_{S1} layer) and then filling up with soft sludge dredged from the Bay (Ac_1 layer). Since in these two areas the sludge layers that required improvement were relatively thin (AP –10 to –20m), the paper drain and pack drain methods were mainly employed for the improvement. By contrast, the Phase III area, which is closest to the shoreline, is grounds recently reclaimed in a short time by first dumping sludge onto the existing alluvial clay layer (Ac_2 layer) about 25 m in thickness and then filling with construction waste soil. Because of this, it was expected that the settlement due to consolidation would be as much as 5 to 9 m. In the Phase III area, therefore, it was necessary to improve even the alluvial clay layer in order to reduce the residual settlement, whereas in the Phase I and Phase II areas, only the sludge layer had to be improved. In the ground improvement of the Phase III area, the composite vertical drain method (the sand drain method supplemented with the paper drain method) and the cylindrical fabric-packed sand drain method were employed to give the same level of consolidation to the sludge layer (Ac_1 layer) and

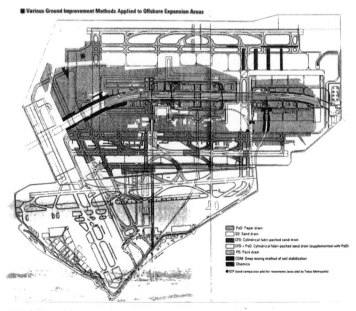

Fig.4 Ground improvement method used in TIA project

alluvial clay layer (Ac₂) in a prescribed period of time.

On the other hand, the restricted ground surface around the airport set a limit to the height of drain machines that could be used in this area. As a result, the maximum depth of ground improvement was limited to AP -28 m. This means that the entire ground could not be improved, hence a certain amount of residual settlement was unavoidable. Therefore, in constructing the new C-Runway in this area, it was necessary to provide allowances against the possible residual settlement.

2.2 Ground improvement by vertical drain

Different ground improvement works carried out on the site were: plastic board drains (PBD), small-caliber fabri-packed sand drains (PD) with a diameter of 12 cm, sand drains (SD) with a diameter of 50 cm, and sand drains partially sheathed with geotextile (FPD) with a diameter of 50 cm. These types of improvements were conducted individually or in combination, depending on the ground condition that varied from place to place. Fig.4 shows the ground improvement method used in the TIA project.

In the sites of Stage-I and Stage-II only Ac_1 layer was improved mainly by plastic board drains. This type of drain is easy to drive into the ground and also cost-effective. The thickness of the Ac1 layer was uneven, and small-caliber fabri-packed sand drains were used in the locations where deep ground improvement was required. This type of drain has relatively high reliability once driven into the ground.

In the site of Stage-III, both the Ac_1 and Ac_2 layers had to be improved. The depth of the ground to be improved was considerable, and the Ac_1 layer was extremely soft. Accordingly, mainly sand drains were used and the Ac_1 layer was sheathed with geotextile. Under the airspace restrictions, it was difficult to drive sand drains through to the bottom of the Ac_2 layer. Accordingly, the drains were driven partially, leaving the lower part of the Ac_2 layer unimproved. To balance the consolidation rate of the Ac_1 layer with that of the Ac_2 layer, additional plastic board drains were driven into the Ac_1 layer (between the ordinary drains) of which the coefficient of consolidation was smaller than that of the Ac_2 layer. For this project the combined use of auxiliary plastic board drains and sand drains partially sheathed with geotextile was applied and a new design method for this "composite vertical drainage" was developed.

Large scale ground improvement works were conducted using vertical drains in an area of about 570 ha over the period of 10 years and 7 months starting August 1984. The depth of improved ground was 28

m at the maximum under the airspace restrictions. The number of drains driven into the ground was as many as 4 millions, their length adding up to a distance of about 80,000 km. At the peak time of the ground improvement works, about 50 units, or 70% of the drain machines available in the country, grouped in Haneda (Fig.5). Accordingly, the settlement of the ground was brought to an end in six months to one year, which otherwise would take 1,000 years. About 50% of allocated budget of the project was invested in the ground improvement works.

Fig.5 A forest of drain machines

a) Plastic board drains

To reduce the residual settlement of the ground, belt-shaped drains of an artificial, high-permeability material (20 cm wide and 3 to 6 mm thick) were driven vertically into the ground (Fig.6). This method was applied to the improvement of clay layer up to 20m depth. Because driving machines of plastic board drains are relatively lightweight, the plastic board drains were driven into the super-soft layer as auxiliary drains to attain a primary level of stability of the ground, and then partially sheathed sand drains, for which the heavier machines were used, were driven into the ground.

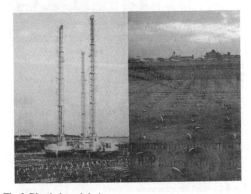

Fig.6 Plastic board drains

b) Sand drain

Sand piles of a large diameter, about 50 cm in diameter, are used as the drainage material to accelerate the rate of consolidation of soft ground. This method is applied for the improvement of soft clay whose depth were beyond 20 m, because, in the case of small diameter drain such as plastic board drain or small-caliber fabri-packed sand drain, the effect of the drain may be damaged by the drainage resistance (or well resistance) which increases with the drain length.

In the Stage-III site, plastic board drains were driven into the dredged clay layer (Ac_1) as auxiliary drains, and sand drains were driven into the alluvial clay layer (Ac_2) through the Ac_1 layer. The Ac_1 layer portions of the sand drains were sheathed with geotextile to ensure integrity and stability of the drain column.

Fig.7 Sand drain

c) Small-caliber fabri-packed sand drains

In this method the sand-filled long and tubular geotextile, about 12 cm in diameter, are driven into the ground to accelerate ground consolidation. Since sand is packed in bags, the sand piles are able to preserve continuity in the ground even if the large deformation may take place during driving of drain or

Fig. 8 Small-caliber fabri-packed sand drains

the further consolidation.

As the diameter of fabri-packed sand drain is larger than that of plastic board drain, the well resistance is smaller. Accordingly, it was used for the improvement of the dredged clay layer Ac_1 whose thickness is relatively larger.

d) Composite vertical drain

Different techniques such as plastic board drains, partially sheathed sand drains, non-through sand drains, supplementary intermediate drains, were used individually or in combination to accelerate the consolidation process of soft ground. Fig.9 shows an example of a composite vertical drain, which consists of plastic board drain, partially sheathed sand drain and non-sheathed sand drain. In some sections, the ground was over-consolidated by the preloading method. Using these techniques individually and in combined with other techniques, the consolidation settlement of the ground was halted in six months to one year, which otherwise would take 1,000 year.

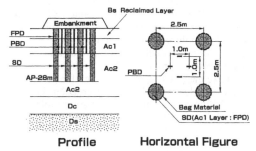

Profile   Horizontal Figure

Fig.9 Composite vertical drain

2.3 Construction of seismic resistant runway

In the Offshore Expansion Project, new runways, taxiways, and aprons are constructed in three phases.

In constructing such facilities, various measures are implemented to provide against the differential and residual settlement of the ground that can occur even after the ground improvement.

a) New A-Runway (3,000 m x 60 m)

The new A-Runway was completed in Phase I of the project and put into service in July of 1988. About 450 m away from former C-Runway toward the shoreline, it has a sandwich pavement structure since the subgrade is soft.

b) New C-Runway (3,000 m x 60 m)

The new C-Runway was completed in the first half of Phase III and put into service in March of 1997. It runs parallel with the new A-Runway at a center to center distance of 1,700 m, allowing the two runways to be used without any interference with each other. Of the three new runways, the new C-Runway is located farthest from the residential areas and can be used 24 hours a day without causing the problems of aircraft noise.

The site of the new C-Runway is a recently-reclaimed land located at the part of the project site that is nearest to the sea. Therefore, the level of groundwater is high and the amount of residual settlement after the ground improvement was expected to reach 1.5 m in 10 years after the opening of the runway. Under those conditions, the subgrade drainage layer was improved using waste materials, etc. and cambering was implemented to provide against the possible residual settlement.

c) New B-Runways (2,500 m x 60 m)

Construction of the new B-Runway was started in April of 1997 as part of the second half of Phase III of the project. Running parallel with the existing B-Runway at a distance of 380 m offshore, the new B-Runway is scheduled to be put into service by the end of fiscal 1999.

Having a length of 2,500 m and a width of 60 m, the new B-Runway is located about 380 m away from the existing B-Runway toward Tokyo Bay. Smooth coordination between the new A-Runway under construction will increase the runway capacity and dissolve the problems of aircraft noise.

Unlike the former airport area and the grounds reclaimed in phases I to III, the construction site of the new B-Runway has a non-uniform soil profile. Similarly, the construction site of the parallel taxiway on the west side of the new A-Runway has nonuniform deposition of soil layers from the construction waste soil layer (Bs layer) to the alluvial sand layer (As$_1$ layer). At both construction sites, the groundwater level is high. At the construction sites, the possibility of ground liquefaction and ground settlement was estimated. As a result, it was found that liquefaction of the construction waste soil layer (Bs layer), dredged sand layer (As$_0$ layer), and alluvial sand layer (As$_1$) layer could occur.

Several new attempts are made in the construction of the new B-Runway. It is being constructed as Japan's first seismic resistant runway provided with measures against liquefaction of the ground. These methods are described below.

To provide safety and stability against the possible liquefaction, the grounds are improved by the sand compaction pile method, which gives vibrations to the ground to compact it and make it hard to liquefy. The sand compaction pile method (SCP method) is a ground improvement method used mainly to increase the bearing capacity, reduce the settlement, and promote the consolidation of sand layer, etc. This method is applied to the construction site down to the alluvial sand layer (As$_1$ layer). Namely, 40 cm diameter casing pipes are driven into the ground (arranged in squares at 1.7 m pitches) and sand is fed

through the pipes and compacted into 70 cm diameter sand piles to increase the soil density, thereby reinforcing the ground and preventing liquefaction.

At and around the intersection with the new A-Runway, it is necessary to provide the measures against liquefaction without causing any adverse effect on the pavement of the new A-Runway already in service. Therefore, a new method has been developed and applied.

A portion of the new B-Runway, about 450 m in length, intersects with the new A-Runway that is already in service. In order to increase the seismic resistance capacity of the intersection without interfering with the airport operation, the ground right under the intersection was improved with the pavement of the new A-Runway kept intact. To that end, methods for preventing liquefaction of the ground right under the existing pavement were developed.

In Japan, there had been no precedent for the reinforcement of seismic resistance of the ground right under a runway in service. Therefore, applicability of various methods was studied. Five different methods were tested and subjected to overall evaluation of workability, economy, etc. As a result, it was decided to adopt the CPG method – a static compaction method – for the ground in the runway intersection area, and the chemical solution injection method– for the ground in the neighborhood of underground structures.

In CPG (compaction grouting) method, a mixture of mortar and special aggregate is injected into holes in the ground to compact the ground by expanding the hole diameters. On the other hand, in chemical solution injection method, colloidal silica solution is injected into holes in the ground to consolidate the ground. Use of colloidal silica as grout material permanently increases the strength of soil. Research works and results related to the development of this method are given at the end of this section.

d) Aprons (taking care of 640 departures and arrivals a day)

The Offshore Expansion Project provides not only loading aprons in the passenger, cargo, and maintenance terminal areas but also night-stay aprons for parking planes during nighttime, maintenance aprons for inspecting and cleaning planes, a compass swing apron for inspecting aircraft's compasses, etc. to secure the functions of Haneda Airport as the hub of the domestic aviation network.

The optimum type of pavement, basically concrete pavement, is employed for each of different aprons taking into consideration the function of the apron, the characteristic of ground settlement at the site, the method of repair of the apron, etc. In particular, for some of the aprons on the east and west sides of the project site, where differential settlement can occur even after the ground improvement, pre-stressed concrete (PC) pavement whose PC layer can be lifted by jacks is employed in order to permit damaged aprons to be repaired during nighttime and made available on the next morning.

2.4 Research and Development of New Chemical Grouting Method

A new grouting method was developed to resist earthquake induced liquefaction by strengthening ground by chemicals is injected into the soil directly beneath a structure, as shown in Fig.10. Series of laboratory tests involving the injection of chemicals into a large soil stratum resulted in the formation of cylindrical structures of improved ground with a diameter of 2.6 meters, confirming that the method would provide effective countermeasure against liquefaction.

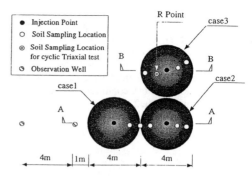

Fig.11 Plan of test facility

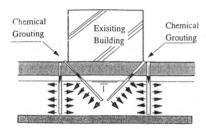

Fig.10 Method of chemical grouting to prevent liquefaction beneath existing structures

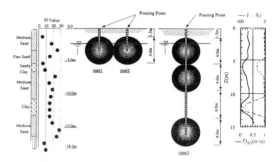

Fig.12 Cross-section of test facility

On the basis of these results, field tests involving chemical injections were carried out at a site in Niigata City. The tests clarified the application and penetration characteristics of the new chemicals. Excavations carried out after the injection process confirmed that solidification had occurred.

a) Outline of Tests

The tests were carried out from October to December 1996 in Niigata City to verify the effects of an injection of the new chemical and to assess the environmental impact. The test items consisted of: (1) grouting volume of acid silica sol (amount: 9.6m³), (2) Low-speed grouting of colloidal silica (amount: 9.6m³), (3) High-speed grouting of colloidal silica (amount: 28.8m³),(4) Soundings to confirm effects, (5) Excavation survey to check penetration area.

A plan of the test area is shown in Fig 11, and a cross-section of the site in Fig. 12. The aim of the tests was to gather data on the following two aspects of the application.

1. Wide-area injection from a single point (diameter of improved structure: 4m)
2. High-speed implementation using rapid injection (20 liters/minute)

The aim of the test was to create a spherical improved structure with a volume of 33.5 m³ through injection from a single point. The chemicals were injected through tubes in three locations to create five improved structures. The total volume of improved ground created through injection of the chemicals was 167m³.

b) Ground Conditions

The ground in the test area consisted of sandy soil interspersed with clay layers containing some silted sand at depths of 6~11 meters. Apart from these layers, the injection area consisted mainly of fine sand with a fine particle content ratio (Fc) of 5% or lower. The physical characteristics of soil at the test site are shown in Table 1.

Table 1 Physical properties of soil at Nigata test site

Depth GL(m)	Density of Soil Particle ρ_s(g/cm³)	Maximam Density ρ_{dmax}(g/cm³)	Minimum Density ρ_{dmax}(g/cm³)	Relative Density Dr(%)	Permeabili Coefficien k(cm/s)
-3~ -4	2.674	1.771	1.366	84	9.40×10⁻⁴
-7~ -8	2.736	1.741	1.313	77	4.80×10⁻⁴
13 -14	2.676	1.758	1.328	63	2.20×10⁻⁴

c) Test Details

Injection was carried out under three sets of conditions using three injection bores. The quantity of chemical injected and the volume of improved ground were the same in Case 1 and Case 2, while in Case 3, three sets of injections were carried out in a vertical direction. The test conditions aimed for each case are shown in Table 2.

Table 2 Test conditions aimed for each case

Case	Grout	Injection Volume m³	Objective Strength(qu) kPa	Improved Volume m³	Injection Rate 1/min	Injection Pressure	Preliminary Washing of Ground Water
1	Asid Silica Sol	9.6	80	33.5	20		none
2	Colloidal Silica	9.6	80	33.5	10	Precedent to Grouting Speed	50 ~ 100% of Injection Volume
3	Colloidal Silica	28.8	80	100.5	20		

The chemicals used for this experiment were superfine silica powder and non-alkaline silica, the permeability and durability of which had been ascertained through laboratory tests. Both are single-solution chemicals. The target improved strength was 80kPa (after curing for seven days). The injection rate for Case 1 and Case 3 was set at 20 liters per minute. The time required for the injection of the chemical in each location was, therefore, 8 hours.

In Case 2 and Case 3, in which superfine silica was used, setting time was substantially influenced by the salt content of the pore water in the ground. For this reason, specific amounts of piped water were injected into the ground prior to the injection of the chemical to remove the salt from the ground.

d) Test Results

One month after the completion of the injection process, ground strength measurements were taken using a tip resistance cone test. In Case 1, the chemical used was non-alkaline silica. Throughout the permeation area, qu was in excess of 200kPa, which is a high level of strength for a situation in which the aim is to prevent liquefaction. In Cases 2 and 3, where the chemical used was colloidal silica, strength was lower overall when compared with the results of the mixture tests carried out in the laboratory. With an average qu of around 50kPa, the improved strength was low in Case 2 and in the middle range of Case 3. However, the average qu in the upper and lower ranges of Case 3 was above the target level of 80kPa.

The improved structures created by filling the pores with silica chemicals were subjected to cyclic triaxial strength tests. In contrast to the unimproved ground, there was no sudden distortion attributable to rising pore water pressure ratios (the liquefaction phenomenon) Some of the samples obtained using the triple tube sampler were subjected to uniaxial compression tests and cyclic triaxial compression tests. As shown in Fig.11, the samples tested were those taken at Point R in Case 3. The results of cy-

clic triaxial tests on undrained soil at various depths are shown in Fig.13. The results of cyclic testing of untreated soil samples from near the base point are also shown for reference purposes.

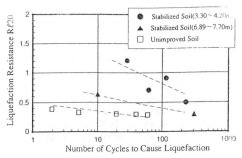

Fig.13 Relation between Rl20 and number of cycles

As is apparent from Fig.13, the results for the improved ground show considerable variation, as was the case with the uniaxial compression test results. However, liquefaction strength in improved soil in which the distortion amplitude (DA) reached 5% was significantly enhanced when compared with the unimproved soil. This shows that the chemical injections can substantially reduce liquefaction in the location where the process is applied.

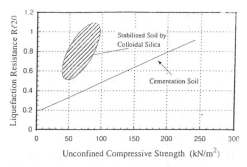

Fig.14 Increase in uniaxial undrained strength due to stabilization

Fig.14 shows the relationship between uniaxial compression strength (q_u) and the liquefaction strength ratio (Rl_{20}), which is the force ratio at which liquefaction is reached after 20 load cycles. There is moderate variation in the case of ground improved with colloidal silica. In the case of ground treated with cement, however, Rl_{20} tends to increase as qu rises. Under the conditions for these tests, the liquefaction strength (Rl_{20}) expected for a given uniaxial compression strength (q_u) is twice as high as with soil treated with cement. This shows that colloidal silica provides a greater strengthening effect.

To measure the extent and form of the improved ground, excavations were carried out 50 days after injection for Case 2 and Case 3. Fig. 15 shows an excavation in progress at GL-2.5m. No consolidation was observed to occur above the groundwater level, since solvent-type chemicals were used, and it was concluded on this basis that consolidation would not occur above this level. In both cases, the solidified ground was in the form of a cylinder with a diameter of 4.5~5.0m with the point of injection at the center.

Fig.15 Photographic view of stabilized soil after excavation

2.5 Construction of parallel taxiway and lightweight soil method

The parallel taxiway on the west side of the new A-Runway is required to serve as a runway in case of emergency. Therefore, like the new B-Runway, it is subjected to ground improvement to prevent liquefaction of the ground.

Having a length of 2,000 m and a width of 30 m, this taxiway guides a plane landed on the new B-Runway efficiently to the passenger terminal. Thus it works in close coordination with the new B-Runway to improve the runway capacity. When it is put into service, a seismic resistant guideway for aircraft from runway to apron will be completed.

Under the parallel taxiway on the west side of the new A-Runway there are large structures, such as access railroad tunnel, access road tunnels, and utility conduits. In the ground improvement, therefore, lightweight treated soil method and deep mixing method (CDM), as well as sand compaction pile (SCP) method, were used paying attention to their

effect on existing underground structures.

In the lightweight treated soil method, artificial lightweight soil having unit weight of 12kN/m^3 (excavated soil mixed with cement and air-foam) is filled to reduce the load on the tunnels and other underground structures, which provides a stable, light overburden, as shown in Fig.16. The CDM method solidifies the ground by mixing cement in the ground.

3 KANSAI INTERNATIONAL AIRPORT 2ND PHASE DEVELOPMENT

3.1 Overview of 2nd Phase Land Development Project

Kansai International Airport (KIA) was constructed on an artificial island in the Osaka Bay , 5 km off the Senshu area southwest of the Osaka City as shown in Fig.17. The airport was inaugurated in September 1994 when the first phase of construction was completed. Since the airport is currently being operated with only one runway, the second phase of construction is planned to focus on the construction of an ad-

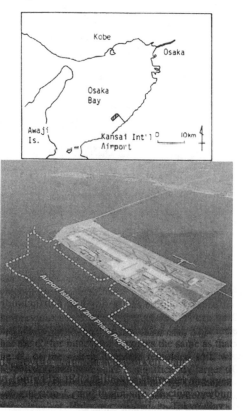

Fig.17 Location and plane view of Kansai International Airport

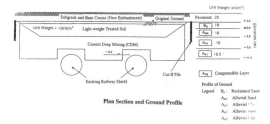

Fig.16 Improvement of parallel taxiway with lightweight treated soil method

ditional runway parallel to the existing one. The construction at the new phase started in summer 1999.

Table 3 shows the comparison of 1st and 2nd phase projects. The reclaimed area in the second phase has an average water depth of 19.5 m, in contrast to 18.0 m in the first phase. The average thickness of the Holocene clay layer in new site location is 24 m (18 m in the first phase), which will increase the overburden pressure on clay deposit by 40% due to reclamation compared to the first phase.

Table 3 Comparison of 1st and 2nd phase projects

	1st phase	2nd phase
Airport area	510ha	545ha
Runway length	3,500m	4,000m
Mean of sea depth	-18.0m	-19.5m
Mean settlement	11.5m	18m
Volume of fill	180mil.m^3	250mil.m^3

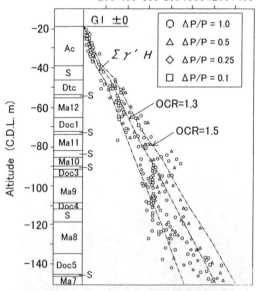

Fig.18 Typical soil profile at the KIA construction site

Fig.18 is the soil profile of the site, showing that the Pleistocene gravel-sand layers and clay layers are accumulated alternately up to 400m depth, and the over-consolidation ratio before the reclamation are 1.0-1.5.

To open Runway B for the planned 2007, the second phase land development project is underway with the schedule shown in Table 4.

The characteristics of construction works of 2nd phase projects are summarized as follows:
1) Restrictions determined by airport operations
   The 2nd phase project site is in an area that can effect aircraft take-offs and landings. However, matters shall be discussed and coordinated with the proper authorities as work progresses in order not to obstruct airport operations.
2) Supplies procurement
   Supplies will be procured with due consideration for local social-economic conditions and in a way that enables work to be completed in as little time as possible, in hope of an early opening of the new runway.
3) Nighttime work
   The scale of work planned for the 2nd phase project is much larger than that of the 1st phase project, but work must still be completed in almost the same amount of time. Therefore, work will proceed at night as well.
4) Work management
   From the success achieved in the 1st phase project, the latest measuring system (GPS) and high speed data processing will be utilized to conduct and manage work efficiently and effectively.

Table 4 Construction Schedule in Phase 2 Project

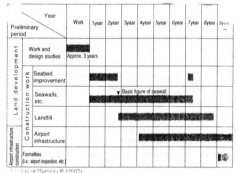

3.2 Seawall structure and improvement of Holocene clay layer by sand drain

The seawall structure in 2nd phase are almost same as those in 1st phase, which are the following 4 types:
a) rubble mound type seawall with wave dissipating block
b) rubble mound type seawall
c) wave-dissipating concrete caisson
d) steel cellular bulkhead type seawall

Rubble mound type seawall with wave-dissipating block

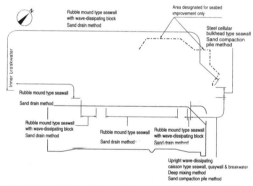

Rubble mound type seawall

Wave-dissipating concrete caisson

Steel cellular bulkhead type seawall

Fig.19(a) cross-sections of seawall structure

Fig.19(b) selection of seawall structure and the ground improvement in 2nd phase Project

Fig.19(a) shows the cross-sections of seawall. In Fig.19(b), the selection of structure and the ground improvement method in 14km length of seawall in 2nd phase project are shown. More than 80% of seawalls are rubble mound type seawall because they are most economical and environment-friendly. A wave-dissipating type concrete caisson will be adopted for quaywalls to keep the enclosed seawater calm. Seawalls connected to quaywalls are also to be constructed as wave-dissipating type concrete caisson or steel cellular bulkhead type seawalls.

In building the airport island, the improvement of seabed under the seawalls is carried out first. The seabed in the reclamation area will be imported after this and, once the basic figure of the seawall ha been built up, land will be reclaimed from the sea by direct dumping and above water placement. Materials will be spread and leveled to complete reclamation.

As shown in Fig.18, the seabed consists of extremely soft alluvial clay of about 25 m of thickness. The consolidation yield stress p_c of the Holocene clay layer is in a slightly overconsolidated state. To improve the Holocene clay layer, the large-scale sand drain method was used in the 1st phase project. The sand drain method is to strengthen the soft clay layer by shortening the drainage path in the layer by installing the sand column. As shown in Fig.20, after laying and spreading a thin sand blanket, sand piles of 40 cm diameter are driven by the special working ship. Under the weight of reclaimed soil, the pore water in the clay is forced out through the sand piles and sand blanket.

The effectiveness of sand drain was clearly shown in the 1st phase project.

Fig.21 shows the comparison of observed settlements and the calculated settlement in the monitoring area at the northern tip of the island, where extensive observations of consolidation settlement were carried out to check the effectiveness of sand drain. As shown in Fig.21, the observed settlement increased in proportion to the reclamation load and

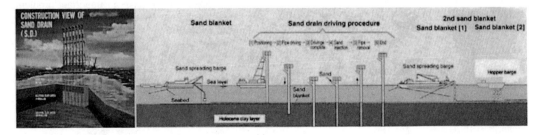

Fig.20 Construction of sand drain by special working ship

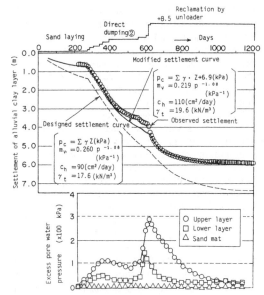

Fig.21 Observed and calculated time-settlement Relationship (Arai, et al., 1991)

ceased 9 months after the reclamation, in which the effect of the sand drain was clearly recognized. However, the observed settlement was smaller than the prediction from the early stage, and the discrepancy looked like widening with the progress of reclamation. Based on the observed data, the set of the soil constants were revised as follows. In the original design, the Holocene clay layer is assumed to be normally consolidated condition i.e. $p_c = \gamma' h$, where γ' is a submerged unit weight of clay and h is a depth. The coefficient of the volume compressibility m_v and the horizontal coefficient of consolidation c_v were determined as follows, by taking the average values of laboratory consolidation test data.

$$m_v = 0.18\, p^{-1.08} \quad (cm^2/kgf)$$
$$c_h = c_v = 90 \quad (cm^2/day)$$

where, p is effective overburden stress and is calculated assuming the unit weight of sand fill of 1.8 tf/m².

To improve the prediction based on Fig.14, the design parameters were modified as follows:

$$p_c = \gamma' h + 0.7 \quad (tf/m^2)$$
$$m_v = 0.15\, p^{-1.08} \quad (cm^2/kgf)$$
$$c_h = 110 \quad (cm^2/day)$$

The unit weight of sand fill was also revised to 2.0 tf/m² in accordance with the field survey.

In Fig.14 is also shown the modified calculated settlement, which actually coincides with the observed one. The modified soil constants were used for the construction control of the whole reclamation work of the airport island.

During the reclamation work, the measurements of the settlements of Holocene layer were carried out at 5 locations. Fig.22 shows a comparison of the observed settlements and those calculated with the modified soil constants, which shows that the modified soil constants gave accurate prediction on the settlements of the alluvial layer improved by the sand drain.

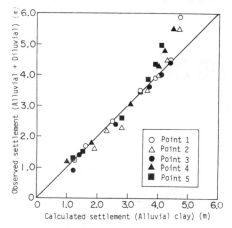

Fig.22 Measured and predicted settlement of Holocene layer

3.3 Settlement of Pleistocene Clay Layer

In 1986, when the airport was designed, the settlements were estimated as shown in Table 5 (Oikawa and Endo, 1990). It was estimated that, after 50years, the total settlement would be 8m, of which 6.5 m of settlement would occur due to compression of Holocene clay and 1.5m would be due to compression of the Pleistocene clay. Before that time, most of engineers hardly expected that the settlements of Pleistocene clay layers might make an important problem, because they did not have experiences of large settlements of Pleistocene layer.

Table 5 Original estimation of settlement at KIA

Layer	Estimated Settlement
Holocene clay	6.5 m
Pleistocene clay	1.5 m
Total	8 m

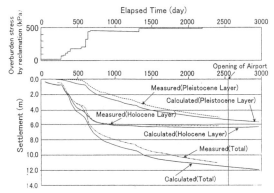

Fig.23 Measure settlement with time in KIA island

Fig.23 shows measured settlement with time at an observation point in KIA, where the largest settlement was observed (Kobayashi,1994). As shown in Fig.16, the consolidation settlement of Holocene clay layer improved by sand drain method ended up within 6 months after reclamation, while the settlement of Pleistocene clay layer commenced when the level of reclaimed land reached almost the sea level, and has continued after opening the airport.

Comparing the observed settlement of Pleistocene layer with the original prediction, the problem difficult to interpret was that the observed rate of settlement was many times faster than the prediction. Finally, it was concluded that the thin sand layers in the Pleistocene layer which had been thought to be discontinuous and incapable of draining the clay, must have been working as drainage layer. Accordingly a revised model to calculate the settlement was developed, and the prediction was modified in 1990 as shown in Table 6. As shown in Fig 23, the modified time-settlement relationship agrees well with the observed settlement.

Table 6 1990 Modified estimation of settlement at KIA

Layer	Estimated Settlement
Holocene clay	5.5 m
Pleistocene clay	5.5 m
Fill	0.6m
Total	11.6 m

The reclaimed area in the second phase has an average water depth of 19.5 m, in comparison with 18.0 m in the first phase. The average thickness of the Holocene clay layer in new site location is 24 m (18 m in the first phase), which will increase the volume of reclaimed sand or gravel by 40%, com-

pared to the first phase. The prediction of settlement in the second phase has been carried out based on the experiences of the first phase and the mean predicted total settlement in the reclaimed island is 18m (Table 6), which is probably the largest consolidation settlement geotechnical engineers have experienced. Therefore, in the second phase project, both the observation of settlement and the version-up of settlement prediction would be the most important works as the construction control.

Table 7 Estimation of settlement in 2nd phase project

Layer	Estimated Settlement
Holocene clay	7.0-8.0 m
Pleistocene clay	9.0-10.0 m
Fill	1.0 m
Total(average)	18.0 m

To explain the large discrepancy of settlement prediction, the mechanical properties of Osaka Bay Pleistocene clay have been studied extensively. Tanaka and Locat (1999) found that there exist abundant microfossils in Osaka Bay Pleistocene clay by a microstructural investigation. It has been reported that microfossils, such as diatoms, have significantly influence the engineering properties of soils (Shiwakoti et el. 1999). Fig.24 show scaning electron micrographs of Osaka Bay Clay, in which microfossils are observed. SEM According to Tanaka and Locat, the microfossils are acting as structural components, which provide the high compressibility of the clay.

Kang and Tsuchida carried out the separated-type consolidometer test of Osaka Bay Pleistocene Clay to study the strain and excess pore pressure distribution during the one-dimensional consolidation. (Kang and Tsuchida, 2000).

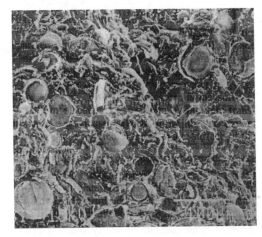

Fig.24 Scaning electron micrographs of Osaka Bay Clay

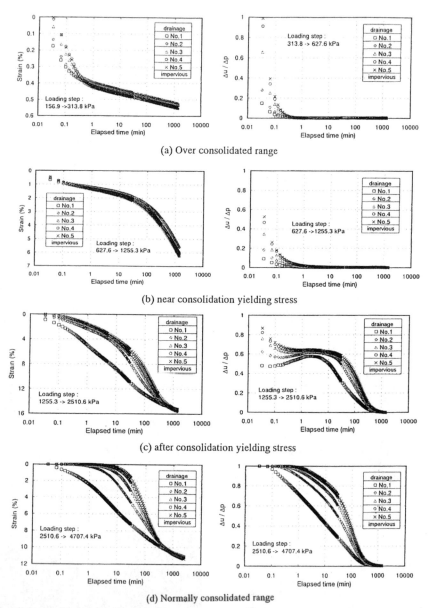

(a) Over consolidated range

(b) near consolidation yielding stress

(c) after consolidation yielding stress

(d) Normally consolidated range

Fig.25 Variation of strain and excess pore pressure as a function of time

Fig.25 shows the variation of strain and excess pore pressure as a function of time for each increment of stage loading. As shown in Fig.25 the consolidation behavior of Osaka Pleistocene clay shows evident contrast between the normally consolidated and the over-consolidated ranges. In the over-consolidated range, primary consolidation is over almost instantly and all subsequent strains for all sub-specimens continue at nearly constant rate of secondary compression. Excess pore pressure also shows instant dissipation within a minute, which means almost no excess pore water pressure is developed in the over consolidated region.

The dissipation characteristics of excess pore pressure display very unique aspect, when the loading stage exceeds the consolidation yielding stress. Particularly it is distinguishable to see the peculiar development of excess pore water pressure at each subspecimen. Once the variation of excess pore pressure shows the decreasing tendency at first, then it increases again to a notable degree after the soil experienced yielding. For the main reason of this, it

140

is considered that excess pore pressure once barely developed in the over-consolidated range is affected by the rapid increase of compressibility progressing in the loading stage shifting over to the normally-consolidated range. This means that the rate of increase in compressibility becomes much higher than the rate of excess pore water pressure dissipation when soil skeletons of Osaka Pleistocene clay with well-developed structure are yielded by a high overburden pressure.

In the normally consolidated range, the consolidation behavior of soil shows delayed compression with distinct delayed dissipation of excess pore water pressure from the start of drainage. Those behaviors in normally consolidated range reveal almost similar variation characteristics to the results for reconstituted sample.

4. CENTRAL JAPAN INTERNATIONAL AIRPORT PROJECT

4.1 Overviews

The Chubu (Central Japan) region has been an industrial center in Japan. The new airport, Central Japan International Airport (CJIA) is being constructed at an artificial island, which is 2km offshore from Tokoname city in Ise Bay and 35km distance from Nagoya city, the largest in Chubu region. The location and the shape of the airport island are shown in Fig.26.

The construction of CJIA started in August 2000, and the construction schedule and the outline of airport facility are shown in Fig.27 and Table 8, respectively. Fig.28 shows the soil profile at the construction site. As shown in Table 8 and Fig.28, the average water depth is 6m and the largest thickness

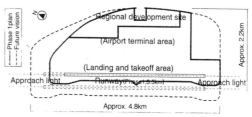

Fig.26 Location and master plan of CIJT

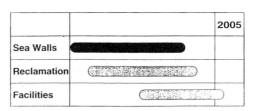

Fig.27 Construction schedule

of soft clayey layer laying on the rigid sedimentary rock is about 30m, which shows that the geotechnical conditions are much better than Kansai International Airport.

An important problem in CJIA project is the sup-

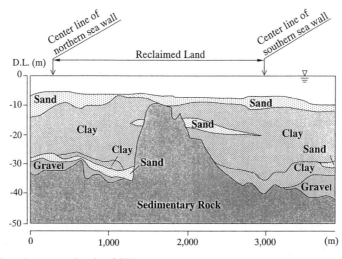

Fig.28 Soil profile at the construction site of CJIA

ply of fills. To construct the artificial island, the total of 56million m³ fill are necessary. Unlike Kansai International Airport, there are not many mountains to supply fills around Tokoname city. In addition, it becomes more difficult to obtain enough amount of soil because of environmental limitations recently. Meanwhile it also becomes difficult to find and construct disposal site for soft clay dredged from navigation channel in Ise Bay. Accordingly, the use of the dredged soft clay as reclamation material have been extensively studied. The island constructed with the dredged clay slurry is usually so week that no structures can be constructed without ground improvement. Although, as seen in Tokyo International Airport Project, the vertical drain method is frequently used for improvement of such a clay layer, the method requires a long time to complete the consolidation.

A new technique, named as Pneumatic Flow Mixing Method, has been developed to reclaim land necessary to construct the CJIA project. In this method, the dredged soft clay is mixed with small amount of stabilizing agent in the pipe during transfer by compressed air pressure and is deposited for sea reclamation. The method requires only stabilizing agent supplier facility to the existing pneumatic facilities. As no mixing blade is required and soil improvement can be performed continuously, this method is found to be suitable in reclaiming sea in relatively short time and in economical way (Kitazume et al 2000).

In CJIA project, a large volume (about 12 million cubic meters) of cement treated soil made by Pneumatic Flow Mixing Method will be used as the fills in the first time. As the clays used for the cement treatment will be taken by dredging in Nagoya Port, the use of pneumatic flow mixing method is contributing to a large scale recycling of waste soils.

Here the result of research and development of Pneumatic Flow Mixing Method are briefly introduced.

Table 8 Outline of Central Japan International Airport

Runway	3,500m × 60m
Airport area	470 ha
Total Investment	768 billion yen
Expected opening	2005
Future vision	Runways :4,000m × 60m × 2
	Airport area 700ha
Water depth	3m – 10m (average 6m)
Volume of fills	56 million m³

4.2 Development of Pipe Mixing Method

In the research project, laboratory mixing tests, centrifuge model tests and field tests were performed.

Here, field construction tests and their results have been described.

Transferring of soft clay in the pipe requires large pressure due to the friction generated on the inner surface of the pipe. When relatively large compressed air is applied into the pipe, however, the clay is divided into small blocks by air pockets in the pipe, as schematically shown in Fig. 29. The formation of the plug flow, which is composed of the clay and the air block, can function to reduce the friction on the pipe surface and in turn can considerably reduce the air pressure required. Generally, the plug flow can best be generated at an air-to-clay ratio of 20 to 400 (Akagawa, 1980).

The injected air pressure is dependent upon many factors such as the properties of clay, the volume of injected air, the pipe diameter, the pipe length and so on. The air pressure of 400 to 500 kN/m² is typically used taking into account the pressure bearing capacity of the pipeline.

Fig.30 shows mixing process with soft clay and stabilizing agent in the pipe during the transferring. Since the clay plugs are transferred at very high speed (of order of about 10 m/sec), the turbulent flow is generated within the plug due to the friction on the inner surface of the pipe, which causes thorough mixing of the clay and the stabilizing agent. It has been established that the thorough mixing can be obtained in the condition of Reynolds number of 500 to 3,000. At least 50 m to 100 m of pipe length is essential to ensure satisfactory mixing.

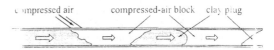

Fig. 29 Schematic view of plug flow

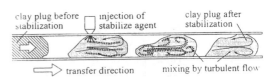

Fig. 30 Sketch of mixing process in the pipe

Fig. 31 illustrates how a typical Pneumatic Flow Mixing Method works. The dredged clay in the barge is loaded into the hopper on the pneumatic vessel at first, and is transferred by the compressed air to the reclamation site. The stabilizing agent is then injected to the clay on the stabilize agent supplier vessel and they are thoroughly mixed during the transferring. There are two types of the method according to where the stabilizing agent is put into; *compressor addition type* and *line addition* type In the former type, the stabilizing agent is put into the clay before the compressed air is injected into the

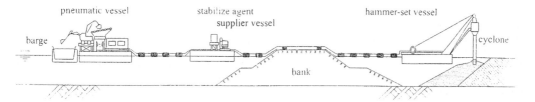

Fig. 31 Schematic flow of pneumatic transfer process

pipe. In the later type, on the other hand, the stabilizing agent is put into after the air injection, as shown in the figure.

The soil mixture is allowed to dump to reclamation site through the cyclone on the hammer-set vessel, which functions to reduce the air pressure transferring the clay plugs. There are several variations available in Japan, in which some equipment are installed in the pipeline to improve the mixture of the clay and agent.

4.3 Field test

Series of field executing tests were performed at Nagoya Port Island in Aichi Prefecture to investigate the effect of the properties of the soil mixture on the transferring.

The tests were performed to determine the optimum air pressure required at the inlet, which depends upon many factors mentioned before.

In the tests, the soil mixture having several combinations of water content and amount of cement and clay, were transferred. The air pressure changes along the pipe were measured in the tests. And also the unconfined compression tests carried out in order to investigate the average strength increase due to the treatment of the in-situ soil.

Vessels used for the field test includes the pneumatic vessel, the stabilizing agent supplier vessel and the hammer-set vessel. A special designed plug detection system is installed in the pipeline to monitor the plug movement and to improve the mixture of the clay and the stabilizing agent.

Soft clays used in the test were marine clay dredged in Nagoya Port. Although the marine clay was dredged at almost same site at Nagoya Port, its properties were found to be different in each test case. The dredged clays were diluted with seawater in the barge to obtain the prescribed water content of $1.3 * W_l$, where W_l is the liquid limit of the clay.

Stabilizing agent used was type-B slag cement, which is one of the most common stabilizing agent in Japan. Seawater was added to the cement in advance to make cement slurry with the water and cement ratio of 100 %, and then the cement slurry was injected directly to the clay plug by the help of the plug detection system mentioned before.

4.4 Results from the Field Test

a) *Characteristics of the clay plug*

Three tests were performed to monitor the plug characteristics by the plug detection system. In the test, the clay mixture having several combinations of the water content and the amount of cement were transferred by the air pressure of about 400 kN/m^2, as summarized in Table 9. Although the measured data have large scatter throughout the tests, it can be found that the clay plugs with an average volume of 0.36 m^3 are transferred at average speed of about 12 m/s and an average interval of about 6 seconds.

Table 9 Test results on plug characteristics

	Case 1	Case 2	Case 3
Test condition			
transferred soil	210 m^3/hr	296 m^3/hr	170 m^3/hr
water content	132.5 %	117.2 %	96.2 %
Liquid limit	70.5	77.6	81.5
cement added	38 kg/m^3	78 kg/m^3	52 kg/m^3
Test results			
plug speed	10.9 m/sec	11.9 m/sec	12.8 m/sec
	(1.6– 25.0)	(1.5 – 25.0)	(1.9 – 25.0)
plug volume	0.41 m^3	0.36 m^3	0.30 m^3
	(0.23–0.52)	(0.25–0.45)	(0.18–0.36)
plug length	4.3 m	3.7 m	3.1 m
	(0.23– 5.4)	(2.6 – 4.7)	(1.9 – 3.8)
plug interval	7.1 sec	4.4 sec	6.4 sec
	(1.3– 30.3)	(0.5 – 18.2)	(0.6 – 29.0)

b) *Air pressure distribution in the pipe*

Another series of tests was performed to investigate the air pressure changes during the transferring, in which the air pressure was measured at five locations along the pipe. In the tests, the water content, the transferred clay volume and the amount of cement were changed to 98 % to 128 %, 182 m^3/hr and 300 m^3/hr and 0 to 80 kg/m^3, respectively.

The measured air pressure changes are plotted in Fig.32 along the transfer distance from the inlet. Relatively large decrease occurred in the pressure between the pneumatic vessel (P_0) and the stabilizing agent supplier vessel (P_1) irrespective of the test condition. These large decreases are probably because the clay in the pipe is still unstable to form the plug flow. After at the point P_1, the air pressures decrease almost linearly to zero at the outlet (P_4) with increase of the transfer distance. These decreases in the air pressure are thought to be attributable to the wall friction on the pipe.

The effect of clay volume transferred on the air pressure decrease is also shown in the figure. The air pressure at the inlet (P_0) becomes large with increase of the clay transferred. As far as the test condition, the required air pressure should be increased to about 100 kN/m^2 when the clay volume transferred increases from 200 m^3/hr to 300 m^3/hr. It is also found that the addition of cement causes increase in the pressure by about 100 kN/m^2 at the inlet (P_0), because the cohesion of the mixture becomes large.

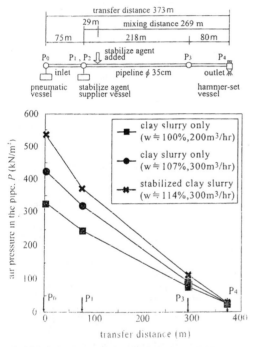

Fig.32 Relation between air pressure and transfer distance

c) *Strength of treated soil*

To investigate the strength profile of the treated soil, other field tests were performed in which the amount of cement added was changed to 38 kg/m^3, 57 kg/m^3 and 68 kg/m^3. In these tests, the clay slurry mixed with the cement was reclaimed into the small ponds excavated in advance. The unconfined compression tests were conducted on the treated soil manufactured in-situ.

Fig.33 shows the strength deviation of the 28 days cured treated soil, which was manufactured with the

initial water content of 125 % and the cement of 57 kg/m^3. In the figure, the strength deviation shows almost similar shape of the standard deviation profile. The coefficient of deviation of the strength is 32.3 %, which is almost same order to that of Deep Mixing Method irrespective of the mixing procedure.

Fig.34 shows test summary of the unconfined compressive strength, in which the average strengths are plotted against the amount of stabilizing agent. The average strengths of the treated soil increase almost linearly with increase of the amount of cement. Other unconfined compression tests were also performed on the field-manufactured specimens whose diameter and height were 50 cm and 100 cm respectively. Their average strengths are also plotted in the figure. It is found that the compressive strength on the large specimen also increases with increase of the amount of cement, but the average strength on the large sized specimen is smaller than that of small

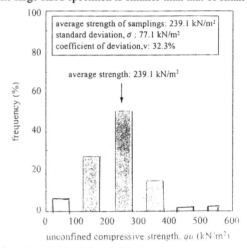

Fig.33 Deviation of treated soil strength

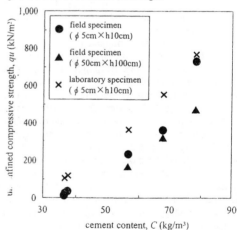

Fig.34 Unconfined compressive strength against the amount of cement

sized specimen. The strength ratio of the large specimen against the small sized specimen is about 0.7, which is almost same as the treated soil manufactured by the Deep Mixing Method. In the figure, test results on the treated soil manufactured in laboratory are also plotted. Laboratory samples found to yield the largest strength among three test specimens.

It was found that the coefficient of deviation on the laboratory samples is relatively small (15 %), and almost constant irrespective of the amount of cement. This means that quite uniform mixing can be obtained in the mixer in a laboratory. The field-manufactured specimens, on the other hand, indicate relatively larger deviation of strength compared to the laboratory specimen. It is found that the field specimen with small size is about 35 % irrespective of the amount of stabilize agent. But the specimen with large size showed almost same coefficient as the small sized specimen as far as the amount of cement exceeds about 50 kg/m^3 but increases in the coefficient to about 60 % when the amount of agent decreases to 38 kg/m^3.

The research, thus, established that the Pneumatic Flow Mixing Method has relatively high applicability for construction of man-made island, sea reclamation and back filling. Fig.35 indicates various types of expected applications of the Pneumatic Flow Mixing Method.

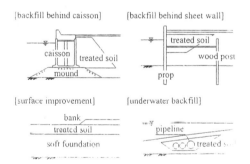

Fig.35 Expected applications of the Pneumatic Flow Mixing Method

5 CONCLUSIONS

Today, coastal airports are increasingly being constructed in many parts of Asia, because of the limited availability of suitable place on land, growing environmental concern, technical feasibility of such projects and consideration of cost- effectiveness. Oftentimes, such coastal land development works possess unique characteristics, inviting new challenges, and opening avenues for further technological advancement. This paper reviews three coastal airports, namely, TIA, KIA and CJIA, each constructed on coastal soft reclaimed land. It highlights exclusive

problem evolved in developing each airport, and describes how they are dealt with.

The TIA was constructed on the reclaimed land, which consisted of dredged slurry and waste soil. Large scale seismic resistant ground improvement works had to be carried out to strengthen the runway against the anticipated strong earthquake motion. To cope with the problem, a new ground improvement technique- involving chemical grouting- was developed and applied in this project. Furthermore, an artificial light-weight treated soil method-consisting of air foam and cement mixture- was developed and applied in constructing new taxi-way, which had to be constructed above existing railway tunnel.

The KIA, which is being developed in an artificial island 5 km off the Osaka Bay, the 1st phase of which was completed in 1994, and whose second phase construction is underway, possesses unprecedented settlement problem. The problems related to the 2nd phase construction project, have been introduced. The improvement of Holocene clay layer and the prediction of settlements of deep Pleistocene cay layers are key the issues in this project.

On the other hand, the CJIA construction project has just been started, for which appropriate fill materials are not available near by, and for which the disposal of dredged waste material has been big problem. Therefore, a new technique for recycling dredged waste material, by treating it with cement, has been developed. Use of large volumes of cement treated waste soils, made by the pneumatic flow mixing method, is planned to be used as the fill material for the first time, making this a unique project to use large scale recycled waste soils.

REFERENCES

Akagawa, K. 1980. Gas-liquid two-phase flow. *Mechanical Engineering 11*: 15. (in Japanese).

Kang, M.-S. & Takashi, T.(2000): Experimental study on the consolidation properties of Osaka Pleistocene clay by separated-type consolidometer test, *IS-Yokohama2000*.

Kitazume, M. (1997): Centrifuge model tests on stability of embankment improved by cement, *Proc. of 32th Annual Conference of Japanese Geotechnical Society*: 2429-2430. (in Japanese).

Kobayashi, M. (1994):"Geotechnical engineering aspects of Kansai International Airport Project", *Proceedings of 39th Geo-engineering Symposium, JSSMFE*, pp.1-6. (in Japanese)

Shiwakoti, D. R., Tanaka, H., Locat, J. and Christine G. (1999). Influence of Microfossils on the Behavior of Cohesive Soil. *11th Asian Regional*

Tanaka, H. and Locat, J. (1999): "A microstructural investigation of Osaka Bay clay: the impact of microfossils on its mechanical behavior." *Canadian Geotechnical Journal*, Vol.36, pp.493-508.

Tsuchida, T.: Evaluation method of structure of aged marine clays, *IS-Yokohama2000*, 2000.9.

Tsuchida, T(2000): "Mechanical properties of Pleistocene clay and evaluation of structure due to aging", Special lecture, *Proceedings of IS-Yokohama 2000*.

Kitazume M., Yoshino N., Shinsha H, Hori R. and Fujino Y. (2000). Field test on pneumatic flow mixing method for sea reclamation, International Symposium on Coastal Geotechnical Engineering in Practice, *IS Yokohama 2000*.

Hyashi K., Yoshikawa R., Hayashi N., Zen K., Yamazaki H. (2000). A field test on new chemical grouting method to improve the liquefaction resistance of sandy layers beneath the existing structures. International Symposium on Coastal Geotechnical Engineering in Practice, *IS Yokohama 2000*.

Special discussion session

Engineering characterization of the Nile Delta clays

D.W.Hight
Geotechnical Consulting Group, London, UK

M.M.Hamza & A.S.El Sayed
Hamza Associates, Cairo, Egypt

ABSTRACT: A ground investigation for a new quay wall at East Port Said, Egypt, is described. The paper focusses on the engineering characterisation of deltaic clays at the site which are of unusually high plasticity and extend to 60m below ground level. The investigation relied on the piezocone and continuous sampling to establish the stratigraphy, soil fabric, and natural variability. High quality samples were taken, using modified Shelby tubes and these were subjected to a limited amount of non-routine testing to measure undrained strengths, compressibility and stiffness. Gas exsolution occurred during sampling and its effects are discussed. Sample quality was evaluated systematically and allowed for in interpretation. This enabled induced variability in the data to be eliminated and gave confidence in the interpreted stress history for the clays, removing a prevailing view that they were underconsolidated.

INTRODUCTION

A new 1.2km long quay wall, container terminal and canal access are planned at East Port Said, Egypt, at the location shown in Figure 1. Extensive deposits of deltaic clays - the Nile Delta clays - are present at the site and extend to depths of 60m below ground level. This paper describes key aspects of the ground investigation that was carried out for the quay wall, its interpretation and the engineering characterisation of the deltaic clays. As with the investigation, the paper focusses on the quality of the samples that were obtained.

AIMS AND PHILOSOPHY OF THE GROUND INVESTIGATION

Previous conventional ground investigations in the vicinity had indicated major variability in undrained strength of the deltaic clays and the possibility of their being underconsolidated, on the basis of 24 hour incremental load (IL) oedometer tests on Shelby tube samples. Results of a typical conventional investigation are shown in Figure 2.

The quay wall has been designed as a novel retaining structure, using deep rectangular barrettes and involving T panel diaphragm walls. Its design,

and the prediction of ground movements during and after dredging in front of the wall, required the use of sophisticated 3D finite difference analyses.

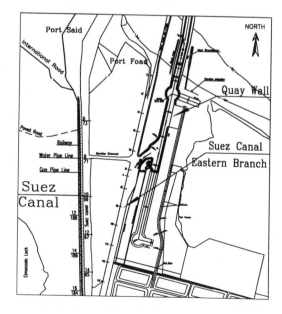

Figure 1. Site location plan

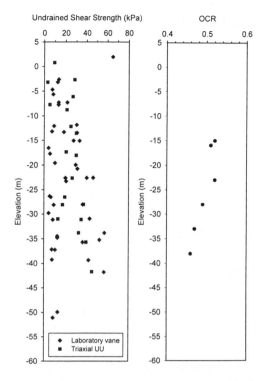

Undrained Shear Strength (kPa)

OCR

* Laboratory vane
■ Triaxial UU

Figure 2. Results of typical conventional investigation of Nile Delta clays

The aims of the ground investigation were to provide:

(i) details of the stratigraphy along the line of the wall,
(ii) the best estimate of engineering properties, from which suitable model parameters could be selected for the 3D analyses,
(iii) conventional design parameters for check calculations to be made,
(iv) sufficient information to assess construction risks.

The philosophy that was adopted to achieve these aims was:

(i) to determine lateral and vertical variability using the piezocone (CPTU) as a profiling tool,
(ii) to take the highest quality samples and assess the quality of individual samples,
(iii) to determine engineering properties in the minimum number of high quality laboratory tests,
(iv) to record the detailed fabric of the soils, including photographs, in continuously sampled boreholes,
(v) to establish site specific correlations between

CPTU parameters, fabric and engineering properties.

The layout of the ground investigation is shown in Figure 3 and comprised:

(i) four boreholes (BH1,6,6A and 10), in which fixed piston, pushed open tube and Mazier sampling was carried out,
(ii) three borings (SPT1, SPT6, SPT10), alongside BH1, 6 and 10, in which standard penetration tests (SPTs) and in situ vane tests were carried out,
(iii) ten cone penetration tests with measurement of penetration pore pressures (QCPTU 1 to 10).

To obtain high quality tube samples, 700mm long stainless steel Shelby tubes were used, on which the cutting edge was sharpened to approximately 5°. The tubes had no inside clearance, outside diameters of 105mm or 86mm, and wall thicknesses of 3mm. The sampling tubes were used either in conjunction with hydraulic Osterberg piston samplers down to 25m b.g.l. and as pushed open tubes at depths below 25m b.g.l., where the piston sampler would not function satisfactorily. 105mm and 86mm OD pushed open tubes were used for the full depth in BH6A. The Mazier rotary coring system utilised a triple tube core

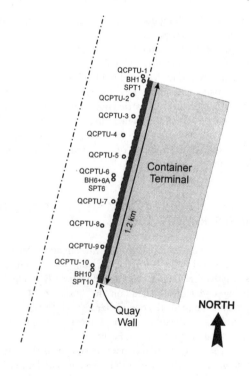

Figure 3. Layout of ground investigation

barrel, the inner tube of which was rigid plastic. Cores of 101mm and 86mm diameter were obtained.

ORGANISATION

The tender documents, prepared by the designers, Hamza Associates, called for an investigation which was to be sponsored jointly by the tenderers, who were asked to ensure that it met all their requirements and provided the parameters necessary for the selection of their particular construction methods and for their temporary works design. To satisfy different tenderers, Menard pressuremeter and Marchetti dilatometer tests were included, but their results are not discussed in this paper. Continuous technical supervision was provided.

The investigation was carried out under normal commercial pressures, with restricted time and budget. The fieldwork was performed by Geotechnical, of Egypt, and the laboratory testing was split between the Norwegian Geotechnical Institute (NGI) and Hamza Associates (HA). Four carefully selected sample tubes from BH6A were carried to NGI for non-routine tests.

STRATIGRAPHY

The stratigraphy determined on the basis of the CPTU records and borehole logs is shown in Figure 4. The following units were identified:

A   Hydraulic fill/dredge arisings which vary in composition along the length of the wall from silty sand at BH1 to plastic clay at BH10. The layer varies between 2.5m and 5m thick.

B   Sand with broken shells, representing the beach before unit A was placed; the sand is described as slightly micaceous. The layer is between 5m and 6m thick, with a relative density that reduces from 90% at the top to 50% at the base.

C   A transition layer, 5.2m to 7m thick, between units B and D which comprises interbedded sandy silts, clayey silts and clay; clayey silt layers thicken towards the base.

D   Plastic organic clay with occasional sand and silt seams.

E   Very plastic organic clay.

The overall thickness of units D and E reduces from approximately 45m in the north (BH1) to 40m in the south (BH10)

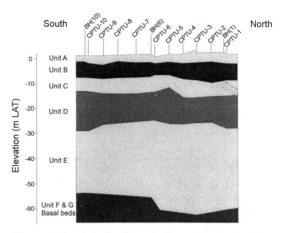

Figure 4. Stratigraphy along the line of the quay wall

F & G Basal beds which comprise sand layers (unit F) and laminated clay layers (unit G). The sands are occasionally cemented by carbonates.

Additional information obtained during construction has confirmed a reasonably consistent top elevation to the basal beds of between -53.2 and -57.6m LAT.

Units B to E are Holocene deposits which coarsen upwards and are thought to have been sedimented at a rate of approximately 500cm/1000 years, with short degradational episodes. Units D and E are regarded as deltaic, deposited in a saline environment; they are distinguished on the basis of the more frequent presence of sand/silt seams in unit D c.f unit E and of a small change in slope of the qc profile. The basal beds are Pleistocene, and are thought to have accumulated in the last interglacial period, either sub-aerially or in a marine environment. Two relatively recent ground lowerings are thought to have occurred.

Unit A was placed as hydraulic fill up to 1977 and has generated excess pore pressures in units D and E as a result of their undrained loading.

INDEX PROPERTIES

The variation in clay content, plasticity index, liquidity index and bulk unit weight with depth are shown in Figure 5. The clay content increases monotonically with depth throughout units D and E. A similar pattern of increasing clay content with depth appears to apply to unit G. Clay contents are high, particularly towards the base of units E and G.

The mineralogy of the clays forming units D and E has been investigated using X ray diffraction. Smectite is the dominant clay mineral and there is a tendency

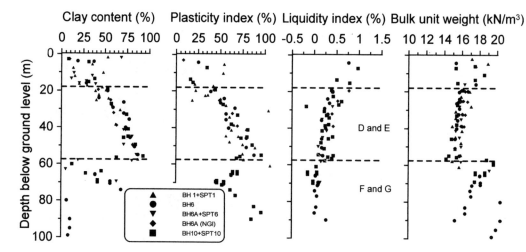

Figure 5. Profiles of index properties

for the smectite content to increase with depth through the units, from approximately 30% to 50%, probably reflecting the overall increase in clay content with depth.

The clays have a significant organic content, between 5.3% and 8.9%, which may influence to some extent compressibility, water content and measurement of Atterberg limits. Chloride contents can be expected to be high, especially in the upper parts, because of sabhka development and the saline environment of deposition.

Plasticity index, Ip, increases almost linearly with depth through units D and E, reaching values as high as 95% at the base of unit E. A similar increase in Ip with depth appears to occur in unit G, from 45% at its top to 95% at its base. In both cases the increase in plasticity index reflects the increase in clay fraction. There is, therefore, a reasonably constant activity (plasticity index/clay content) with depth of between 0.7 and 1.1.

Liquidity index, I_l, reduces with depth through units D and E, from approximately 0.5 to 0.1. I_l in unit G appears to be almost constant, just above zero, although this value may have been affected by swelling of the laminated clay samples on stress relief.

Bulk unit weight, γ, appears to reduce with depth through units D and E, reflecting the increasing clay content and water content. Reasonable averages are $16kN/m^3$ for unit D and $15.5kN/m^3$ for unit E. These are low values, consistent with the high plasticity and clay content. There is a pronounced step in bulk unit weight on entering the basal beds, although the same trend of reducing γ with depth can be seen.

CPTU RECORDS AND NATURAL VARIABILITY

A typical CPTU record showing the interpreted boundaries between units A, B, C, D and E is shown in Figure 6. Profiles of cone tip resistance, q_c, and penetration pore pressures measured at the cone shoulder, u_2, in units C, D and E in QCPTU 1 to 10 are superimposed in Figure 7. These indicate that the materials between -12.0m LAT and -37.5m LAT are remarkably consistent along the line of the quay wall. The greater density of silt seams in unit D c.f. unit E can be seen in Figures 6 and 7.

SAMPLE QUALITY

The particular sample tube geometry should have minimised the strains imposed during the penetration of the tube. Because of the clay's high plasticity, the soil should have been tolerant of these levels of strain, and high quality samples should have been obtained.

Sample quality has been evaluated on the basis of:
(i) the initial effective stress, p_i', measured in UU triaxial compression tests (Figure 8(a));
(ii) values of G_{max} measured on samples reconsolidated to in situ stresses in the simple shear, DSS, apparatus and compared to in situ values (Figure 8(b));
(iii) the magnitude of the volumetric strains during reconsolidation to in situ effective stresses in oedometer and CAU and CIU triaxial tests, expressed as the ratio de/e_o, where de is the

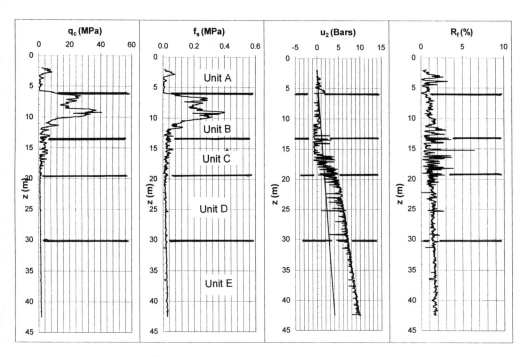

Figure 6. Typical CPTU profile

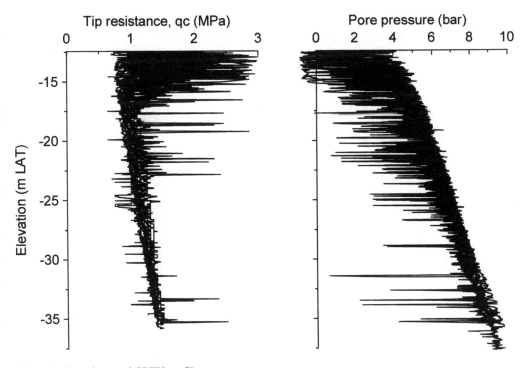

Figure 7. Superimposed CPTU profiles

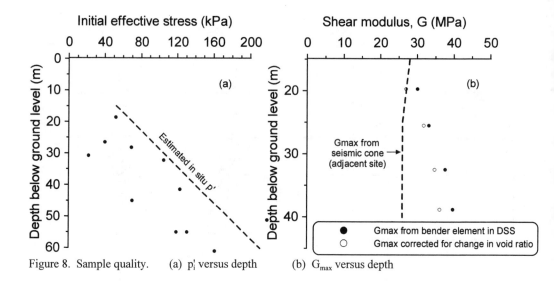

Figure 8. Sample quality.   (a) p'_i versus depth   (b) G_{max} versus depth

change in void ratio and e_o is the initial void ratio, following Lunne et al (1997) (Figures 9(a) and (b)).

Measurements of p'_i were, with one or two exceptions, reasonably close to the values that might be expected after perfect sampling in these plastic clays, i.e. reasonably close to the in situ values of mean effective stress, p'_o, indicating that sample quality was likely to be good. The values of G_{max} measured in the DSS tests were, surprisingly, higher than values measured in situ with the seismic cone at an adjacent site, even after allowing for the effects of reduction in void ratio after reconsolidation of the DSS samples. This discrepancy,

which may be related to differences in frequency between the two types of measurement, requires further investigation.

Lunne et al (1997) have proposed the four categories of sample quality shown in Figure 9 and based on de/e$_o$. On the basis of measured compressibility curves, and for this investigation only, we have modified the boundary between good to fair and poor samples to be de/e$_o$ of 0.081.

It can be seen from Figures 9(a) and (b) that the majority of the samples are very good to fair. The quality of each sub-sample taken from the four tubes tested at NGI was assessed and the results are shown

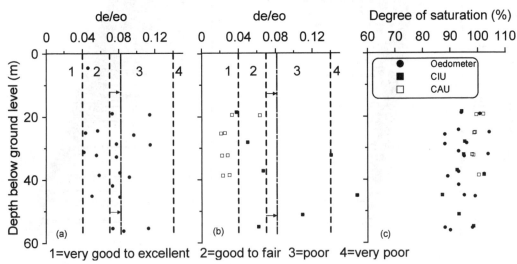

1=very good to excellent 2=good to fair 3=poor 4=very poor

Figure 9. Sample quality on basis of (a) and (b) volumetric strains during reconsolidation to in situ stresses, and (c) degree of saturation

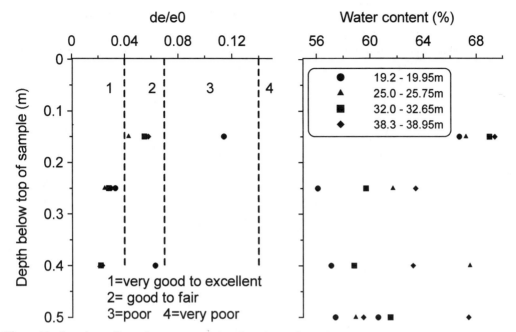

Figure 10. Sample quality and water content in selected sampling tubes

in Figure 10. These reveal:

(i) the lower quality of the sample from 19.2m, which was traced to a distorted sampling tube, and

(ii) the lower quality of the sub-samples taken towards the top of each tube - this is consistent with these sub-samples being of higher water content and with the samples having to be taken as open tube rather than piston samples.

It became evident that several of the poor samples were associated either with the presence of silt laminae or layers, or with low degrees of saturation. In the case of the former, the silt laminae were not able to sustain the required suction and water content redistribution and swelling of the clay occurred; the potential for swelling was high because of the high plasticity of the clays.

Gas exsolution was suspected, not only on the basis of measured degrees of saturation (Figure 9(c)), but also on the basis of one-dimensional compressibility (Figure 11) and of effective stress paths during undrained shear (Figure 12). Gas exsolution was not unexpected, bearing in mind the depth from which the samples were being taken, their high in situ void ratios, and their significant organic content.

The effect of gas exsolution on compressibility in conventional oedometer tests, i.e. without back

pressure, is illustrated in Figure 11 as causing increased compressibility, damage to structure and reduced values of vertical yield stress. The effect of gas exsolution on the results of undrained shear in CIU triaxial compression tests is illustrated in Figure 12 as causing a discontinuity in the effective stress path, when pore pressures rapidly increase, presumably as pores enlarged by gas exsolution now collapse.

ONE DIMENSIONAL COMPRESSIBILITY

Pattern for intact samples

The results of four incremental load oedometer tests performed by NGI are superimposed in Figure 13(a). These reveal the effect of increasing plasticity with depth, both in terms of increasing initial void ratio, increasing compressibility and increasing importance of creep. The damage to the sample from a depth of 19.5m, resulting from the distorted sampling tube, is also evident.

After eliminating the results of tests on samples affected by gas exsolution and by other forms of disturbance, two patterns of one-dimensional compressibility emerged from the remaining IL oedometer tests - these are illustrated in Figure 13(b).

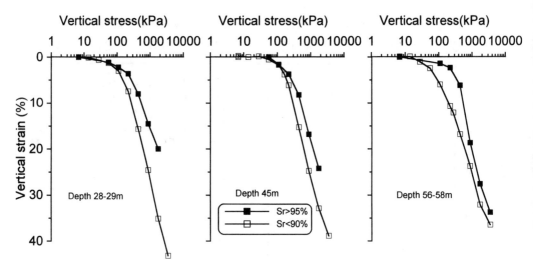

Figure 11. Effect of gas exsolution on compressibility

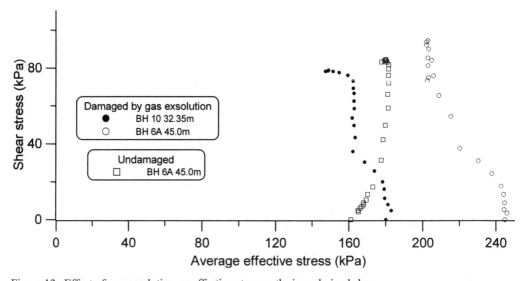

Figure 12. Effect of gas exsolution on effective stress paths in undrained shear

Samples showing lower overall compressibility are almost certainly those having a layered fabric.

Pattern for intact versus reconstituted samples
Oedometer tests were run on samples which had been reconstituted at a water content equal to their liquid limit. Typical comparisons between intact and reconstituted samples from the same depth are shown in Figure 14. The intrinsic parameters e_{100}^* and C_c^*, defined by Burland (1990) and measured on the reconstituted samples, are shown in comparison to

published correlations in Figure 15. It is evident that:

(i) The void ratio of the reconstituted soil at a vertical effective stress of 100 kPa, e_{100}^*, is, in most cases, appreciably higher than for other clays. This would suggest the presence, for example, of diatoms which increase water content without increasing compressibility.

(ii) In one case (samples from 29.0m), the intact sample shows the classic response of a mildly structured soil in relation to that of the reconstituted soil.

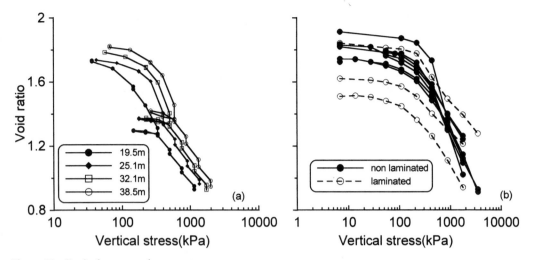

Figure 13. Typical compression curves

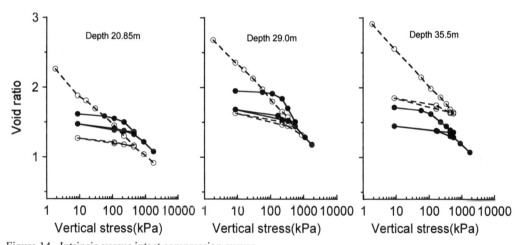

Figure 14. Intrinsic versus intact compression curves

(ii) In the case of samples from 35.5m, the compression curve for the intact sample lies below that of the reconstituted sample, suggesting the presence of a layered fabric in the intact soil that was not reproduced in the reconstituted soil. The importance of laminae was seen in comparisons of intact and reconstituted soil behaviour in the laminated facies of the Bothkennar Clay (Hight et al, 1992).

Overconsolidation ratio

Overconsolidation ratio, OCR, has been determined from incremental load oedometer tests by HA on the basis of Casagrande constructions. The compression curves used for estimating OCR are those applying to

the end-of-primary consolidation (eop). OCR has been determined from the NGI incremental load oedometer tests using both the Casagrande construction and a method due to Janbu (1963) - the latter values have been plotted. In the NGI tests, load durations varied from 60 minutes for loading to σ'_{vy}, 150 minutes for loading from σ'_{vy} to $2\sigma'_{vy}$ and 24 hours for increments beyond $2\sigma'_{vy}$, where σ'_{vy} is the vertical yield stress.

The estimated values of OCR are plotted versus depth in Figure 16(a). Interpretation of this data requires that account is taken of the quality of each sample and the degree of saturation, S_r, in each. Eliminating results from poor quality samples, having de/e_o greater than 0.081 and $S_r < 0.9$, and from

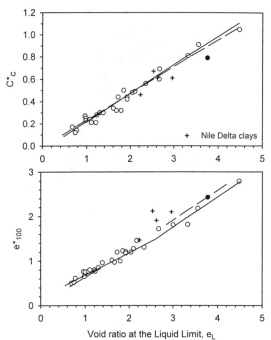

Figure 15. Intrinsic parameters, e* and C_c*

layers D and E, from approximately 1.4 to 1.2; the precise value of OCR depends on the method of interpretation and the test duration. The conclusion is quite different to that from a conventional investigation (Figure 2).

The interpreted values of $(\sigma'_{vy}-\sigma'_{vo})$ have been plotted against depth in Figure 16(b). Again, a reasonably clear picture emerges after eliminating data from poor quality samples, with $(\sigma'_{vy}-\sigma'_{vo})$ approximately constant with depth in units D and E, between 60 and 80 kPa. This would be consistent with the effects of a previous, but recent, reduction in groundwater level, and subsequent rise.

A similar exercise has been carried out for the basal beds which indicates that OCRs are 1.9 to 2.0 and that $(\sigma'_{vy} - \sigma'_{vo})$ reduces with depth in each main clay layer, which would be consistent with the effects of drying from the surface of each layer and with the sub-aerial depositional environment.

Compressibility parameters

An interpretation of the oedometer compression curves has been made to determine:

(i)   the compression index, Cc, i.e. the slope of e-log σ'_v curve in the normally consolidated region - the virgin compression line,

(ii) the recompression index, Cr, i.e. the average slope of the unloading-reloading e-log σ'_v curve,

(iii) the void ratio at unit pressure, e_1, found by extrapolation of the virgin consolidation line to a pressure of 1 kPa.

laminated samples, refer Figure 16(a), a clear picture emerges, with OCR reducing with depth through

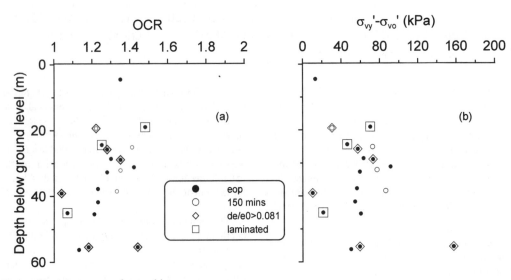

Figure 16. Assessment of stress history

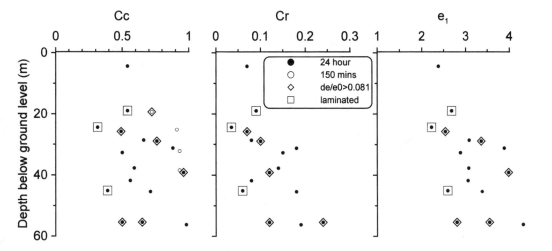

Figure 17. Assessment of compressibility parameters

Variations of Cc, Cr and e_1 with depth are presented in Figure 17. Again, the data has been reviewed in the light of sample quality and the problem of gas exsolution. Low values of Cc and e_1 were associated with damaged, desaturated samples. The values selected for analysis are shown in Table 1.

TABLE 1

Unit	Cc	Cr	e_1	$C_{v \, oc}$	$C_{v \, nc}$
				m²/yr	
D	0.65	0.08	3.0	4.5-9	0.9-2
E	0.75	0.15	3.5	3-8	0.9-2

Coefficients of consolidation
Data from the oedometer tests suggested the values shown in Table 1 for C_v for vertical drainage.

The presence of the sand seams within unit D will lead to higher values for C_h.

UNDRAINED STRENGTH

Undrained strengths of units D and E have been measured in:
(i) unconsolidated undrained triaxial compression tests, with pore pressure measurement (UU) - these tests were run partly to provide conventional design parameters,

(ii) isotropically consolidated undrained triaxial compression tests, with pore pressure measurement, involving consolidation to either:
- in situ mean effective stress, po' (CIU(po')), or
- twice po' (CIU(2po')),
(iii) undrained triaxial compression and extension tests on samples anisotropically consolidated to estimated in situ stresses (CAUTC and CAUTE),
(iv) undrained (constant volume) simple shear (DSS) tests on samples one dimensionally consolidated to estimated in situ vertical effective stress.

The test programme was designed so that the UU, CIU (p_0') and CIU ($2p_0'$) tests were run on sets of three specimens from the same tube and the CAUTC, CAUTE and DSS tests were also run on sets of three specimens from the same tube.

The strengths measured in the triaxial tests have been corrected for the effects of the restraint imposed by both the membrane and filter paper drains

The results of the CAUTC, CAUTE and DSS tests are summarised in Figures 18 and 19 as effective stress paths and stress-strain curves. Undrained strength profiles are presented in Figure 20. Figure 20(a) shows only the results of the CAUTC, CAUTE, DSS and CIU (po') tests. Figure 20(b) shows only the results of the UU tests. Figure 20(c) presents the strength ratios.

The following may be noted:
(i) there is significant anisotropy of undrained strength in the clays; the ratio of undrained strength in triaxial extension to that in triaxial compression is approximately 0.7;

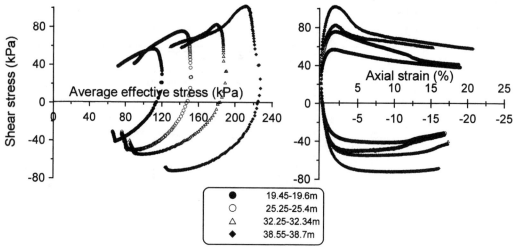

Figure 18. Undrained shear in triaxial compression and extension

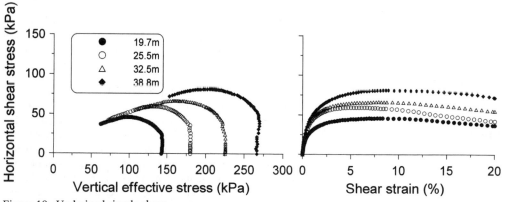

Figure 19. Undrained simple shear

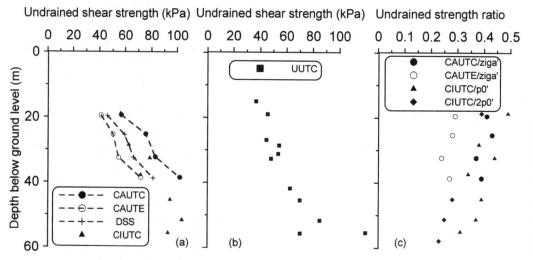

Figure 20. Undrained strength profiles

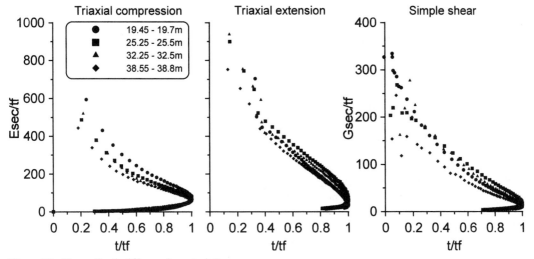

Figure 21. Normalised stiffness characteristics

(ii) there is undrained brittleness in triaxial compression;

(iii) strengths increase more or less uniformly with depth; the pattern identified on the basis of the CAU tests is confirmed and extended by the results of the CIU tests;

(iv) the variability evident in previous investigations, even in UU tests, is largely eliminated and the pattern of an increase in strength with depth is consistent with the CPTU records;

(v) strength ratios, c_u/σ_{vo}', are consistent with a lightly overconsolidated clay; strength ratios from CIU (po') tests show an overall trend to decrease with depth, reflecting the increasing plasticity and reducing OCR with depth;

(vi) strength ratios for the clay in a normally consolidated state, i.e. from the CIU (2po') tests, show a clear trend to reduce with depth, and, therefore, with increasing plasticity.

The CAU tests were run with an axial strain rate of 1.4%/hour; the UU and CIU tests were run at rates of between 0.6 and 1.2%/hour; the DSS tests were run at 5%/hour. Axial strains at failure in the CAU and CIU triaxial compression tests were generally between 1% and 2%, giving times to failure of 1 to 2 hours. In DSS tests, failure strains were approximately 6% to 9%, giving times to failure of 1 to 2 hours. Larger strains at failure (approximately 5 to 9%) and longer times to failure (approximately 5 to 6.4 hours) were involved in the CAU triaxial extension tests.

Significant undrained rate effects can be expected in these high plasticity clays. Shearing rates in the field will be considerably lower than those used in the laboratory tests, so that a correction to the laboratory measured strengths was made.

The sensitivity of the clay has been measured in fall cone tests and in in-situ vane tests and these show that the sensitivity of the clays is low (approximately 2-4), which is consistent with the relatively low liquidity index.

EFFECTIVE STRESS STRENGTH PARAMETERS

Effective stress strength parameters, c' and ϕ', for triaxial compression, have been determined for the clays in a normally consolidated state on the basis of the CIU ($2p_o$') tests. These tests indicate the following:

Unit	c'_{nc}	ϕ'_{nc}
D	0	24.5°
E	0	20.5°

The lower values for unit E again reflect its higher plasticity. At high stresses, values for ϕ'_{nc} as low as 15.5° to 16.7° have been measured on samples from the base of layer E.

161

STIFFNESS

Stiffnesses for units D and E have been evaluated on the basis of NGI's triaxial and simple shear tests. Values of secant Young's Modulus, E_{sec}, and of secant shear modulus, G_{sec}, have been calculated at different mobilised values of shear stress, t, ($=(\sigma_1-\sigma_3)/2$ in triaxial tests and $= \tau_h$ in simple shear tests). The variation in E_{sec}/t_f and G_{sec}/t_f versus t/t_f are plotted in Figure 21, where t_f is the value of either $(\sigma_1-\sigma_3)/2$ or τ_h at failure.

There is similarity in stiffness values for the different shearing modes suggesting isotropy at these values of mobilised strength. It may be noted that the normalised stiffness values for triaxial compression decrease with depth, again consistent with the increasing plasticity of the clay with depth.

CONCLUDING REMARKS

The investigation has demonstrated the benefits of taking high quality samples and of taking into account sample quality in a systematic way in the interpretation of data. The benefits of profiling with the piezocone, using this to assess natural variability, and of relying on a limited amount of non-routine laboratory testing have also been demonstrated.

It has been shown that the deltaic clays forming units D and E are lightly overconsolidated, with OCR reducing with depth. They have anisotropic strengths. The clays have unusually high plasticity indices at depth and a significant organic content which is probably responsible for the exsolution of gas on stress release. Methods to prevent gas exsolution and to allow deep piston sampling need to be introduced.

ACKNOWLEDGEMENTS

The authors wish to thank the Port Said Port Authority and the Maritime Transport Sector of the Ministry of Transport of the Government of Egypt for permission to publish the paper. The tenderers subscribing to the investigation were: Soletanche Bachy, Rodio, Trevi, Kvaerner Cementation, Spie Foundations, Bouygues, Necso, Sefi Fontec, Ballast Nedam International, Campenon Bernard, Bauer International, Archirodon and Hyundai. Geotechnical (Egypt) carried out the fieldwork. Philippe Bousquet-Jacq of Terrasol was responsible for technical supervision.

REFERENCES

- Burland, J. B. (1990). On the compressibility and shear strength of natural clays. *Géotechnique,* 40, 3, 329-378.
- Hight, D. W., Bond, A. J. & Legge, J. D. (1992). Characterisation of the Bothkennar clay: an overview. *Géotechnique*, 42, 2, 303-347.
- Janbu, N. (1963). Soil compressibility as determined by oedometer and triaxial tests. *European Conf. on Soil Mechanics and Foundation Engineering,* Wiesbaden, Vol. 1, 19-25.
- Lunne, T., Berre, T. & Strandvik, S. (1997). Sample disturbance effects in soft low plastic Norwegian clay. *Symp. on Recent Developments in Soil and Pavement Mechanics,* Rio de Janeiro, 81-102.

Coastal Geotechnical Engineering in Practice, Nakase & Tsuchida (eds)
© 2002 Swets & Zeitlinger, Lisse, ISBN 90 5809 151 1

Glacial and non-glacial clays: An overview of their nature and microstructure

J.Locat
Department of Geology and Geological Engineering, Laval University, Québec, Canada

H.Tanaka
Port and Airport Research Institute, Yokosuka, Japan

ABSTRACT: The nature and microstructure of most clayey sediments is largely influenced by the characteristics of the sedimentary environment : it source, sedimentation rate, temperature and diagenesis. This work will illustrate the difference between glacial and non-glacial environments and will show how the nature and the origin of the sediment can influence its nature, structure and ultimately its mechanical behavior, this is particularly true for fossilifereous soils.

1 INRODUCTION

The strong development of geotechnical engineering in Asia has provided an opportunity to confront the behavior of the Asian clays with those of well known Canadian or British clays (see Tanaka, this volume). Over the last few years, major efforts were put to understand the nature of selected Asian clays (Locat *et al.* 1996, Tanaka and Locat 1999, Tanaka *et al.* 2000) which are mostly being deposited under non-glacial conditions.

Many of the clayey sediments used to develop the major concepts in soil mechanics were formed under sedimentary conditions often dominated by glacial environments. We consider here that a glacial clay as such when most of the sediment originated from glacial erosion whereas non-glacial clays would be mostly the product of weathering. There could be another division which could include rapidly growing mountain ranges which also yield significant amount of sediments to the ocean. These region are either found in both cold and warm environments and, to some extent may not provide enough time for sufficient in situ weathering to develop. Including this subdivision is outside the scope of this work.

2 SEDIMENTARY ENVIRONMENTS

In nature, sedimentary environments can be divided into three: fluvial, lacustrine and marine (Figure 1). According to local conditions there could be some intermediate milieu. For example, estuaries would fall as a transition from fluvial to marine.

A schematic description of the inter-relation between the sedimentary environments and the final deposit is shown in Figure 1. The sedimentary envi-

ronments are characterized by sedimentation source and rate, temperature and finally diagenetic processes. For clay minerals, Carroll (1970) indicated that they are essentially the weathering product or chemical decomposition of igneous or metamorphic rocks. When clay minerals are found in sedimentary rocks they are mostly of detrital origin. Therefore, the type of clay mineral will often reflect on the regional geology and sedimentological conditions.

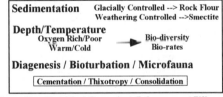

Figure 1. Elements of the sedimentary environments and their relation to microstructure.

Post-depositional changes resulting from leaching of salts (case of some glacial clays, Locat 1996), or aging, will modified the mechanical and physicochemical properties of the soil (Locat and Lefebvre 1986). In the context of non glacial clays, the transformation of organic matter or the dissolution of some microfossils with burial, will also bring about changes in physico-chemical properties by changing surface properties of the sediments. Good parameters

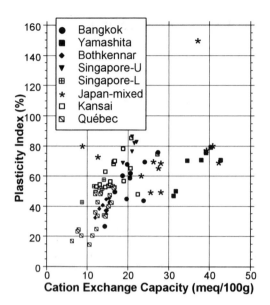

Figure 2. Relationship between the plasticity index and the CEC of selected clays (Japan mixed clays are from Locat *et al.* 1996, and Kansai clays from Tanaka and Locat 1999).

Table 1. Comparing glacial and non-glacial sediments or soils.

	Glacial	Non-Glacial
Source Material	Rock flour	Weathering
Environment	Inland seas, Fjords	Estuaries, Seas
Age	Postglacial	Pleistocene / Holocene
Water Temp.	Cold	Warm, Cool
Clay Formation	Detritic	Diagenetic / Authigenic
Swelling Clays	Rare	Abundant
Amorphous	Traces	Abundant
Organic Matter	Low	Low to High
Activity	Low	Medium to High
Sensitivy	High	Low

to evaluate the global physico-chemical properties of soils or sediments are the plasticity index and the cation exchange capacity (CEC, see Figure 2).

The sedimentary basin itself will influence the final characteristics of the sediment by it parameters such as temperature of the water, salinity and availability of organic matter. These factors will influence the biological diversity and its impact on the sediments, *i.e.* bioturbation. Other processes active during and after sedimentation are related to phenomena such as thixotropy, cementation and consolidation (Silva 1974, Locat and Lefebvre 1986). The deposit itself will be qualified as either detritic (organic or inorganic), calcareous (rich in calcareous debris, *e.g.* coccolith) or siliceous (rich in siliceous debris, *e.g.* diatoms).

Therefore, for sedimentary environments under the influence of glaciers, most of the sediment produced will originate from glacial erosion and little clay minerals would be derived from weathering. In the case of non-glacial source, the sediments originate from soils or rocks that have been weathered and the clay mineral are usually reflecting the weathering process taking place on land with granitic terrain yielding mostly kaolinite and illite while basaltic terrain will provide more smectite. A comparison between glacial and non-glacial environments is proposed in Table 1. In addition to difference related to how the original sediment is produced, non-glacial environments differs significantly by a higher organic activity and warmer temperatures. Non-glacial clays also tend to have a higher activity and lower sensitivity.

Sedimentary basin like lakes can also accumulate various types of sediments, but because of the low salinity, the do not develop sediments with high sensitivity.

3 NATURE AND MICROSTRUCTURE

3.1 *Approach*

The clay mineralogy has been carried out on an oriented mount of the clay fraction (obtained by centrifugation) and was obtained using the X-ray diffraction analysis (XRD) on oriented (N), glycolated (G) and heated (550°C, H) specimens using a Siemens diffractometer. Clay minerals are identified as follows: I: illite, S: smectite, K: kaolinite, C: chlorite, V: vermiculite. Other mineral can be observed, even in the clay fraction, like quartz (Qz) and hornblende (Ho). The glycolated treatment is used to distinguish the swelling clay minerals like smectite and mixed-layers clay minerals (*e.g.* chlorite-vermiculite). If present, the first order peak (usually between 5° and 8° 2θ for the natural, N, oriented mount) of the clay swelling clay mineral will shift to the left. Heating will help distinguish if kaolinite and/or chlorite are present. Kaolinite peaks are destroyed by heating. Also, for swelling clays and vermiculite, when heated at 550°, their first order peak will collapse to a value close to that of the first order illite peak, at about 9° 2θ. When this happens, the illite peak increases in intensity.

Scanning electron microscopic (SEM) observations were carried out on freeze-dried specimens for which the exposed surface was cut while the sample was still frozen.

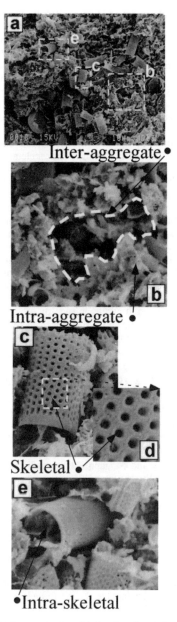

Figure 3. Classical model with pore space as inter and inter-aggregate (sample from Port-Cartier, Québec). Dashed square in (a) (scale bar at 10 μm) is shown in (b) (scale bar at 1 μm) and dashed square in (c) (scale bar at 10 μm) is shown in (d) (scale bar at 1 μm).

Figure 4. Microstructural model showing all pore families (inter-aggregate and inter-particle is considered the same here, scale bar in (a) is at 10 μm).

In most of the literature, the pore network has been described according to the example shown in Figure 3 (a clayey silt from Port-Cartier, Québec) where the pore network is descibed in terms of inter-aggregate and intra-aggretate porosity (Delage and Lefebvre 1984).

Tanaka and Locat (1999) have introduced the terms skeletal and intra-skeletal to describe the pore network associated with the presence of various types of microfossils.

The skeletal pore diameter can be as low as 1 μm and as high as 100 μm for the intra-skeletal pore diameter of diatoms and up to 200 μm for foraminifera.

The intra-aggregate pore are usually much less than 0.5 μm, and as shown by Locat (1996), the inter-aggregate (or particle) pore space is close when the water content (or void ratio) is close to that of the plastic limit. The terminology used herein will consist of the following (see also Figure 4): inter-aggregate (or inter-particle), intra-aggregate, skeletal and intra-skeletal. An illustration of the terminology is provided in Figure 4 obtained on a clay sample from Hachirogata (Japan). The skeletal pores are those found on the shell of the fossil remnants. The intra-skeletal pores are those inside the microfossils. In some case, as we will see later, fossils are broken so that the intra-skeletal pore space becomes part of the inter-aggregate pore space. In some cases, when aggregates are not common, the pore family can reduce simply to inter-particle porosity and skeletal porosity if some porous microfossils are present. In our description of the microstructure, we will also mention the size and shape of aggregates or particle, like the very angular shape of the quartz particle shown in Figure 3c and 3d.

Because of the above mentioned factors, a simple relationship cannot be expected for all the sediments. Still, illitic clays will fall in the lower part of the dispersion while high smectitic clays, like some Japanese soils (e.g. Hachirogata) will generally have a higher CEC and plasticity index. CEC values can also be influenced by the amount of microfossils presents, in particular if they have a significant intra-skeletal pore space (Tanaka and Locat 1999). Clays with limited amounts of smectite, like Québec, Bothkennar and Singapore clays will show a clear proportionality between CEC and Ip.

The specific physico-chemical characteristics of the samples used for SEM presentation in this work are compiled in Table 2 (presented at the end).

3.2 Ariake, Kobe and Kansai clays, Japan

The nature of Ariake clay, geochemistry and mineralogy in particular, has been studied in details by Othsubo et al. (1995) and by Tanaka et al. (2000).

The Ariake clay site is located on the western part of Kiushu Island (Japan) in the Saga Prefecture. The upper part of the Ariake profile is believed to show evidence of leaching with the resulting development of a sensitivity often found for glacial clays in Québec and Scandinavia (Othsubo et al. 1995). The above mentioned studies have recognized the dominance of smectite (low swelling type) in the clay along with the presence of other clay minerals such as kaolinite, illite and vermiculite (Figure 5a). Recent investigation (Tanaka et al. 2001) has also revealed the presence of chlorite at all depth. In most

Japanese clays, smectite is the dominant clay mineral. This is largely due to the presence of basaltic volcanic rocks (Kimura et al. 1991), or soil developed under warm and humid conditions. The volcanic activity also provides a large source of silica which is supporting the development of diatomaceous ooses.

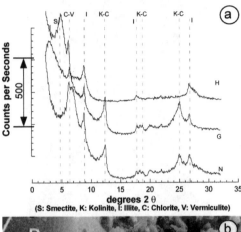

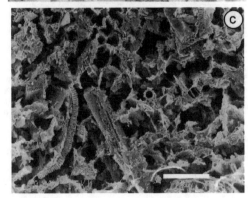

Figure 5. XRD and SEM of Ariake clay at 3m (scale bars at 10μm)

The SEM pictures of Ariake clay, at 3 m (most likely in the zone of sensitive clay (Othsubo *et al.* 1995), are shown in Figure 5b and 5c, both being at about the same scale. They reveal a well flocculated structure and the presence of abundant fossil debris. The image in Figure 5b shows centric diatoms that have been preserved more or less intact (D in Figure 5b). With the presence of large diatoms (about 30 μm in diameter, the texture of the sample in Fig. 5b is coarser than in Fig. 5c. In Figure 5c, the aggregates are made of angular particles and have about the same size and shape, i.e. about 10 μm long and about 2 to 3 μm thick with short contact bridges often made of single particle. The high natural water content of 147% is well evidenced in Figure 5c where large inter-aggregate pores, about 5 μm in diameter, are seen with little intra-skeletal pore space. In fact, the fossils are broken so that the intra-skeletal pore space coincides with the inter-aggregate pore space. In Figure 5b, the aggregates are replaced mostly by single flaky particles with a less well open structure than in Figure 5c.

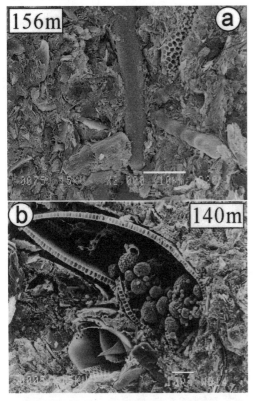

Figure 6. SEM pictures of (a) Kobe and (b) Kansai clays (scale bars at 10 μm).

As depth increases in a sedimentary environment, the inter-aggregate pore space should gradually van-

ish. This is well shown by the example of SEM picture from Kobe (Figure 6a) and Kansai (6b), both sites located in Osaka Bay. The Kobe sample is taken from a core located closer to the mouth of the Osaka River than the Kansai sample (near Kansai Airport, Tanaka and Locat 1999). These deep samples show that the inter-aggregate pore space is closed with some intra-skeletal pore space still open for the case of the Kansai sample. Kobe sample is coarser than Kansai sample with some angular silt particles and debris of diatom microfossils evidenced by the row of small chambers making the skeletal pore space.

The Kansai sample presents spectacular diatom microfossils filled with framboïdal pyrite (Fig. 6b) surrounded by a fine and compacted clay matrix which is molded around the fossils. It is interesting to see that some of the intra-skeletal pore spaces of the diatoms in Kansai samples remain quite large. Not well known is the evolution of the pyrite which can have been formed soon after deposition therefore providing increased resistance against the compaction of the fossil. This would mean that if the chemical equilibrium was such as to bring about the dissolution of pyrite, it could also translate an increased compressibility. The potential impact of microfossils on the behavior of some Japanese clays have been demonstrated by Tanaka and Locat (1999) and by Shiwakoti *et al.* (2001) and will be briefly illustrated later.

3.3 *Yangsan clay, Pusan, Korea*

The Pusan clay samples investigated were from the southern part of Korea, near the town of Pusan. These sediments were deposited during glacial period under high sedimentation rate. The actual sedimentary environment is very much like that of an estuary. There is an ongoing debate about the degree of over-consolidation for this deposit. Tanaka *et al.* (2001), using high quality samples, found that the Yangsan clay was normally consolidated but with still a low OCR value compared to clays of similar age in Japan *(i.e.* about 10 000 years old). The Yangsan clay is also quite sensitive to sample disturbance (Tanaka *et al.* 2001). The Yangsan clay profile consists of two layers: upper and lower. The upper layer is slightly coarser than the lower layer and has an OCR value between 1 and 2.

The mineralogy of the clay fraction of the Yangsan clay is mostly composed of illite with significant amount of chlorite and vermiculite (Figure 7a). Traces of kaolinite and detectable amounts of quartz and hornblende were found. It is interesting to note here the mineralogical changes which occurred after heating at 550°. We can see that the chlorite peak has remained but that another "shoulder" has developed at an angle of about 7-8° 2θ. This may result from the transformation of some irregular mixed-layered minerals in regular ones, probably of chlorite/vermiculite type. Note here also that the intensity of the first order illite peak has increased sharply, in part because of the collapse of the vermiculite. Apart from the presence of mixed-layered minerals, such a mineralogical composition is quite similar to that of sediments derived from glaciated granitic terrain as also seen in Eastern Canada (Locat 1996).

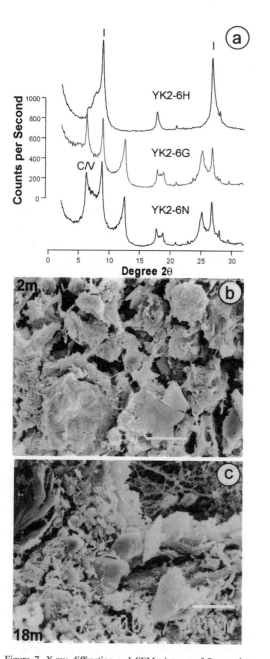

Figure 7. X-ray diffraction and SEM pictures of Pusan clay (scale bars at 10 μm).

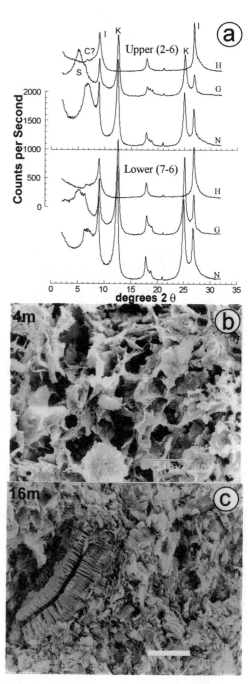

Figure 8. X-ray diffraction and SEM images of Singapore clay (scale bars at 10 μm).

The SEM images obtained from two samples are shown in Figure 7b (upper layer) and 7c (lower layer). The upper layer is well flocculated, with building blocks consisting primarily of rounded aggregates (about 10 to 30 μm in diameter) linked to each other by short bridges about 1 μm thin to form an inter-aggregate pores about 5 to 10 μm in diameter.

This type of bounding between aggregates could explain the high sensitivity to sample disturbance reported for this clay (Tanaka *et al.* 2001). Although not shown here, some fossil remains were found in this soil profile and were more abundant in the lower layer (Tanaka *et al.* 2001, Locat and Tanaka 1999). The main difference between the two layers is that the lower layer contains more single particles and shows much less bounding. Individual angular silt particles are also abundant. In Figure 7c most of the picture shows the shell of a pellet (upper right corner) with its inner web of organic matter.

3.4 *Singapore clay, Singapore*

The Singapore site consists of two layers: an upper layer formed during the Holocene (less than 2m at this site) and a lower layer formed more than 10000 years ago (Tanaka *et al.* 2000) with a dessicated crust in between (Pitts 1984). Work reported hereafter is restricted to the lower layer.

Amongst all the soil tested, the Singapore clay showed the most intense kaolinite peaks (Figure 8a) as also revealed by the presence of large kaolinite crystals (Figure 8c). Apart from kaolinite, smectite and illite are clearly present. The weak peak of kaolinite/chlorite, left after heating, indicates that some chlorite is present.

The SEM pictures shown in Figures 8b reveals a well flocculated structure. The aggregates are mostly composed of agglomerated flaky clay particles. Found throughout the sample were pieces of kaolinite clay minerals (see the typical book-like stack of sheets, Fig. 8c) which must have formed in place or have been carried over only a very limited distance. It is interesting here to compare both SEM pictures. In Figure 8b, the sample is coarser (clay fraction at 46%) and it is 70% in Figure 8c. However, the water content for both samples is 95.9% and 49.9% respectively. This difference is due to burial, and, as indicated above, the inter-aggregate pore space is already quite smaller in the sample taken at the depth of 16m (Figure 8c).

3.5 *Bangkok clay, Thailand*

The Bangkok clay studied here is from a 18m deep profile which also exhibits, from a strength point of view, two layers: upper down to about 12m and than the lower layer which contains larger amounts of silt and sand. The Bangkok clay has been deposited in

what look like an estuary with a salinity sufficient to enable foramifera to live.

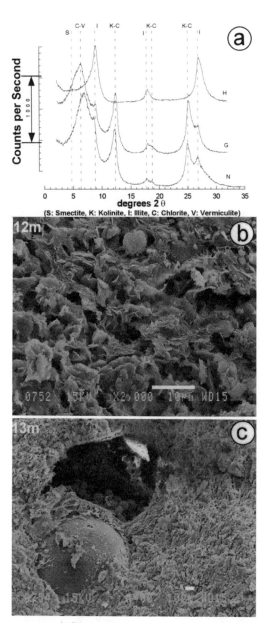

Figure 9. X-ray diffraction and SEM images of Bangkok clay (scale bars at 10 μm).

The clay minerals (Figure 9a) are mostly smectite (evidenced by he very significant shift of the smectite peak upon glycolation) and illite with of kaolinite. There is no evidence here of chlorite and the increase in the intensity of the first order illite peak result mostly from the collapsing of smectite and

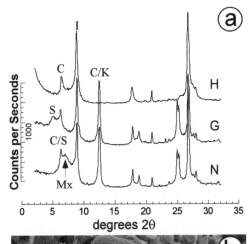

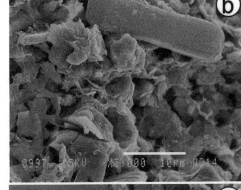

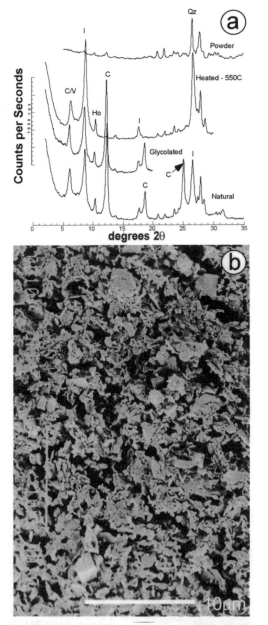

Figure 10. X-ray diffraction and SEM images for Louiseville clay.

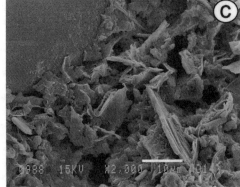

Figure 11. X-ray diffraction and SEM images of Bothkennar clay (scale bars at 10 μm)

possibly vermiculite. Primary minerals such has quartz were also observed.

The SEM images show a flocculated structure composed of aggregates consisting of bounded fine particles. The aggregates are composed of agglomerated flaky particles (most likely of smectite) about 5 to 10 μm in diameter. The porous network consist primarily of inter-aggregate pores with occasional

intra-skeletal pore space when foraminifera are encountered (Fig. 9c). The microfossils are quite large (about 1mm) and are composed either of silica or calcite (here it consists of calcite).

An interesting observation which has been made for many sample of Bangkok clay is that the particles or aggregates often appears draped with a translucent material (coating ?) which could act as a week cementing agent. This is particularly true for the example shown in Figure 9a.

3.6 *Louiseville and Port Cartier clays, Québec*

Louiseville and Port-Cartier clays are both part of the sensitive clays found in Eastern Canada but their physico-chemical properties are quite different, although their nature is not so different their main source sediments remains the pre-Cambrian shield to the north. The Louiseville site being located in the middle of the postglacial Champlain Sea basin (Lapierre *et al.* 1988, Locat 1996) has received significant sediment supplies from the Appalachian Mountains which are rich in illite and chlorite. The Port-Cartier site is located on the north flank of the postglacial Goldtwaith Sea, about 400 km to the northeast of Québec City, and is just inside of the Laurentian Plateau. It is therefore much closer to the sediment source with little contribution from the Appalachian rocks to the south. In addition, the porewater salinity of the Louiseville clay is still at about 7 g/L, while for the Port-Cartier clay, the leaching as removed most of the salts in the porewater. The sensitivity of the Port-Cartier clay is more than 500 (!) and, once remolded, it flows just like the example shown in Figure 13b.

The clay mineralogy of the Louiseville clay is presented in Figure 10a. It consists mostly of illite with some chlorite and traces of kaolinite and vermiculite with significant amounts of quartz and feldspar in the clay fraction. The large amount of chlorite is suggested here by the high intensity of the 14 nm peak of the heated specimen. Such a mineralogy is typical of glacially derived sediments (Locat 1996).

The SEM image of the Louiseville clay, in Figure 8b, shows a well flocculated structure with angular clay-size particles forming aggregates about 5 to 15 μm in diameter. The pore network consist of interaggregate and intra-aggregate pores. Contrary to the samples seen before, in the case of Louiseville clay, the aggregate still exhibit a visible intra-aggregate porosity. This may be the result of the process of aggregate formation which may have taken place during sedimentation with little cementation or bounding strength. Microfossils are usually scarce in most post-glacial clays in Québec, except for the youngest ones.

The Port-Cartier clay, shown in Figure 3, is mostly composed of silt grains with very short and small surface contact between the particles. There are only very few aggregates so that the pore network consists primarily of inter-particle pore. The water content of this clay is quite low (35%) although its sensitivity is more than 500 and the liquidity index greater than 4. The large angular grains are mostly composed of quartz (Figure 3c) . The non-flaky small particles seen in Figure 3b and 3d are feldspar minerals which often, along with quartz, are the dominant minerals, even in the clay fraction (Locat 1996).

3.7 *Bothkennar clay, United Kingdom*

Bothkennar clay site is located along the Fourth Valley in Scotland and has been a site investigated in great details (Paul *et al.* 1992). The sediment was deposited soon after glaciation, and like Louiseville clay, it results from glacial erosion. In this area, the source of the sediments is from metamorphic, igneous and sedimentary rocks with a large content in chloritic and mica schist. The clay fraction of most samples recovered from this site varies between 19 and 31% with a plasticity index varying from 32 to 45% and a liquid limit between 55 and 75%. The activity of the Bothkennar clay is at about 1.5.

The clay mineralogy (Figure 11a), reflecting the source rock, is composed of illite (or micas), significant amounts of chlorite and mixed-layered clays minerals of the swelling type. Compare to the sensitive clays from Québec, the relatively higher activity for the Bothkennar clay is most likely due to the presence of some swelling minerals which are absent, for example in Louiseville and Port-Cartier clays.

The microstructure is shown in Figure 11b and c. The soil is flocculated with elementary particles either made of microfossils or large mica or chlorite crystals often more than 10 μm in diameter (see Fig. 11c). The aggregates are about 5 to 20 μm in diameter.

The pore network consists primarily of two families: intra and inter-aggregate. The skeletal and intra-skeletal is visible but not very abundant. The inter-aggregate pores are about 1 to 2μm in diameter.

4 IMPACT OF MICROFOSSILS ON SOME PHYSICAL PROPERTIES

One of the major characteristics soil from which most of the soil mechanics concept has been developed is the absence of microfossils (if present, they were not considered). This section will briefly look into how they can influence some of the basic engineering properties. It is beyond the scope of this paper to go in great details about this topic and the reader is referred to papers of Tanaka and Locat (1999), Shiwakoti *et al.* (2001) and Locat and Tanaka (2001) for more insights.

The study of the Kansai Airport sediments has provided a unique occasion to look at the effect of diatoms on the physical properties of sediments. An example is shown in Figure 12 where the Osaka Bay soil, at a depth of about 363m below the sea floor, contains large amounts of diatoms (Fig. 12a) to a point that it directly controls both the grain size distribution (Fig. 12b) and the pore network (Fig. 12c). The SEM picture in Figure 12a is quite revealing about the potential role of microfossils. Here, the inter-aggregate pore network is almost closed while the intra-skeletal pores are still quite open!

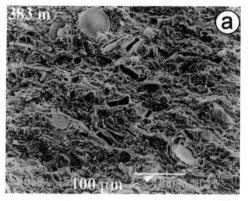

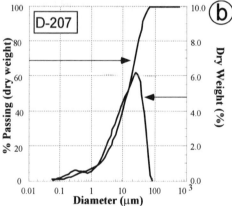

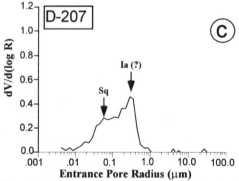

Figure 12. SEM , grain size and pore size distribution of a fossilifereous clay at Kansai airport (modified after Tanaka and Locat 1999). Ia: inter-aggregate, Sq: skeletal.

The sediment is largely composed of diatoms which make more than one third of the soil. The grain size distribution, given in Figure 12b, shows

that the mode is that of diatoms, at about 30 μm. These microfossils have a fine skeletal porosity which is picked up by mercury porosimetry measurements (Tanaka and Locat 1999), where the main peak (Sq), shown in Figure 12c, represent the 1 μm pores on the surface of the skeleton (*i.e.* the skeletal porosity). So, here, the microfossils are part of the soil particles and do influence the grain size distribution. Here the inter-aggregate pore space approaches that of the intra-agregate pore space.

The matrix (Figure 12a) is mostly composed of smectite and is very much compacted. So, the actual water content of this sample is largely influenced by the open intra-skeletal porosity so that although the overall liquidity index may be high, that of the matrix could be much lower. This available space for compaction in the pore network could be translated into a kind of stored strain energy which could become available if the yield stress is exceeded.

The influence of diatoms on the geotechnical properties of soils has been illustrated by Tanaka and Locat (1999) and in greater details by Shiwakoti *et al.* (2001). Locat and Tanaka (2001) have also reviewed the potential impact and influence of some microfossils (like foraminefera and diatoms) and have proposed to consider that, in addition to organic and inorganic soils we also consider fossilifereous soils (or sediments).

5 ENVIRONMENT AND STRUCTURATION

The above sequence of clayey sediments or soil has shown the diversity of mineralogy and microstructure style of various selected clays. The source rocks or sediments, the sedimentary environment and the regional climatic conditions will influence on the final nature and microstructure of clay sediments.

5.1 *Glacial clays*

For the case of emerged glacial sediments, Locat and Lefebvre (1986) proposed a general framework for explaining the development of structuration in sensitive clays. This model, also presented by Locat (1996), is shown in Figure 13. This model is presented into two parts: one representing the changes in void ratio as a function of time and stress, and the second showing the changes in the liquidity index also as a function of stress or time.

The void ratio-stress relationships are provided for two reference curves: firstly when a sediment deposited in marine conditions (salinity of 30 g/L) and, secondly, for a sediment deposited in fresh water (salinity at 0 g/L).

The third curve represent the evolution of the void ratio in a sediment element which has been cemented soon during the consolidation process. The equilibrium void ratio, controlled by cementation, is shown up to about σ'_{vo} which marks the end of sedimentation and stress increase but also the beginning of the leaching process resulting from glacio-isostatic rebound. Then, upon loading, the void ratio remains constant until the yield stress, corresponding to the bounding strength, is reached. After, the compressibility is high and the relationship migrates toward the equilibrium curve for the leached clay, *i.e.* at about 0 g/L. The void ratio-stress relationship can explain the high compressibility but not the high sensitivity. Using the liquidity index makes it possible to show the impact of leaching on the development of the very low remolded shear strength (or high liquidity index) as illustrated in Figure 13b. The gray zone in the right diagram of Figure 13a corresponds more or less to the strain energy available upon remolding.

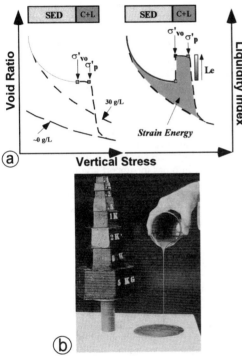

Figure 13. (a) Development of structuration in sensitive clay as a result of cementation and leaching (adapted from Locat 1996, (b): a very sensitive clay, NRC archive). SED: sedimentation, C+L: cementation plus leaching.

In the above example, the structuration is the result of the following processes: (1) sedimentation (usually high rate, low bioturbation, see also Perret *et al.* 1995), (2) cementation (weak cement due to silica or calcium precipitates), and (3) leaching (for more details, see Quigley 1980, Torrance 1983 and Locat 1996).

One element to remember here is that for most of the sensitive clays, the amount of swelling clays is quite small. This is significant since swelling clays, like smectite, respond in the opposite direction when leached, *i.e.* the liquidity index would decrease, and the remolded shear strength would increase so that the sensitivity would be reduced. One exception to this is the low swelling smectite of Ariake clay (Othsubo *et al.* 1995) which, because of its nature, does respond like a non-swelling clay (*e.g.* illite) to the reduction in the salinity (leaching).

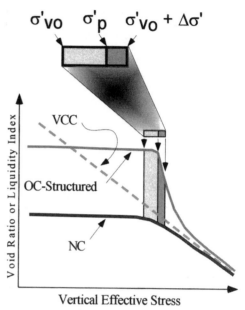

Figure 14. Structuration of a non-glacial clay caused by the presence of microfossil particles.

5.2 Fossiliferous clays (the case of diatoms)

Some clays, like the Kansai clay, do contain significant amounts of diatoms, at least in some layers (Tanaka and Locat 1999). In such a case, we would like to propose that the presence of significant amount (still remain to be determined) of diatoms could provide the clay with structural element, the diatoms, which will resist compaction during the early stage so that the clay matrix will be compacted around them and some arching could develop. This is illustrated by the conceptual model shown in Figure 14, in relation to a virgin compression curves (VCC), two compression curve are shown for one sample which is normally consolidated (NC) and another one which is over-consolidated (OC). The NC behavior could represent a soil which has no significant amounts of diatoms or one where the *in situ* confining stress has exceeded the yield strength of the diatoms embedded in the clay matrix (Fig. 12a). The OC sample would result from the presence of building block or structural elements composed mostly of intra-aggregate pore space due to the pres-

Table 2. Physico-chemical characteristics of various samples used in this work.

Sample	CF (%)	w_n(%)	w_p(%)	w_L(%)	I_p(%)	OM(%)	CEC meq/100g	SSA m^2/g	G_s
Ariake 3m	140	147	47	112.7	65.8	3.16	27.4	105	2.615
Hachirogata 6m	15	179	71	221.2	150.2	7	36.5	225	2.488
Kobe 156m	44	37.7	23.6	66.6	43.0				
Kansai 140m	29	61.1	38.7	108.4	69.7	1.7	16.4	126	2.685
Kansai 180	36	77.7	44.3	122.6	78.3	1.4	18.7	150	2.567
Kansai 383	19	43.3	42.2	116.8	74.6	1.6	27.2	120	2.712
Pusan 5.4m	33	46.5	22.2	53.0	30.8	-	9.2	44	2.722
Pusan 21.4m	40	55.9	24.9	57.9	33	-	16.7	84	2.726
Singapore 4m	46	95.9	30.2	116.5	86.3	1.95	21.0	167	2.775
Singapore 16m	70	49.9	23.2	65.7	42.5	0.57	8.8	106	2.759
Bangkok 12m	56.8	59.3	24	84.5	60.5	2.68	18.8	151	2.736
Bangkok 13m	59.1	58.5	25	87	62	2.72	20.6	178	2.730
Bothkennar 15.7m	-	47.4	-	-	-	2.43	12.8	76	-
Louiseville	80	72	27	72	45	1.0	13.0	80	2.780
Port-Cartier	-	35	17	22	5	<1.0	-	-	-

CF: clay fraction; OM: organic matter; CEC: cation exchange capacity; SSA: specific surface area

ence of large amounts of more or less intact diatoms shells.

Therefore, in a given soil profile, like under the Kansai Airport (Tanaka and Locat 1999), one could find alternating layers with variable yield stress values which could be caused by the presence of more or less abundant fossilifereous layers. In this case, the structuration would result from the following processes: (1) environmental conditions prone to the presence of large concentration of diatoms fossils: source of silica, low sedimentation rate and bioturbation, (2) consolidation of clay matrix and development of arching around resistant microfossils.

6 CONCLUSIONS

We have put together a series of XRD and SEM microphotographs of various clays, mostly in Asia, to illustrate the diversity of composition and microstructure. We believe that we have shown that the role of the source of the sediment, the sedimentary environment, and the post-depositional evolution are significant in the development of the macrostructure of soils and sediments and could be use to explain some of their behavior. We have illustrated the potential role of microfossils on basic physical properties of soils or sediments. Our observations on the effect of diatoms microfossils on the development of shear strength has led us to propose a conceptual model to explain the variability of observed values

of pre-consolidation pressures (σ'_p) observed at the site of the Kansai Airport.

Such a model can be put in contrast to that of sensitive clays in order to support argumentation to the effect that some of the conceptual models developed from experiences with glacial clays do not necessarily apply directly to non-glacial clays, at least those rich in microfossils.

7 ACKNOWLEDGEMENTS

The Authors would like to thank the Port and Harbor Research Institute of Japan, the Ministry of Education (Québec), the National Science and Engineering Research Council of Canada and Fonds F.C.A.R. for their financial support in these studies. We also thanks the many contributors to this work: Shiwakoti (PHRI, Japan), C. Goulet and M. Choquette of Laval University. H. Tremblay (Laval University) helped us to edit the final version of the manuscript.

8 REFERENCES

Carroll, D., 1970. Clay minerals: a guide to their X-ray diffraction. Geological Society of America, Special Paper 126, 80p.

Delage, P., and Lefebvre, G., 1984. Study of the structure of a sensitive Champlain clay and its evolution during consolidation. *Canadian Geotechnical Journal*, 21: 21-35.

Kimura, T., Hayani, I., and Yoshida, S., 1991. Geology of Japan. University of Tokyo Press, 287p.

Locat, J., 1996. On the development of microstructure in collabsible soils: lessons from the study of recent sediments and artificial cementation. *In: Genesis and Properties of Collapsible Soils* by E. Derbyshire, T. Dijstra, and I.J. Smalley,. Klewer Academic Publisher, Dordrecht, The Netherlands, pp. 93-128.

Locat, J., and Lefebvre, G., 1986. Origin of the structuration of the Grande-Baleine marine sediments. *Quarterly Journal of Engineering Geology*, 19: 365-374.

Locat, J. and Tanaka, H., 1999. Microstructure, mineralogy, and physical properties: techniques and application to the Pusan clay. *In: Proceeedings of the 99 Dredging and Geo-enviromental Conference*, Seoul, Korea, Korea Geotechnical Society, pp. 15-31.

Locat, J., and Tanaka, H., 2001. A new class of soils: fossilifereous soils ? *In: Proceedings of the ISSMGE Conference*, Istambul, in press.

Locat, J., Tremblay, H., Leroueil, S., Tanaka, H., and Oka, F., 1996. Japan and Québec clays: their nature and related environmental issues. *In: Proceedings of the 2nd International Congress on Environmental Geotechnics*, Osaka, Japan, pp. 127-132.

Ohtsubo, M., Egashira, K., and Kashima, K., 1995. Depositional and post-depositional geochemistry, and its correlation with geotechnical properties of marine clays in Ariake Bay, Japan. *Géotechnique*, 45: 509-523.

Perret, D., Locat, J., and Leroueil, S., 1995. Strength development with burial in fine-grained sediments from the Saguenay Fjord, Québec. *Canadian Geotechnical Journal*, 32: 247-262.

Quigley, R.M, 1980. Geology, mineralogy, and geochemistry of Canadian soft soils: a geotechnical perspective. *Canadian Geotechnical Journal*, 17: 261-285.

Paul, M.A., Peacock, J.D., Wood, B.F., 1992. The engineering geology of the Carse Clay of the National Soft Clay Research Site, Bothkennar. *Géotechnique*, 42: 183-198.

Pitts, J., 1984. A review of the geology and engineering geology in Singapore. *Quarterly Journal of Engineering Geology*, 17: 93-101.

Shiwakoti, D.R., Tanaka, H., Tanaka, M., and Locat, J., 2001. Influence of diatom microfossils on engineering properties of soils. Submitted to *Soil and Foundations*.

Silva, A.J., 1974. Marine geomechanics: overview and projections. *In: Deep Sea Sediments, Physical and Mechanical Properties*, A.L. Inderbitzen, (ed.), Plenum Press, , New York, pp.: 45-76.

Tanaka, H., and Locat, J., 1999. A microstructural investigation of Osaka Bay clay: the impact of microfossils on tis mechanical behaviour. *Canadian Geotechnical Journal*, 36: 493-508.

Tanaka, H., Locat, J., Shibuya, S., Thiam Soon, T., and Shiwakoti, D.R., 2000. Characterizaiton of Singapore, Bangkok and Ariake Clays. *Canadian Geotechnical Journal*, vol. 37, in press.

Tanaka, H., Mishima, O., Tanaka, M., Park S.-Z., Jeaong, G.-H., and Locat, J., 2001. Characterization of Yangsan clay, Pusan, Korea. Submitted to *Soils and Foundations*.

Torrance, 1983. Towards a general model of quick clay development. *Sedimentology*, 30: 547-555.

Torrance, K.J., 1984. A comparison of marine clays from Ariake Bay, Japan, and the South Nation River landslide site, Canada. *Soils and Foundations*, 24: 75-81.

Engineering properties of lean Lierstranda clay

T.Lunne
Norwegian Geotechnical Institute, Oslo, Norway

ABSTRACT: This paper describes the engineering properties of lean clay from the Lierstranda test site near the city of Drammen, Norway. This glacial clay was deposited in salt water and is therefore fairly homogeneous and of low plasticity. The clay deposit is believed to never have been subjected to loads greater than the present overburden, but due to secondary consolidation and possibly other processes the clay exhibits an apparent overconsolidation. Strength and deformation characteristics are given based on tests on high quality block samples. The significant effects of sample disturbance are demonstrated by comparison with tests on standard piston tube samples. Correlations to results from in situ tests are also included.

1 INTRODUCTION

Since the early 1950s, the Norwegian Geotechnical Institute (NGI) has been studying the engineering geology and geotechnical characteristics of the Drammen clay (e.g. Bjerrum, 1967, 1972). Most of the research has been concentrated on the clay in the center of the city of Drammen; however, since about 1985, the test site at Lierstranda just a few km outside Drammen has been the location for research on model piles, in situ testing and sample disturbance. Japanese researchers have also tested clay from Drammen and Lierstranda; e.g. Nutt et al. (1995), Tanaka (1997), Tanaka et al. (1996), Dam et al. (1999), & Watabe (1999).

Although the soil classification parameters of the clays in the center of Drammen and in Lierstranda are quite similar, advanced laboratory testing has recently shown that there are some differences in the soil parameters between the two sites. Lunne and Lacasse (1999) give an overview of the in situ testing, sampling and laboratory testing that has been done at the two sites.

This paper concentrates on the engineering properties of the Lierstranda clay. Emphasis is placed on the low plasticity (or lean) clay below about 12 m. However, as the testing included sampling and laboratory testing on the upper more plastic clay, some results from this layer are also included on the boring profile and other composite plots.

2 SITE DESCRIPTION AND GEOLOGY

The city of Drammen is located about 40 km south – west of Oslo, Norway, and the test site at Lierstranda is just a few km north of Drammen's center, see Fig. 1.

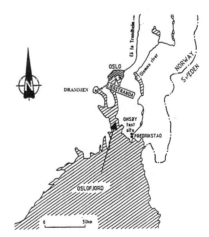

Figure 1. Location of Lierstranda test site outside Drammen

About 20,000 year ago, the climate became milder in northern Europe, and Norway started to emerge from the huge ice sheets which covered the Scandinavian peninsula during the last Pleistocene glaciation. Isostatatic crustal rebound of the previously glaciated areas took place. As the glaciers retreated, rivers of melt-water discharged large quantities of sand, silt and clay into the fjord at the front of the ice sheet. Due to the saltwater deposition, the late-glacial clay which filled the bottom of the valley was fairly homogeneous and of relatively low plasticity (Bjerrum, 1967).

As the country gradually rose above the sea level, the valley took the character of a narrow fjord with only limited communication with the ocean. The climate became milder again, and finer material such as silt and clay were transported down as the glaciers retreated. The clays deposited post-glacially to the lower part of the valley showed more segregation and somewhat higher plasticity than the previously deposited underlying clay. The character of the landscape gradually changed from that of a fjord to that of a valley. The final stage of deposition, to which the uppermost sand, silt and clay in the Drammen area belong, dates back to about 1,000 to 3,000 years ago.

The clays deposited in the latter part were never subjected to loads greater than the present overburden.

3 MINERALOGY GEOCHEMISTRY AND CLASSIFICATION DATA

X-ray diffraction analyses on a sample from 13.1 m depth resulted in 65 % of illite/muscovite minerals , 25 % chlorite and 0-5 % of each of the minerals : quartz, K-feldpar and plagioclase.

The salinity at Lierstranda, 6-16 m below ground level is between 29-32 g/l, decreasing slightly with depth.

Geochemical tests showed that the organic carbon content was about 5000 mg/kg.

Figure 2 shows photos of SEM micrographs of an undisturbed sample from 13.2 m depth. A relatively open structure can be observed. No evidence of cementing can be seen.

The boring profile in Fig. 3 includes soil classification data. In the lean Lierstranda clay below about 12 m water content ranges from 31 to 38 %, the liquid limit between 33 and 40 % and the plasticity index between 14 and 18.5 %. Sensitivity, as measured by the vane test, ranges from 8 to 15 in the lean clay.

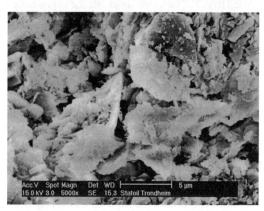

Figure 2. SEM pictures, 13.2 m depth

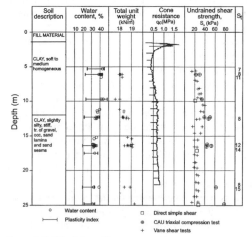

Figure 3. Soil boring profile

178

4 IN SITU STRESSES AND STRESS HISTORY

Figure 4 gives the in situ stresses and apparent pre-consolidation stress with depth. A slight excess pore water pressure has been included in the calculation of the effective vertical stress. It is believed that this excess pore water pressure is caused by a recently placed 1.5 m thick fill.

The effective horizontal stress has been estimated based on results of in situ vane tests and CAUC (anisotropically consolidated undrained, sheared in compression) triaxial tests following a procedure described by Aas et al. (1986).

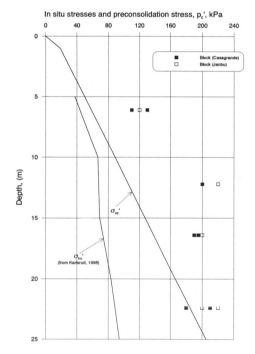

Figure 4. In situ stresses

The values of the preconsolidation stresses, p_c', have been obtained by interpretation of CRSC (constant rate of strain consolidation) tests on block samples, following the methods proposed by Casagrande (1936) and Janbu (1969). As outlined in the geological description, the Lierstranda clays are not believed to be mechanically overconsolidated yet they exhibit a yield stress or apparent preconsolidation stress when tested in the oedometer apparatus. According to Bjerrum (1967, 1972) this apparent preconsolidation is caused by delayed consolidation and possibly some chemical processes. It should also be pointed out that the p_c' values obtained from CRSC tests are at NGI termed as 'rapid' p_c'. Several studies both in

Norway and other countries have shown that the p_c' values obtained from CRSC tests are higher than those found from conventional incremental loading oedometer tests with 24 h load steps. For instance Leroueil et al. (1983) reported, for Champain Sea clays, an average ratio of 1.28 between the p_c' measured in CRSC tests performed at strain rates of 2 to $5*10^{-6}$ s^{-1} and in conventional 24 h oedometer tests. For the Lierstranda clay it was found that the 'rapid' p_c' (rate of strain $2*10^{-6}s^{-1}$) is 15 –20 % higher than the 24 h p_c'. At NGI it is common practice to keep the load constant for 24 h before the start of the unloading/reloading loop. In this way it is possible to construct the 24 h oedometer curve from the CRSC test (see also Sandbækken et al., 1986).

5 SHEAR STRENGTH

5.1 Triaxial compression strength

Figure 5 shows summaries of the stress strain and stress path curves from CAUC triaxial tests following the procedures given by Berre (1982). It can be observed that the behaviour of the clay is characterized by peak shear stress occurring at small axial strains and by significant strain softening. Further, the inclination of the stress path up to peak is very close to 1:3 which would indicate that it behaves in this respect. like an isotropic elastic material.

Figure 6 shows normalized values of undrained shear strength (s_u/σ_{vo}') vs depth. At 6.1 m; s_u/σ_{vo} = 0.55 which reduces to 0.31 at a depth of 22.4 m. This is consistent with the decrease in p_c' and OCR with depth as indicated in Fig. 4.

5.2 Strength anisotropy

Two consolidated triaxial tests were sheared in extension (CAUE) with the results included in Figures 5 and 6. For the two tests the anisotropy ratio, s_u^{CAUE}/s_u^{CAUC}, is 0.42 (plastic clay) and 0.37 (lean clay). These values are somewhat higher than those reported for Drammen clay by Bjerrum (1972): 0.38 and 0.26 for the plastic and lean clays respectively.

In addition three direct simple shear tests (DSS) were carried out with results shown in Fig. 6. The anisotropy ratio, s_u^{DSS}/s_u^{CAUC}, for these three tests are 0.72 (plastic clay) and 0.72/0.89 (lean clay). Bjerrum reported 0.75 and 0.65 respectively for the plastic and lean Drammen clay. The reasons for the differences in the anisotropy values for the present study on Lierstranda clay and those reported for Drammen clay by Bjerrum (1972) are thought to be: (i) the tests Bjerrum reported were on piston samples which are somewhat more disturbed than the block samples used for the Lierstranda clay; (ii) as mentioned in the introduction there are some differences in the behaviour of the two clays.

179

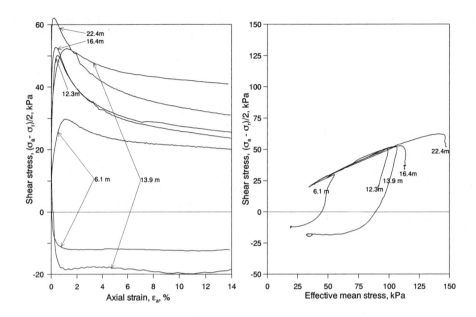

Figure 5. Summary plots of stress-strain curves and stress paths for CAU tests on block samples

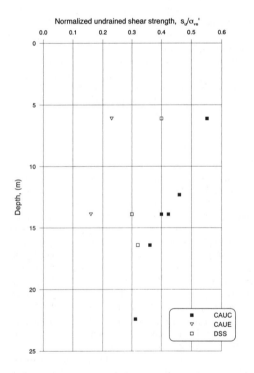

Figure 6. s_u/σ_{vo}' vs depth for CAUC, CAUE and DSS tests on block samples

5.3 Effect of rate of strain on undrained shear strength

The effects of rate of shear strain on shear strength has only been investigated by two sets of DSS tests for the lean Lierstranda clay. The results from these two tests are summarised in Table 5.1 below.

At NGI the shear strain rate normally used during DSS tests is 4.5 %/hour. It can be seen that by reducing the strain rate by a factor of 10 the undrained shear strength reduces by 6 and 12 %. This is within the range reported by Berre and Bjerrum (1972) for CAUC tests for the plastic Drammen clay.

5.4 Results of SHANSEP tests

Two sets of SHANSEP tests were carried out according to the principles outlined by Ladd and Foott (1974). Table 5.2 gives the results of these tests:

Referring to the table the OCR=1 samples were consolidated into the normally consolidated range following K_o- consolidation. The values of σ_{ac}' and σ_{rc}' for the OCR=2 tests represent the final consolidation stage after unloading. The OCR=2 tests were also K_o- consolidated. The values of s_u/σ_{ac}' for the OCR=1 of 0.301 and 0.316 fit well with values reported by Ladd (1991). The m-values for the SHANSEP equation :

$$(s_u/\sigma_{ac}')_{oc}=(s_u/\sigma_{ac}')_{nc}*OCR^m$$

180

Table 5.1 Results of DSS with two strain rates

Depth m	w %	w_l %	I_p %	σ_{vo}' kPa	σ_{max}' kPa	σ_{min}' kPa	s_u kPa	γ_f %	Rate of strain,%/h
13.9	33.3	39.2	18.0	124	170	123.7	37.7	3.0	4.5
14.0	34.0	39.2	18.0	124	170	124.2	35.6	2.6	0.45
16.4	29.0	34.4	14.6	146	170	145.3	47.0	2.5	4.5
16.4	27.7	34.4	14.6	146	170	145.7	41.1	3.3	0.45

Table 5.2 Results of SHANSEP tests on lean Lierstranda clay

Depth m	Sample type	σ_{vo}' kPa	σ_{ac}' kPa	σ_{rc}' kPa	OCR	s_u kPa	s_u/σ_{ac}'	m
12.3	75mm	109.0	211.9	110.0	1	67.1	0.316	
12.6	54mm	109.5	106.1	73.9	2	55.1	0.519	0.72
15.2	54mm	175.0	175.0	122.0	2	88.1	0.503	0.74
15.4	54mm	175.5	349.8	181.9	1	105.4	0.301	

included in Table 5.2 of 0.72 and 0.74 are somewhat lower than the average value of 0.8 given by Ladd et al (1977).

5.5 Effective stress strength parameters

From the stress path plots in Fig. 5 it is possible to define the following effective stress parameters:
- internal friction angle, $\phi' = 28°$
- cohesion, c = 5 kPa
These values are well within the general experience of Norwegian marine clays.

6 DEFORMATION CHARACTERISTICS

6.1 Small strain shear moduli

Small strain shear moduli have been calculated from measurements of shear wave velocities using bender element tests (Dyvik and Madshus, 1985). Fig. 7 shows the results of such measurements carried out on both triaxial and CRSC tests on block samples. It can be observed that the CRSC G_{max} measurements consistently give somewhat lower values than the CAUC tests. Although this is contradictory to findings reported by Dyvik and Olsen (1989), it is believed that this difference is due to differences in measurements and interpretation procedures for the two types of tests.

For the measurements in the triaxial tests the ratio G_{max}/s_u^{CAUC} varies between 940 and 1030 which is quite a lot higher than the 4-500 reported for Bangkok clay by Shibuya and Tamrakar (1999).

Figure 7 also includes G_{max} values interpreted from Raleigh wave measurements in the field (BRE, 1990). The G_{max} from the field tests at Lierstranda are significantly higher than the laboratory values. It is believed that this difference is not associated with the problem of sample disturbance, but rather with differences in measurements and interpretation procedures for the two types of tests.

6.2 Constrained modulus as measured by CRSC tests

Constant rate of strain oedometer tests (CRSC) were carried out, one typical example is shown in Fig. 8, which is included in Chapter 8 on sample disturbance. Table 6.2 summarises the results of the CRSC tests on block samples.

With reference to Table 6.2 the following definitions need to be made:

In general the constrained modulus is defined as $M = \Delta\sigma_v'/\Delta\varepsilon_v$.

M_0 is the constrained modulus at the in situ vertical stress, σ_{vo}'. M_1 is the average constrained modulus in the stress range σ_{vo}' to p_c', where p_c' is the apparent preconsolidation stress.

When the vertical effective stress is above σ_v' the constrained modulus is linearly increasing with σ_v' according to the following formula proposed by Janbu (1969): $M = m (\sigma_v' - p_r)$;

m is a dimensionless constant and p_r is a constant with the same units as σ_v'.

The values for m and p_r reported in Table 6.2 are within the normal range for Norwegian soft clays according to Janbu (1969).

Table 6.2 Results of CRSC tests on block samples

Depth m	w_i %	e_o at σ_{v0}'	σ_{v0}' kPa	M_0 MPa	M_1 Mpa	m	P_r Kpa	c_{v0} m²/year	c_{v1} m²/year	$\Delta e/1+e_0$
12.3	38.6	1.03	109	8.0	5.7	17.7	120	56.0	35.6	0.038
16.4	32.6	0.86	146	8.3	5.5	21.6	110	32.6	21.2	0.049
22.4	33.1	0.86	200	5.2	4.5	21.4	60	33.4	29.0	0.072

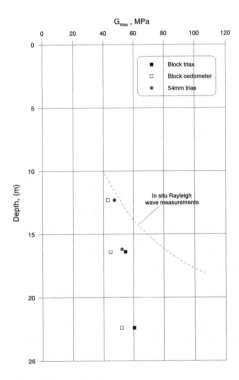

Figure 7. G_{max} vs depth

An interesting note is that in Norway a plot like that shown in the lower diagrams in Fig. 8 is used to determine the preconsolidation stress (or yield stress). We have found that it can in many cases give a better definition than the classical Casagrande method.

The M and c_v values are very much influenced by sample disturbance. See Chapter 8.

7 COEFFICIENTS OF CONSOLIDATION AND PERMEABILITY

7.1 Coefficient of consolidation, c_v

Table 6.2 included two values for c_v. c_{v0} is the value at the in situ vertical effective stress, while c_{v1} is the average value in the stress range from σ_{v0}' to p_c'.

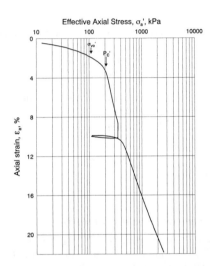

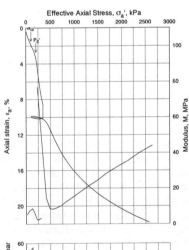

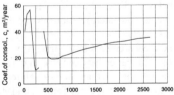

Figure 8. Results from CRSC test on block sample; 12.3 m depth

7.2 Coefficient of permeability, k

The coefficient of permeability, k, has been computed from pore water pressure measurements in the bottom of the sample during CRSC testing as described by Sandbækken et al. (1986). Some direct measurements of permeability (falling head tests) confirmed the computed ones. For the three block samples at 12.3; 16.2 and 22.4 m the k – values range from 0.4 to 0.65 m^2/year (or 1.5 to 2.5 $*10^{-9}$ m/sec). These values are actually quite similar to values of k reported by Tavenas et al. (1986) for eastern Canada clays.

8 EFFECT OF SAMPLE DISTURBANCE ON MEASURED SOIL PROPERTIES

As mentioned most of the strength and deformation parameters referred to above are based on laboratory measurements on high quality block samples. It is a well known fact that sample disturbance is especially pronounced in clays of low plasticity. Lunne et al. (1997a) have shown that this is indeed also the case for the lean Lierstranda clay. Lunne et al gives a detailed account of the effects of sample disturbance on measurements carried out in the CAUC triaxial and the CRSC tests. Only some of the results will be included herein to highlight the significant effects of sample disturbance.

The samplers used, in addition to the Sherbrooke block sampler, were the NGI 54 mm composite piston sampler (Andresen and Kolstad), 1979 and the Japanese 75 mm hydraulic piston sampler (Tanaka et al. 1996).

Figure 9 shows the results of CAUC tests carried out on samples from 12.3 m in terms of stress strain curves and also in terms of stress paths. Fig.10 shows measured s_u as function of depth. It can be observed that the undrained shear strength is reduced by 22 to 36 % for the piston samples relative to the block samples. Fig. 11 shows the results of CRSC tests on the three samples from 12.3 m depth as blown up plots of the constrained modulus, M, and the coefficient of consolidation, c_v, vs effective vertical stress.

It can be observed that the M and c_v values in the stress range σ_{vo}' to p_c' measured on piston samples are reduced as much as 42 % and 55 % respectively when compared with the results of the tests on block samples. It is worth mentioning that the samples taken by the Japanese 75 mm sampler were generally of higher quality compared to those taken by the 54 mm NGI sampler.

Regarding the effect of sample disturbance on other soil parameters the reader is referred to Lunne et al. (1997a).

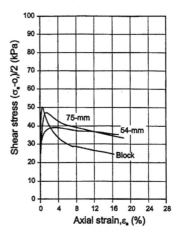

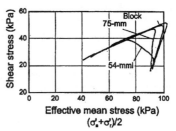

Figure 9. Effects of sample disturbance on CAUC tests at 12.3 m depth

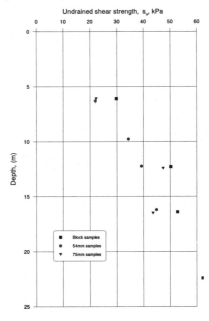

Figure 10. Effects of sample disturbance on s_u from CAUC tests

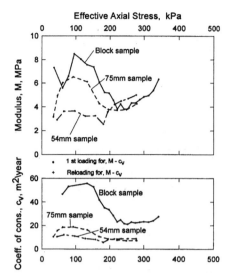

Figure 11. Effects of sample disturbance on CRSC test results; 16.4 m depth

9 CORRELATIONS BETWEEN IN SITU TEST AND LABORATORY TEST MEASUREMENTS

9.1 Vane tests

The boring profile in Fig. 3 shows the results of 2 vane profiles. It can be observed that the vane shear strength is systematically lower than the CAUC shear strength. In fact the vane shear strength is comparable to the three values of s_u measured by DSS tests. This fits well with general observations in Norwegian low plastic clays as reported by Aas et al. (1986).

9.2 Piezocone tests (CPTU)

The boring profile in Fig. 3 included the cone resistance as measured with a piezocone.

Table 9.1 summarises results of the CPTU at the 3 depths where CAUC and CRSC laboratory tests have been performed on block samples.

The three last columns in the table below relate to correlations between stress history and undrained shear strength. Leroueil et al. (1995) defined $N_{\sigma T}$ from the following equation:

$$q_{net} = N_{\sigma T} * p_c'$$

The $N_{\sigma T}$ values for the lean Lierstranda clay range from 2.6 to 2.7 whereas Leroueil et al. found $N_{\sigma T}$ to be 3.6 on average for eastern Canada clays.

N_{kt} is the 'classical' cone factor relating the corrected cone resistance q_t (e.g. see Lunne et al. 1997b) to the undrained shear strength (as measured in a CAUC test) as follows:

$$q_{net} = (q_t - \sigma_{v0}) = N_{kt} * s_u^{CAUC}$$

As can be seen from Table 9.1 N_{kt} varies from 9.7 to 12.6. These values compare well with results from other soft Norwegian clays (e.g. Lunne et al. 1997b). Leroueil (1999) reported N_{kt}-values in the range 11 to 18 ; but these were based on vane shear strength and are not directly comparable.

In soft clays where the CPTU gives small q_c values compared to the range of the load cell it is in many cases possible to get more accurate measures of the penetration pore pressure. With measurement of u_2 (just behind the cone) the excess pore pressure, Δu_2, can be used to define a pore pressure cone factor, $N_{\Delta u}$, from the following formula:

$$\Delta u_2 = N_{\Delta u} * s_u^{CAUC}$$

For other soft Norwegian clays this factor lies in the range of 5 to 9. At Lierstranda the range is small, from 7.1 to 7.7. In practical soil investigations in soft clays in Norway the undrained shear strength profile to be used in design is based on interpretation of the CPTU results using both the net cone resistance and the excess pore pressure as indicated above.

Figure 12 includes also ranges for N_{KT} and $N_{\Delta u}$ found by Lunne et al. (1985) using the triaxial compression test on conventional 72 mm diameter samples. It can be observed that when comparing the new and the 'old' data: 1) There is less scatter in the new data, 2) The new N-values are lower than the 'old'. Use of the new correlations therefore result in higher s_u-values.

10 SUMMARY

The low plasticity Lierstranda clay is believed to never have been subjected to loads greater than the present overburden, but due to aging effects the clay exhibits an apparent overconsolidation. Triaxial compression tests on high quality block samples show peak shear stress at very small strain and significant strain softening. Strength anisotropy is found to be similar to other soft clay. Parallel testing on piston tube samples have shown that the effects of sample disturbance can be very significant for the lean Lierstranda clay.

ACKNOWLEDGEMENT

The author would like to thank several colleagues at NGI who has contributed to the high quality laboratory testing carried out in this project. Also the valuable discussions with and input from Stein Strandvik, Toralv Berre and Knut Andersen are very much appreciated.

Table 9.1 Correlations between CPTU and laboratory test results

Depth m	σ_{vo}' kPa	q_{net} kPa	Δu_2 kPa	B_q	R_f %	OCR	s_uCAUC kPa	$N_{\sigma T}$	N_{kt}	$N_{\Delta u}$
12.3	106	630	383	0.61	1.3	2.3	50	2.6	12.6	7.7
16.4	138	535	409	0.76	1.3	1.5	53	2.6	10.1	7.7
22.4	184	603	443	0.73	1.3	1.2	62	2.7	9.7	7.1

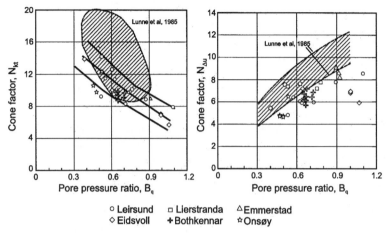

○ Leirsund　　□ Lierstranda　　△ Emmerstad
◇ Eidsvoll　　+ Bothkennar　　☆ Onsøy

Figure 12. Cone factors N_{KT} and $N_{\Delta u}$ vs pore pressure ratio, B_q

11 REFERENCES

Aas, G., Lacasse, S., Lunne, T. & Høeg, K. 1986. Use of *in situ* tests for foundation design on clay. *In situ 1986. ASCE Speciality Conf.* Blacksburg Va, USA, Vol. 1:1-36.

Andresen, A. & Kolstad, P. 1979. The NGI 54 mm samplers for undisturbed sampling of clays and representative sampling of coarser materials. *Proceedings of International symposium on Soil Sampling. Singapore 1979.* Also published in Norwegian Geotechnical Institute, Oslo. Publication No. 130,1980.

Berre, T. 1982. Triaxial testing at the Norwegian Geotechnical Institute. *Geotechnical Testing Journal*, Vol. 5, No. 1/2:3-17.

Berre, T. & Bjerrum, L. 1973. Shear strength of normally consolidated clays. *Int. Conf. On Soil Mechanics and Foundation Engineer,* 8 Moscow 1973. *Proceedings*, Vol. 1.1, pp. 39-49. Also publ. in: Norwegian Geotechnical Institute, Oslo. *Publication, 99.*

Bjerrum, L. 1967. Engineering geology of Norwegian normally consolidated marine clays as related to settlements of buildings; 7th Rankine lecture. In *Géotechnique*, b. 17, No. 2, 1967: 83-118.

Bjerrum, L. 1972. Embankments on soft ground, state-of-the-art report, *Proceedings of the Specialty Conference on the Performance of Earth and Earth-Supported Structures*, Lafayette, Indiana, Vol. 2:1-54.

BRE 1990. Rayleigh wave measurements at Lierstranda, Holmen and Museum Park in Drammen, Norway- factual report. Building Research Establishment. Report No. G/GP/9015.

Dam, T.K.L., Yamane, N. & Hanzawa, H. 1999. Direct shear results on four soft marine clay. Proc. *Bothkennar, Drammen, Quebec and Ariake clays.* Yokosoka, Japan, Feb. 1997, pp. 229-240.

Casagrande, A. 1936. The determination of the pre-consolidation load and its practical significance. *Int. Conf. On Soil Mechanics and Foundation Engineering*, 1. Cambridge, Mass. 1936. *Proceedings*, Vol 3, pp. 60-64.

Dyvik, R. & Madshus, C. 1985. Lab measurements of G_{max} using bender elements. *Advances in the art of testing soils under cyclic conditions, proceedings of a session in conjunction with the ASCE Convention* in Detroit, Michigan 1985, pp. 186-196. N.Y., American Society of Civil Engineers.

Dyvik, R. & Olsen, T.S. 1989. G_{max} measured in oedometer and DSS tests using bender elements. *Int. conf. on Soil Mechanics and Foundation Engi-*

neering, 12. *Rio de Janeiro 1989. Proc.*, Vol. 1, pp. 39-42. Also publ. in: Norwegian Geotechnical Institute, Oslo. *Publication, 181*, 1991.

Ihler, H. 1996. Effekt av prøveforstyrrelse ved bestemmelse av geotekniske egenskaper for leire fra Lierstranda. Cand.Scient. Thesis, Oslo University.

Janbu, N. 1969. The resistance concept applied to deformation of soils. *Seventh Int. Conf. On Soil Mechanics and Foundation Engineering,*, Mexico, *Proceedings*, Vol. 1, pp. 191-196.

Karlsrud, K., Lunne. T. & Brattlien, K. 1996. Improved CPTU interpretations based on block samples. *Nordic Geotechnical Conference, 12. Reykjavik 1996. Proc.* Vol. 1, pp. 195-201.

Ladd, C.C. 1991. Stability Evaluation During Stage Construction. *Journal of Geotechnical Engineering,* ASCE, Vol. 117, No. 4, April. pp. 540-615.

Ladd, C.C. & Foott, R. 1974. New Design procedure for Stability of Soft Clays. *JGED, ASCE*, Vol. 100, No. GT7, pp. 763-786.

Ladd, C.C. and Foott, R. 1977. *Foundation Design of Embankments Constructed on Varved Clays*. United States Depatment of Transportation, FHWA, Contract No. DOT-FH-11-8973.

Ladd, C.C., Foott, R., Ishihara, K., Schlosser, F. & Poulos, H.G. 1977. Stress-Deformation and Strength Characteristics. *Proc., 9th International Conference on Soil Mechanics and Foundation Engineering*, Vol. 2, Tokyo, pp. 421-494.

Lefebvre, G. & Poulin, C. 1979. A new method of sampling in sensitive clay. *Canadian Geotechnical Journal*, Vol. 16:226-233.

Leroueil, S. 1999. Geotechnical characteristics of Eastern Canadian clay. *Proc. of Int. Symposium on characterization of soft marine clays – Bothkennar, Drammen, Quebec and Ariake clays*. Yokosoka, Japan, Feb. 1997, pp 3-32.

Leroueil, S., Demers, D., La Rochelle, P., Martel, G., & Virel, D. 1995. Practical use of the piezocone in eastern Canada clays. *Int. Symp. on Cone Penetration Testing*, Linkoping, 2, 515-522.

Leroueil, S., Tavenas, F. & Le Bihan, J.P. 1983. Propriétés caractéristiques des argiles de l'est du Canada. *Canadian Geotechnical J.*, 20(4): 681-705.

Lunne, T. & Lacasse, S. 1999. Geotechnical characteristics of low plasticity Drammen clay. *Proc. of Int. Symposium on characterization of soft marine clays – Bothkennar, Drammen, Quebec and Ariake clays*. Yokosoka, Japan, Feb. 1997, pp 33-56.

Lunne,T.,Christophersen,H.P. and Tjelta,T.I.1985 Engineering use of piezocone data in North Sea clays. Proc.ICSMFE , San Francisco 2, 907-912.

Lunne, T., Berre, T, & Strandvik, S. 1997a. Sample disturbance effects in soft low plastic Norwegian clay. *Conference on Recent Developments in Soil and Pavement Mechanics, Rio de Janeiro, June 1997. Proc.* pp. 81-102.

Lunne, T., Robertson, P.K. & Powell, J.J.M. 1997b. Cone penetration testing in geotechnical practice. Published by E & FN Spon, London, 1997.

Norwegian Geotechnical Institute. 1996. Optimal Use of Soil Data from Deep Water Sites. Results of Laboratory Tests. NGI report 521676-4, March 1996.

Norwegian Geotechnical Institute. 1998. Deep Water Sampling and the Influence of Gas. Analysis and recommendation report. NGI report 521676-7, 15 November 1998.

Nutt, N., Lunne, T., Hanzawa, H. & Tang, Y.X. 1995. A comparative study between the NGI direct shear apparatus. NGI Report No. 521666-1, March 1995.

Sandbækken, G., Berre, G. & Lacasse, S. 1986. Oedometer testing at the Norwegian Geotechnical Institute. *Consolidation of soils: testing and evaluation: a symposium*. Fort Lauderdale, Fla. 1985. American Society for Testing and Materials. Special Technical Publication, 892:329-353.

Shibuya, S. &Tamrakar, S.B. 1999. In situ and laboratory investigations into engineering properties of Bangkok clay. *Proc. of Int. Symposium on characterization of soft marine clays – Bothkennar, Drammen, Quebec and Ariake clays*. Yokosoka, Japan, Feb. 1997, pp. 107-132.

Tanaka, H. 1997. Characteristics of Drammen clay measured by PHRI. Port and Harbour Research Institute, Yokasuka. Report.

Tanaka, H., Oka, F. & Yashima, A. 1996. Sampling of soft soil in Japan. *Marine Georesources and Geotechnology*, 14:283-295.

Tavenas, F., Tremblay, M., Larouche, G. & Leroueil, S. 1986. In situ measurement of permeability in soft clays. *ASCE Special Conf. on Use of In Situ Tests in Geotechnical Engineering*, Blacksburg: 1034-1048.

Watabe,Y. 1999. Mechanical properties of K_0 – consolidation and shearing behaviour observed in triaxial tests for five world wide clays- Drammen, Louiseville, Singapore, Kansai and Ariake clays. *Proc. of Int. Symposium on characterization of soft marine clays – Bothkennar, Drammen, Quebec and Ariake clays*. Yokosoka , Japan, Feb. 1997, pp 241-254.

Coastal Geotechnical Engineering in Practice, Nakase & Tsuchida (eds)
© 2002 Swets & Zeitlinger, Lisse, ISBN 90 5809 151 1

The mechanical behavior of different European clays within a general framework of clay behavior

F.Cotecchia
Department of Civil Engineering, Technical University of Bari, Italy

ABSTRACT: The behaviour of clays depends on their structure, which is the result of their composition and geological history. However, clays of different structure are shown to follow similar patterns of behaviour, controlled by the same parameters, within a general framework of clay behaviour, although the values of the parameters depend on the structure of the material. A parameter for structure is proposed, that is the stress sensitivity S_σ. It is shown that it is possible to relate the differences in size of the mechanical properties of different clays to their differences in stress sensitivity. The applicability of these concepts and relationships is shown for 4 italian marine clays, of different origin and geological history, and for one british clay. The composition, history and structure of the materials is briefly discussed. It is shown how the behaviour of clays, from soft to stiff, can be represented within a single general picture, which can be of use to the engineer to interpret the behavioural data of the material and choose the values of the mechanical parameters to be used in design.

1. INTRODUCTION

The mechanical behaviour of clays is influenced by their structure, which depends on the physical and chemical conditions applying during deposition (e.g. water chemistry, pressure, temperature and organic content), consolidation and unloading, as well as on geological processes such as ageing, diagenesis, weathering and tectonics. At the micro-level, the clay *structure* is the combination of *fabric*, that is the geometric arrangement of the particles, and *bonding*, that is the result of the inter-particle forces not of a purely frictional nature (Lambe & Whitman 1969). Therefore bonding is not necessarily a solid link, particularly with clays, but it is the consequence of electrostatic, electromagnetic or other forces connecting the particles, which have developed in the soil geological life.

The structure of a clay whose response is not affected by disturbance may be considered to be *stable*. Generally clays sedimented rapidly, such as the reconstituted clays consolidated in the laboratory from slurry (Burland 1990), possess stable structures and exhibit similar mechanical behaviour between each other. This behaviour is interpreted well by the critical state framework (Schofield & Wroth 1968). Natural clays may have structures different from those of the same clays reconstituted in the laboratory and can be sensitive to mechanical disturbance, in which case the clay structure is not stable and the clay exhibits differences in mechanical behaviour from the corresponding reconstituted clay. However, since the basic elements of structure are similar for all clays, clays of structure exhibit also

Fig.1 Location of the italian clays studied.

similarities in mechanical behaviour. Following this approach to the influence of structure on clay behaviour, Cotecchia & Chandler (2000) have proposed a general framework for the mechanical behaviour of clays which shows that the differences in mechanics between clays of different structure concern principally the magnitude of the clay mechanical properties rather than the patterns of mechanical response. The authors also show that the differences in size of response for clays of different structure can be related to differences in a single structure parameter, the *stress sensitivity* S_σ. Clays of different structure are seen to exhibit differences in behaviour only if they have different stress sensitivity S_σ and

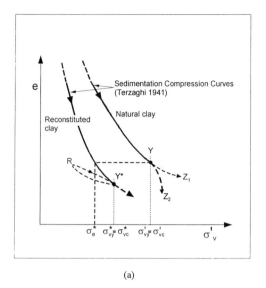

(a)

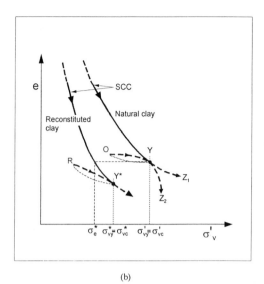

(b)

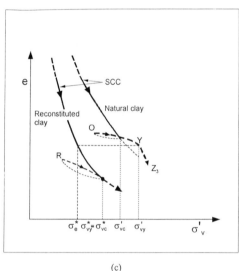

(c)

Fig. 2 Compression behaviour of clays: (a) with sedimentation structure, (b) overconsolidated solely due to unloading, (c) with post-sedimentation structure (Cotecchia & Chandler 2000).

their differences in response are shown to be related to the differences in S_σ within the framework.

The present paper shows the applicability of the general behavioural framework from Cotecchia & Chandler to four marine clays from Italy (Figure 1). Pappadai (Cotecchia 1996; Cotecchia & Chandler 1995; 1997; 1998) and Montemesola (Cafaro & Cotecchia 2001) clays are stiff blue clays sampled close to Taranto, on the Ionian coast, the Sibari clays (Coop & Cotecchia 1995 and 1996) are firm intertidal clays located on the other side of the Ionian coast, whereas the scaly Variegated clays from Senerchia (Santaloia et al. 2000) are stiff heavily

tectonized clays located in the Apennines. In addition, the paper will compare the behaviour of Bothkennar clay, which is a soft clay located in Scotland (Hight et al. 1992), with the behaviour of the stiffer clays from Italy mentioned above. The composition, structure and geological history of the clays will be discussed in some detail and related to their mechanical behaviour.

2. A GENERAL FRAMEWORK FOR THE MECHANICAL BEHAVIOUR OF CLAYS

Despite differences in mineralogy, pore water chemistry, deposition environment and rate, all clays follow, during deposition and normalconsolidation, a *sedimentation compression curve* (SC curve) in the e-σ_v' plane (Terzaghi 1941), which is of the shape shown in Figure 2. The figure shows the SC curves for a natural clay and for the same clay reconstituted in the laboratory, which differ as result of the differences in structure of the two clays, which are of the same mineralogy, but were formed in different environments and conditions. Thus the state of a clay depends on the structure developing in it since the start of its geological history. During normal consolidation in situ, that is when moving along the SC curve, the clay is continuously at gross yield (Y in Figure 2; Hight et al. 1992), where *gross yield* is intended as a state of volume (in the specific volume, v, deviatoric stress, q, mean effective stress, p', space) which separates two domains characterized by significantly different stiffness properties and rate of structure degradation with straining.

If the structure of a clay is the result solely of normalconsolidation (Figure 2a), Cotecchia & Chandler propose to call it a *sedimentation structure*. If some geological processes subsequent to normalconsolidation intervene to modify the clay structure, such as simple mechanical unloading, or creep, thixotropy, diagenesis, weathering or tectonic shearing, then the clay structure may be defined as a *post-sedimentation structure*. If the clay has either a sedimentation structure (Figure 2a) or a post-sedimentation structure due to simple mechanical unloading (Figure 2b), it will exhibit gross yield on the SC curve and will either follow it or fall below it with further compression. In fact, if the clay is over-consolidated for mechanical unloading, it recovers its original sedimentation structure at gross yield. If the clay has a post-sedimentation structure due to geological processes which strengthen the clay bonding permanently, such as diagenesis, according to Cotecchia & Chandler (2000) it will exhibit gross yield to the right of the SC curve, as shown in Figure 2(c). Other studies after Cotecchia & Chandler (2000) have shown that other geological processes, such as weathering (Cafaro & Cotecchia 2001) and tectonic shearing (Santaloia et al. 2000), may give rise to post-sedimentation structures which make the clay exhibit gross yield to the left of the SC curve. Thus, in general it is useful to divide clays into the two classes proposed by Cotecchia & Chandler (2000) since, although they include large varieties of structures, they differentiate the clays whose gross yield is fundamentally controlled by stress history from those for which also geological processes other than stress history control the clay mechanical behaviour and gross yielding.

In their work Cotecchia & Chandler examine the behaviour of clays from either classes. The main aspects of the general behavioural framework that they propose are briefly reviewed in the following and the reader can refer to the original paper for more explanations. Since Cotecchia & Chandler do not discuss both weathered and tectonized clays, these will not be considered in the following review. However it will be shown later that also these clays fit the framework.

In general, clays of either sedimentation or post-sedimentation structure are similarly controlled by the yield stress ratio, expressed either as YSR, the ratio of the gross yield vertical effective stress σ_y' in one-dimensional compression to the current vertical effective stress, σ_{v0}', or YSR_{is}, the ratio of the mean effective stress p_{iy}' in isotropic compression to the current p', or YSR_{K0}, the ratio of the mean effective stress p_{K0y}' in one-dimensional compression to the current p'. YSR is different from the geological OCR (the ratio of the preconsolidation stress σ_{vc}' to the current vertical effective stress σ_{v0}') only for post-sedimentation structure clays (Figure 2c).

Burland (1990) has shown how the void index: I_v =$(e-e_{100}*)/(e_{100}*-e_{1000}*)$ normalizes the compression behaviour of clays for composition, where $e_{100}*$ and $e_{1000}*$ are the void ratios of the clay reconstituted and normally consolidated in the laboratory to a vertical effective stress of 100 and 1000 kPa respectively. The one-dimensional normal consolidation line of reconstituted clays in the I_v-σ_v' plane is unique (the intrinsic compression line ICL; Burland 1990), and represents the SC curve of all reconstituted clays. For all types of either natural sedimentation or post-sedimentation structure clays (if tectonized clays are

excluded), gross yield Y lies to the right of the ICL in the I_v-σ_v' plane as result of differences in structure between the natural and the reconstituted clay. Pre-gross yield all clays exhibits a compression index (C_r) which is lower than that post-gross yield (C_c) and the clays with gross yield Y to the right of the ICL have a post-gross yield compression index C_c either equal to or higher than that of the reconstituted clay, C_c*. The * is used here to refer to reconstituted clay properties. The distance between the gross yield point Y and the ICL is indicative of how much different is the natural clay structure from that of the reconstituted clay. Thus Cotecchia & Chandler propose to use the ratio of the gross yield stress σ_y' to the equivalent stress σ_e* (Figure 2), taken on the ICL for the same void ratio as that of the natural clay at gross yield, as parameter of the clay structural strength, which they define as *stress sensitivity* $S_\sigma = \sigma_y'/\sigma_e*$.

For a clay of stress sensitivity S_σ, the gross yield states in both one-dimensional and isotropic compression are shifted to the right of the corresponding compression curves of the reconstituted clay in the v-p' plane, and the ratios of either the isotropic, p_{iy}', or the one-dimensional gross yield mean effective stresses, p_{K0y}', to the corresponding equivalent stresses: $p_{ie}*$ and $p_{K0e}*$ respectively, equal the stress sensitivity, so that: $S_\sigma = \sigma_y'/\sigma_e*= p_{iy}'/p_{ie}*= p_{K0y}'/p_{K0e}*$. Therefore the differences in structure cause the offset of the whole compression framework of the natural clay with respect to that of the reconstituted clay. The post-gross yield compression curves of the natural clay, in either one-dimensional or isotropic conditions, are parallel to each other, as for reconstituted clays, and their slope is generally higher than that of the reconstituted clay compression curves. Thus the patterns of response to compression of clays of different structure are similar, but the compression properties (i.e. parameters which locate the compression curves in the v-p' plane) are different in magnitude.

Clays of different stress sensitivity exhibit differences also in the magnitude of the shear properties. Figure 3(a) shows a scheme representing the typical pattern of response to shear of natural clay specimens one-dimensionally consolidated pre-gross yield to different values of YSR_{K0} along the swelling line shown in Figure 3(b). The scheme in the figure is seen to apply generally to clays, from soft to stiff (Cotecchia & Chandler 1997; Kavvadas & Amorosi 2000; Lerouil 1997 and 1998). The stress paths are normalized for volume by the equivalent pressure p_e*, which is shown in Figure 3(b). They are bounded by an arch-shaped state boundary envelope. Generally the stress paths of either one-dimensionally or isotropically normally consolidated specimens, rise to the apex of this state boundary envelope, as shown in Figure 4. This figure also shows that the normalized state boundary envelope of the natural clay lies outside the normalized state boundary surface of the reconstituted clay. This is the main effect on shear behaviour of the higher structural strength of the natural clay ($S_\sigma > 1$) with

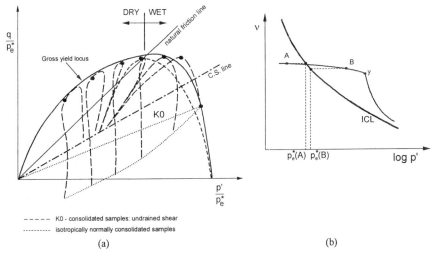

Fig.3: (a) Shear behaviour of natural clay specimens consolidated along the swelling line in Figure 3(b).

respect to the reconstituted clay ($S_\sigma= 1$) (Cotecchia & Chandler 1997; Burland 1990; Lerouil 1997). Terzaghi defined, in 1944, the *sensitivity* of clays as the ratio of the undrained strength of the undisturbed clay to that of the same clay after remoulding at the same water content. Therefore the sensitivity of normally consolidated specimens of natural clay, S, may be expressed as the ratio :

$$S = q_{max}(v_0)/q_{max}*(v_0) = (q_{max}/p_e*)/(q_{max}*/p_e*) \quad (1)$$

v_0 being the clay specific volume. S also equals the ratio between the maximum vertical size of the normalized state boundary envelope of the natural clay, (q_{max}/p_e*), to that of the reconstituted clay $(q_{max}*/p_e*)$. Cotecchia & Chandler propose to define *strength sensitivity*, S_t, of a clay, either normally or over-consolidated, the ratio :

$$S_t = (q_{max}/p_e*)/(q_{max}*/p_e*) \quad (2)$$

that is the sensitivity S after compression of the clay to gross yield (YSR=1).

The authors then show that for both sedimentation and post-sedimentation structure clays, the stress sensitivity S_σ equals the strength sensitivity S_t:

$$S_\sigma = S_t \quad (3)$$

The authors refer, for example, to gross yield and sensitivity data (S values) reported by Skempton & Northey (1952) for normally consolidated - sedimentation structure clays, which show that for all clays the stress sensitivity S_σ equals the measured sensitivity S, that in turn equals the above defined strength sensitivity S_t, being the clays examined normally consolidated. In addition the authors show that compression data for clays from soft (Bothkennar clay) to very stiff (Gault clay), which are reported in Figure 5, are indicative of strength sensi-

tivities (Table 1) which equal the stress sensitivity of the clay, confirming the equality (3). The data from Skempton & Northey as well as those in Figure 5 indicate that in general soft clays are of higher sensitivity than stiff clays, the latter being of low to medium sensitivity. Relation (3) implies that the schemes of behaviour in Figure 6 apply to the general behaviour of clays. Figure 6(a) shows the loci of gross yield states, in the I_v-σ_v' plane, for clays of equal strength sensitivity S_t. According to this scheme, clays of equal strength sensitivity have gross yield states along a single curve, parallel to the ICL, although the clays have different origin, geological history and structure. For sedimentation structure clays the gross yield locus coincides with the sedimentation compression curve. Consequently Figure 6 shows that clays of different origin and deposition environment follow the same SC curve in normal consolidation in situ if developing structures of equal strength sensitivity. Figure 6(a) is the base of the *sensitivity framework*, as defined by Cotecchia & Chandler (2000) and Chandler (2000).

According to the framework, clays of equal stress sensitivity have the same state boundary surface in the I_v-q/M-p' space (Figure 6b), where I_v normalizes the clay behaviour for the influence of composition on specific volume and M* (the stress ratio at critical state for the reconstituted clay) normalizes the clay strength for the influence of composition (Graham et al. 1988; Coop et al. 1995). Also, the framework implies that the state boundary envelopes of specimens of equal stress sensitivity, but different specific volume, are similar between each other and to the state boundary envelopes of the reconstituted clay. The size of these envelopes is related to stress sensitivity. Consequently clay specimens having the same stress sensitivity are confined by a SBS which can be normalized for volume as for the SBS of reconstituted clays (SBS*).

Post-gross yield the clay structure undergoes sig-

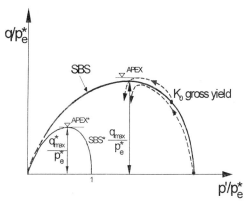

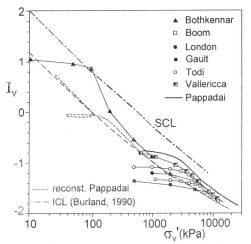

Fig.4 State boundary envelopes of a natural clay and of the same clay reconstituted in the laboratory (Cotecchia & Chandler 2000, modified).

Fig.5 One-dimensional compression curves of soft to stiff natural clays (Cotecchia & Chandler 2000).

nificant degradation with straining and, consequently, the stress sensitivity reduces, as shown in Figure 7. The clay states A and B in the figure lie post-gross yield and plot on gross yield loci relating to a stress sensitivity value lower than that at gross yield (Y). According to the sensitivity framework, the state boundary envelopes corresponding to the gross yield points A and B should be smaller than the original one, as shown in Figure 7(b). It follows that the state boundary surface and the gross yield loci of natural clays compressing post-gross yield and having a compression index C_c higher than C_c* cannot be normalized solely for volume, as it is instead the case for reconstituted clays. In fact the essential difference between natural sensitive and reconstituted clays is that the latter have constant sensitivity, of 1, both before and post-gross yield, whereas the natural clays may reduce in stress sensitivity with post-gross yield straining. Since

$(q_{max}*/p_e*)$ is constant for reconstituted clays according to the critical state framework, from equation (2) it follows that :

$$S_\sigma = S_t = (q_{max}/p_e*)/\text{constant} \qquad (4)$$

and consequently

$$q_{max}/(p_e* \, S_\sigma) = \text{constant} \qquad (5)$$

Therefore Cotecchia & Chandler propose p_e* as the normalizing factor for the effects of volume on the clay strength and S_σ as the normalizing parameter for the effects of structure. They suggest that a single normalized gross yield curve should therefore be found, for all clays, both pre and post-gross yield, in the normalized stress plane $q/(p_e* \, S_\sigma)$ – $p'/(p_e* \, S_\sigma)$, where S_σ may either equal σ_y'/σ_e* or $p_{iy}'/p_{ie}*$ or $p_{K0y}'/p_{K0e}*$, as shown in Figure 8. The validity of this framework is shown in the following with reference to some of the clays indicated in Figure 1.

3. THE STIFF BLUE CLAYS AT PAPPADAI, MONTEMESOLA AND TARANTO

The clays at Pappadai, Montemesola and Taranto (Figure 1) are part of the sub-Apennine Blue Clays, a geological formation deposited, in the Upper Pliocene and Lower Pleistocene, in the Bradano Trough and in Apulia, that currently outcrops in large areas of Southern Italy. Since the sub-Apennine blue clays were deposited initially in isolated basins, they exhibit some geotechnical variability.

The apulian Blue clays overlie either a transgressive Pliocene calcarenite or a Cretaceous calcareous bedrock. In particular the Pappadai and the Montemesola clays were deposited in early Pleistocene (1200000 years ago) in the Montemesola Basin (Figure 9), which was initially separated from the Taranto Basin, as reported by Cotecchia & Chandler (1995). Palaeontological analyses indicate that the sea was protected at the time of deposition of both these clays, and that the deposition environment was characterized by reducing conditions with little water circulation, comparatively more still than for the Taranto Basin, characterized by more open conditions. The deposition conditions for the clays at Pappadai were even quieter than for those at Montemesola, being Pappadai in the centre and Montemesola closer to the north-western margin of the basin. This difference in deposition conditions is mildly reflected in the index properties of the soils, which are reported in Table 2. The average liquid and plastic limits at Pappadai are 55% and 30% respectively, with average clay fraction of 50%. The clays closer to the margin of the basin, at Montemesola, appear to have a similar granulometry to Pappadai clay, and are only slightly less plastic. However, the Montemesola clay reflects the vicinity to the sea coast at the time of deposition in the fact that it has more frequent coarse inclusions. Also, the shallower clays at Pappadai, which were deposited in a stage when the sea was more open than for the clays at depth,

Table 1 Stress sensitivity and strength sensitivity of the clays in Figure 5.

CLAY	$S_\sigma = \sigma_{vy}'/\sigma_e^*$	p_{iy}'/p_{iy}^*	S_t
Bothkennar (Smith 1992; Burland 1990)	6	7	7.3
Todi (Burland et al. 1996)	2.25		2.3
Boom (Coop et al. 1995)	1.5		1.5
Vallericca (Burland et al. 1996)	2.47		2.5
Pappadai (Cotecchia 1996; Cotecchia & Chandler 1997)	3.5	3.2	3.2
London (Burland 1990; Cotecchia 1990)	> 2		2.1

have more coarse inclusions than those at depth, which instead sometimes (e.g. 25m depth) are laminated. It can be presumed that for horizontal laminations to occur a quiet deposition environment existed, with little water circulation. Cotecchia & Chandler (1995) examine the geotechnical properties of the whole clay deposit at Pappadai, discussing the disuniformities present in it and relating them to the deposition stages, therefore the reader can refer to their work for details.

Mineralogical analyses show that the dominant clay mineral, both at Pappadai and Montemesola, is illite, with subordinate kaolinite, chlorite and smectite. Therefore the clays at different locations in the basin are not significantly different in either their mineralogy or grading or plasticity, but they have different void ratios, which are smaller about the margins of the deposit than in the central part (Table 2). The differences in void ratio reflect differences in fabric of the two clays: the fabric is more densely packed at Montemesola than at Pappadai, although in both the clays it has the typical features of a compacted bookhouse (Cotecchia & Chandler 1995 and 1998; Cafaro 1998; Cafaro et al. 2000).

The Blue clays of the Montemesola Basin have generally high and highly variable carbonate contents: 27%-40%. These high carbonate contents and their large variability are due to the occurrence of significant quantities of calcareous fragments in the clay, which derive either from the erosion and transport of the limestone surrounding the basin, or to the deposition of foraminifera and nannofossils, densely present in both clays, as observed by means of the scanning electron microscope (Cotecchia & Chandler 1995; Cafaro 1998).

The Taranto clays appear to be coarser than both the Pappadai and the Montemesola clays due to the more open sea where they were deposited. Their index properties are also reported in Table 2. These clays are still mainly illitic, but have lower clay fraction, which is reflected into their lower liquid

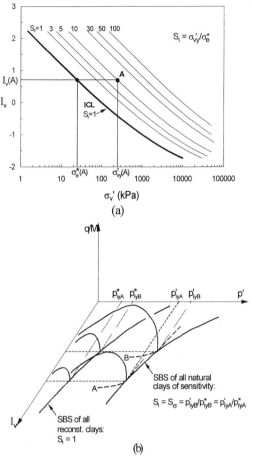

Fig.6 Sensitivity framework: (a) gross yield loci; (b) state boundary surfaces of clays of given stress sensitivity (Cotecchia & Chandler 2000).

limit and plasticity index, although the clays are slightly more active due to a higher percentage of smectite (Cafaro et al. 2000). Their void ratios are closer to those at Montemesola than those at Pappadai, that is consistent with the more turbulent deposition conditions.

In both the Montemesola and the Taranto Basin, erosion has given rise to the overconsolidation of the clays. Higher values of OCR apply to the Pappadai clays (OCR=3 at 25m depth) by comparison with the Montemesola ones (OCR=2 at 25-30m depth). The clays at Taranto are even more overconsolidated. However the gross yield pressures in one-dimensional compression for all these clays have been affected by diagenesis, which occurred when the clays were buried at depth before erosion and which has strengthened the clay bonding, making the YSR of the clay higher than the OCR (Figure 2c). Cotecchia & Chandler (1997) show that Pappadai clay possesses a diagenetic carbonate bonding due to the presence of a carbonate film covering the clay

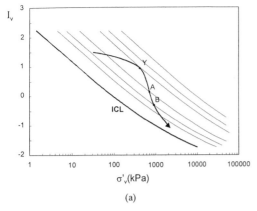

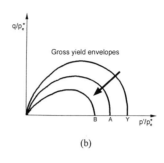

(a)

(b)

Fig.7 (a) Change in stress sensitivity with post-gross yield compression; (b) gross yield envelopes for reducing S_σ.

particles and inter-particle contacts, which they identify by means of chemical micro-probing of the clay samples. The results of oedometer tests on both natural undisturbed and reconstituted Pappadai clay specimens are plotted in Figure 10. The gross yield of the natural clay plots to the right of the ICL, at a distance corresponding to a stress sensitivity S_σ of 3.5. The gross yield state plots also to the right of the original preconsolidation state as effect of diagenesis, and is indicative of a post-sedimentation structure (Figure 2c). Diagenesis not only has doubled the YSR of the clay, but it has also made its swelling index decrease, as suggested by the high swell sensitivity (Schmertmann 1969) C_s^*/C_s, which is equal to 2.5. Post-gross yield the natural clay exhibits a compression index C_c of 0.5, higher than C_c^*, although the high pressure tests show that the natural clay state tends to superimpose to those of the reconstituted clay only at very high vertical effective stresses σ_v' (above 13MPa). However, Cotecchia & Chandler (1998) show that, even at very high stresses, the natural clay fabric is very different from that of the reconstituted clay for the same specific volume.

The swell sensitivity of the Montemesola clays, 4.3 (Cafaro et al. 2000), is even higher than that at Pappadai, being indicative of a bonding which inhibits significantly the natural clay swelling capacity, although for these clays the stress sensitvity S_σ is lower than for the Pappadai clay (S_σ =2.4).

The shear stress paths followed by natural Pappadai clay specimens consolidated pre-gross yield are shown in the plane of stresses normalized for volume by p_e^* in Figure 11. These stress paths appear to reach an arch-shaped state boundary envelope, that is also the gross yield locus of the clay, according to the scheme in Figure 3a. The stress paths abandon the state boundary envelope after following it for a while, and they tend to a critical state ($dq/d\varepsilon_s=0$, $dp'/d\varepsilon_s=0$, $d\varepsilon_v/d\varepsilon_s=0$) inside it. Figure 11 shows that the clay specimens isotropically consolidated to YSR of 1 (i.e. normally consolidated) exhibit a peak strength about the apex of the gross yield locus (or state boundary envelope) according to

the scheme in Figure 4. Similar behaviour has been observed for the Montemesola clay (Cafaro & Cotecchia 2001) and for other clays (Kavvadas & Amorosi 2000; Leroueil 1997). Figure 12 shows, in the plane of stresses normalized for volume, the gross yield locus of undisturbed Pappadai clay, the state boundary surface of reconstituted Pappadai clay and the shear stress paths of the natural clay after compression post-gross yield. The strength sensitivity S_t of the undisturbed clay computed from these data is 3.2, that is very close to the stress sensitivity strength sensitivity and the stress sensitivity to 2.7. Therefore the data support the sensitivity framework previously discussed. Pappadai clay appears to undergo significant structure degradation and weakening in post-gross yield compression. Consequently, its shear behaviour and boundary envelopes are not normalized solely by p_e^*, according to Figure 7(b).

Figure 13 shows that the normalization for both volume and structure by the use of p_e^* and S_σ succeeds in reducing the gross yield data of natural Pappadai clay, the state boundary envelope of the reconstituted clay and the shear stress paths of the natural clay compressed post-gross yield to a single gross yield locus, as expected according to Figure 8 and to the sensitivity framework.

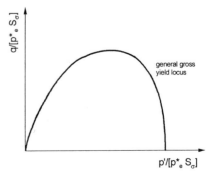

Fig.8 General normalized gross yield locus.

193

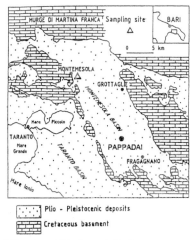

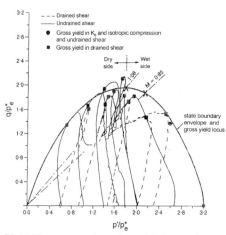

Figure 9 Taranto and Montemesola Basins.

Fig.11 Shear stress paths on Pappadai clay specimens consolidated isotropically pre-gross yield (Cotecchia & Chandler 1997).

3.1 The yellow clays of the Montemesola Basin and the effects of weathering

The top clays in the Montemesola Basin are yellow due to oxidisation and weathering. Thick yellow clay strata, up to 25 m thickness, overlie the unweathered clays which are of grey colour and have been discussed previously. Cafaro & Cotecchia (2001) have reported the causes and the mechanical effects of this weathering, as resulting from a comparative study of both the composition and the mechanical properties of both the yellow and the grey clays. They have found that the main cause of weathering in the region has been the occurrence of shrinking-swelling cycles within the top clays in recent Quaternary, which have resulted from the drying due to a semiarid climate and the rising and draw-down of the water-table. Weathering has affected the structure of the clay, making it more

densely packed and weakening its bonding, and has reduced the clay void ratio. The stress sensitivity of the clay has been consequently reduced. Cafaro & Cotecchia (2001) find that at present the yellow clays have a stress sensitivity S_σ=1.5, that is lower than that of the grey clays, 2.4, which was also the stress sensitivity of the top clays before weathering. Thus weathering shifts the gross yield state of the clay to the left in the e-σ_v' plane. According to the sensitivity framework the strength sensitivity should also be reduced by weathering. The shear data from tests on both grey and yellow clay specimens confirm this, as shown in Figure 14, where the pre-gross yield state boundary envelopes of both the clays are shown normalized for volume by means of p_e^* (Cafaro & Cotecchia 2001). The strength sensitivities S_t computed according to the envelopes in the figure are about 2.4 for the grey clay and 1.5 for the yellow clay, therefore equal to the above mentioned stress sensitivities S_σ.

4. THE FIRM CLAYS OF THE SIBARI PLAIN

The Sibari Plain is located within a graben, trending ENE-WSW, enclosed by several faults bordering the calcareous-dolomitic Monte Pollino to the north and the crystalline Monte Sila to the south (Figure 15). In the Upper Pliocene and Lower Pleistocene the graben was submerged by the sea and the deposition of sub-Apennine Blue Clays occurred. After emersion, in the Late Pleistocene, the plain has been filled by sub-littoral and alluvial deposits, of about 400 m thickness. Since the Pleistocene, Mediterranean coastlines have been subjected to significant changes. V. Cotecchia et al. (1971) showed that the Flandrian regression lowered the sea level of 100 m below the present sea-level in the northern part of the Ionian sea. Later, in the last 14000 years, the sea-level has generally risen, although small fluctuations have occurred in recent times, as shown

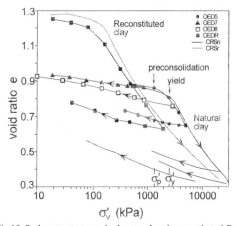

Fig.10 Oedometer tests on both natural and reconstituted Pappadai clay specimens (Cotecchia & Chandler 1997).

in Figure 16. Due to the activity of the extensive fault systems, complex tectonic movements and-downwarping have taken place in the easternmost part of the Sibari plain, which apparently did subside in the Holocene despite the general uplift at slow rates of the Calabrian arc. The above mentioned tectonic movements and glacioeustatic sea - level changes have both contributed to the geomorphological evolution of the plain in the Quaternary, with the formation of the above mentioned alluvial/littoral deposits. Consequently, in the last 2500 years the archaic city of Sybaris, the Hellenistic Thurium and Roman Copiae, which were located in the plain (Figure 15), have disappeared, the shoreline has moved 2 Km to the East and the course of river Coscile has moved to the South. The archaic city of Sybaris, a greek colony lying at about sea level in the plain between 720 and 510 BC, now lies below both ground-water level and sea-level (-2.5 m a.s.l.) and is overlaid by Thurium (-1.26 m a.s.l.) and Copiae (+0.48 m a.s.l.), as shown in Figure 17.

In the last decade a multi-disciplinary research has been developed (Progetto Strategico Beni Culturali Mezzogiorno-CNR) to investigate the subsidence of the plain in the Quaternary and to define the engineering works to access the archaic remains. Three fundamental causes have been recognized for the quaternary subsidence of the plain: the tectonic movements and the glacieustatic sea-level changes mentioned above, as well as the geotechnical compression of the sediments being formed.
The top 100 m sediments located below the archaological area have been thoroughly investigated, as reported by Cotecchia et al. (1994), Coop & Cotecchia (1996), Pagliarulo et al. (1995), Cotecchia et al. (2000); Pagliarulo & Cotecchia (2000). The lithological profiles of four deep boreholes drilled at the archaeological sites (Figure 15) are shown in Figure 18 and the corresponding geotechnical profiles are shown in Figure 19. The deposits are of high variability and are formed of sands, clayey silts

and clays of transitional environments. On the basis of lithological and palaeontological analyses, Pagliarulo et al. (1995) demonstrated that between 30 and 70 m depth the soil deposition environment was lagoonal and continental to the north-west of the archaeological site, becoming intertidal and sublittoral south-east, where the soils are more clayey. Above 30m depth lagoonal and continental deposits are present over the whole archaeological area, which were formed in the last 4000 years. The existence of a stratum of higher clay fraction between 30 and 70 m depth is evident in the clay fraction profile in Figure 19. However the silt-clay stratum is still highly variable due to the interbedding of silt and sand strata within the silty clays typical of the transitional environment. Between 70 and 100 m depth sands and gravels occur, which form an acquifer confined at the top by the silty clays.

Despite the variability of the soils their plastic limit is constant, 22%, while the liquid limits reflect the variations in clay fraction. The activity values suggest that the principal clay minerals present in the soils, down to 70 m depth are illite and chlorite, as also ascertened by mineralogical analyses (Cotecchia et al. 1994). As shown in Figure 18, several ^{14}C datingd have been carried out on both peat samples from the continental alluvial deposits and fossil organic remains (Pagliarulo & Cotecchia 2000).

The datings show that the top 70 m of sediments were deposited in the Holocene, whereas the lower coarser soils are from the Late Pleistocene. Also, the data show that deposition was far slower between 30000 and 12000 years ago, when the sea level was lowered due to the Flandrian regression, than in the last 12000 years, when the sea level has risen again (Figure 16). Pagliarulo & Cotecchia (2000) have related the datings of the sediments to the concentra-

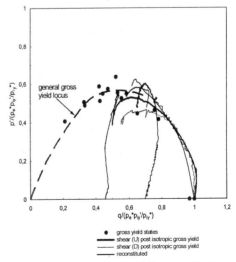

Fig.13 Shear behaviour and gross yield states of both natural and reconstituted Pappadai clay normalized for both volume and structure (Cotecchia & Chandler 2000).

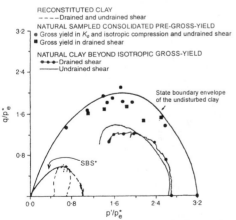

Fig.12 Gross yield data and stress paths of both natural and reconstituted Pappadai clay specimens (Cotecchia & Chandler 1997).

Table 2 Index properties of the sub-Apennine blue clays.

Index Propsrties	PAPADAI CLAY (Cotecchia 1996)	MONTEMESOLA CLAY (Cafaro 1998)	TARANTO CLAY (Cafaro et al. 2000)
G_s	2.69	2.71	2.71
v	1.84	1.65	1.60
γ(kNn/m^3)	18.7	19.6	20.2
γ_d(kNn/m^3)	14.3	16	16.6
CF (%)	50	49	32
LL (%)	65	51	43
PI (%)	35	28	21
A	0.70	0.57	0.67
Carbonate content (%)	28	34	31

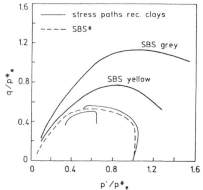

Fig.14 State boundary surfaces of the grey, the yellow and the reconstituted clay (Cafaro & Cotecchia 2001).

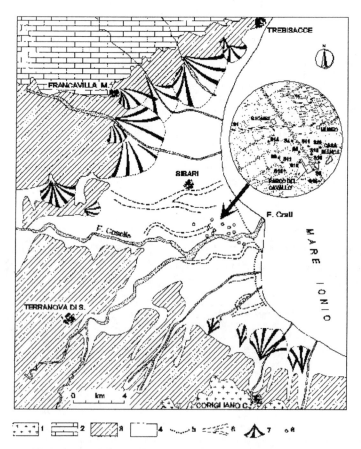

Fig.15 Geological map of the Sibari plain (Pagliarulo & Cotecchia 2000).

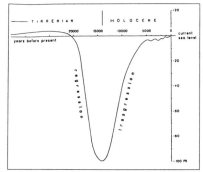

Fig.16 Glacioeustatic sea level changes of the Ionian Sea (Cotecchia et al.1971).

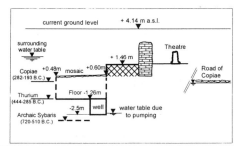

Fig.17 Archaeological section at Parco del Cavallo (Sibari) (Coop & Cotecchia 1996).

tions of the oxhigen isotop $d_{18}O$ reported by Orombelli & Ravazzi (1996); thus they have derived the climatic conditions present at the time of deposition of the various soil strata. Pagliarulo & Cotecchia reconstruct that the deep sands, deposited before 12000 years ago, during the Wurmian glaciation, had slow rates of deposition, little organic content and few fossils due to the cold climate, the sea water level being 60m lower than at present. On the contrary, the higher deposition rates characterizing the formation of the shallower deposits (between 12000 and 4000 years ago) are consistent with a transgressive phase of higher temperatures. Deposition rates have reduced in the last 5000 years due to a new reduction of the temperatures. Coop & Cotecchia (1995) report the results of oedometer and stress path tests on both undisturbed and reconstituted clay specimens. The one-dimensional compression curves of both the natural and the reconstituted specimens are found to be highly dispersed in the e-logσ_v' plane due to the different composition of the soils. However, once normalized for composition, by means of the void index I_v, all the reconstituted normal compression curves reduce to the ICL in Figure 20, and the natural samples have gross yield states and normal compression curves located in a narrow band parallel to and to the right of the ICL. The closeness of the gross yield states to the in-situ states shows that the undisturbed soils, of OCR=1, have YSR=1; therefore they possess a sedimentation structure. The parallelism of the normal compression curves of the natural soils to the ICL suggests that the soils have of constant sensitivity post-gross yield. These observations suggest that soil structures may differ due to the variability of the transitional deposition conditions, but affect similarly the behaviour of the in-situ soils.

Coop & Cotecchia (1995) recognize that the constant stress sensitivity post-gross yield is due to the presence of layering in the specimens. They demonstrate that a constant S_σ with post-gross yield compression is typically due to layering by showing the results of tests on reconstituted layered soils. They also report thin sections of Sibari layered specimens. Therefore, in this case the stress sensitivity of the

soil is not related to a non-stable structure, rather it results from the layering of non-sensitive clay strata. Figure 21 shows the results of triaxial tests, on both natural and reconstituted specimens, after normalization for current volume by means of p_e^*. The deviatoric stresses have been also normalized for the influence of composition on strength by means of the stress ratio at critical state M. The normalized stress paths for all the reconstituted specimens lie on a single Roscoe surface and reach critical state in the same point. The stress paths of the natural specimens, which have been normally consolidated post-gross yield, follow other state boundary envelopes, whose offsets with respect to the SBS of the reconstituted clay are indicative of different stress sensitivities. The natural specimens appear to be mildly strain-softening and to reach critical state to the right of the ICL, as also demonstrated by the stress path of an overconsolidated natural specimen. The stress paths in Figure 21 are further normalized for structure in Figure 22 by means of the stress sensitivity, here expressed as $S_\sigma = p_{K0y}'/p_{K0e}^*$. As result of this normalization the stress paths lie all close to the SBS of the reconstituted clay. This demonstrates that the procedure proposed by Cotecchia & Chandler normalizes for volume, composition and structure, the shear response not only of clays possessing a sensitive non-stable structure, but also of layered clays. The compression data in Figure 20 have been used by Coop & Cotecchia (1996) to model the compression of the sediments at Sibari. The authors have shown that the primary consolidation of the soil strata alone is able to account for the net movements which have occurred at the site with respect to the sea level since the construction of the ancient Greek city. Also in modern time there has been a continued settlement of the site; 11 cm have been measured in the last 50 years. From the geotechnical model Coop & Cotecchia have shown that this is likely to have been the result of a small draw-down of the water table in the upper few tens of metres resulting from water extraction from irrigation wells. In addition, the pumping at the archaeological site (which has lowered the water level from ground level down to 4.3 metres) has resulted in the reduction of the void ratios shown in the I_v – m profile. This profile shows a continuous increase of void index with decreasing

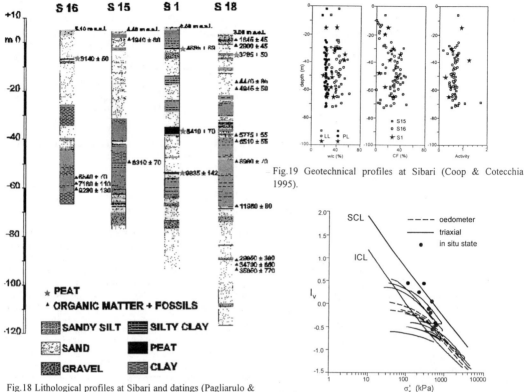

Fig.18 Lithological profiles at Sibari and datings (Pagliarulo & Cotecchia 2000).

Fig.19 Geotechnical profiles at Sibari (Coop & Cotecchia 1995).

Fig.20 Compression curves of Sibari clay specimens normalized for composition (Coop & Cotecchia 1995).

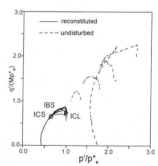

Fig.21 Shear stress paths of the Sibari clays normalized for volume (Coop & Cotecchia 1995).

depth up to about 10 m down, as would be expected for normally consolidated soils in hydrostatic conditions, and much smaller I_v values in the top 5 m of the borehole, due to the compression of the sediments resulting from the pumping.

5. THE SCALY CLAYS OF THE VARIE-GATED CLAY FORMATION

The Variegated Clay Formation is a turbidite formed in Paleogene – Upper Cretaceous and subsequently subjected to tectonic dislocation and intense shearing. The geological formation principally includes scaly clays and calcareous and marly strata. It is widely present in the Apennines, where these soils are still subject to tectonic events (e.g. 1980 Irpiniaearthquake) and are often involved in landsliding and instability processes. Santaloia et al. (2000) discuss the Variegated Clays present at Senerchia (Av.; Figure 1). In particular the authors present data showing the mechanics of the scaly clays which are part of this formation. The main clay minerals are inter-stratified illite-smectite, kaolinite, illite and chlorite. The clay fractions are between 53 and 61%, the corresponding liquid limits vary between 50 and 80% and the plasticity indexes between 23 and 54%.

The structure at the size of the laboratory specimen is scaly, since fissures and shear planes form an intricate network, dividing the material into millimetric hard scales (Picarelli 1998). Generally the clay particles forming these scales are not disaggregated by standard remoulding, and some scales can be present in the clay even after reconstitution in the laboratory.

Santaloia et al. show that the compression curves

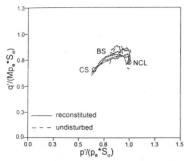

Fig.22 Shear stress paths of the Sibari clays normalized for both volume and structure (Cotecchia et al.1997).

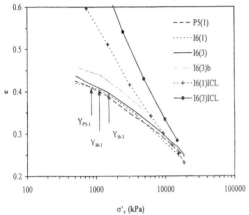

Fig.23 Oedometer tests on both natural and reconstituted specimens of Senerchia scaly clays (Santaloia et al.2000).

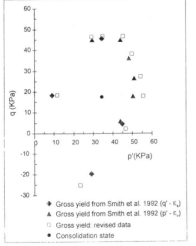

Fig.24 Bothkennar clay: gross yield data (original data from Smith et al.1992).

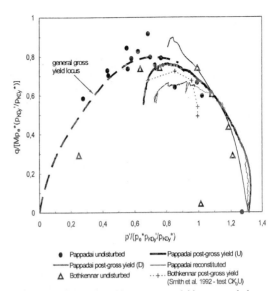

Fig.25 Pappadai and Bothkennar gross yield states and shear stress paths normalized for both volume and structure (Cotecchia & Chandler 2000).

of the natural clay specimens plot, both pre and post-gross yield, to the left of the ICL in the e-σ_v' plane, as shown in Figure 23, where the results of high pressure oedometer tests on both natural and reconstituted clay specimens are shown. The location of the gross yield states to the left of the ICL implies values of stress sensitivity below 1 according to the sensitivity framework. The authors suggest to implement the behaviour of tectonized soils in the general framework of clay behaviour from Cotecchia & Chandler (2000) by considering these soils to possess a *negative structure*, in contrast to the *positive structure* of soils of stress sensitivity above 1

Santaloia et al. (2000) and Fearon (1998) show that these negative-structure clays have strength properties poorer than those of the corresponding reconstituted clays. Both the critical state friction angles (12-15°) and residual friction angles (5-6°) of the Senerchia clays are particularly low and their state boundary envelopes lie within the reconstituted state boundary surface. Thus, also for negative-structure soils the stress sensitivity S_σ appears to equal the strength sensitivity S_t. This finding suggests that the sensitivity framework can be used to interpret also the behaviour of tectonized soils.

6. APPLICATION OF THE SENSITIVITY FRAMEWORK TO SOFT BOTHKENNAR CLAY

A thorough analysis of the characteristics of Bothkennar clay is presented by Hight et al. (1992), therefore the reader can refer to this work for details about the geological features of the soils, their structure and geotechnical features. For the mechanical behaviour of the soil the reader can also refer to Smith (1992) and Smith et al. (1992). The site is located in Scotland, on the South side of the river Forth (Figure 1). The mechanical test results presented by Smith et al. (1992) and used by Cotecchia & Chandler (2000) to check the applicability of the sensitivity framework to this soft clay, refer to clay samples taken within the clayey silts of the Claret Beds (mottled facies; Paul et al.1991), generally referred to as the soft clays of Bothkennar. These are post-glacial, and were deposited in shallow intertidal water between 8500 and 6500 years BP (as for the Sibari clays), when sea level was rising and a temperate climate prevailed. However, differently from the Sibari clays, the deposition of the soft Bothkennar clays occurred in relatively sheltered and quiet, saline or brackish water, 7 to 20 m deep. Post-depositional processes have included some reworking of the original fabric, some leaching and erosion of up to one metre, and ground-water lowering of up to 2.75 m. The principal clay minerals are illite and kaolinite and the clay fabric is defined by Paul et al. (1991) as a boxwork fabric with individual burrows which have cement coating and internal mucous lining. Therefore the bonding of the clay is not due to a general presence of inter-particle cement, and is fundamentally that created during deposition and consolidation along the SC curve. The clay fraction of the samples considered (6m depth) is between 29 and 41%, the liquid limit is 74-86% and the plastic limit is 27-35%.

The one-dimensional compression curve of the clay, shown in Figure 5, is indicative of a sedimentation structure at gross yield and a stress sensitivity S_σ of about 4.6. In Figure 24 the gross yield data from stress path tests on reconsolidated undisturbed Bothkennar clay (Smith 1992) are plotted. In Figure 25 both Bothkennar and Pappadai clay gross yield data (taken from the previous figures) are plotted after normalization for volume (p_e^*), structure (S_σ) and composition (M) (Cotecchia & Chandler 2000). The figure shows that there is a considerable similarity between the locus of gross yield states of both Pappadai and Bothkennar clay. A further check of these normalizations is provided by a test on a specimen of Bothkennar clay one-dimensionally consolidated after gross yield (test CK_0U). Owing to creep during consolidation, the normalized stress path for this test starts within the general SBS, but it soon reaches the general SBS and gross yield locus.

The data appear to confirm that there is a general normalized gross yield locus which applies to many clays and lies close to the state boundary surface. The gross yield locus is arch-shaped, although the comparison between the gross yield locus and the SBS is not very satisfactory for Bothkennar clay, as generally observed for other soft high-sensitivity clays, in the region of the p' axis. Cotecchia & Chandler show that this may be effect of the higher disturbance soft highly-sensitive clays are subjected to when loaded isotropically from undisturbed anisotropic stress states. As a consequence, for soft sensitive clays the gross yield curve lies close to the general SBS in the region where deviatoric stress q is increasing, whereas with reducing values of q the gross yield points are less clearly defined and lie progressively further from the SBS.

7. CONCLUDING REMARKS

The investigation of the geological history, structure and mechanics of different clays has shown that there are many similarities in the patterns of response to loading of clays of different structure, origin and history, irrespective of their being soft, firm or stiff. Though the same clays may have mechanical properties and parameters controlling these patterns of response which differ in value due to soil structural differences. The qualitative similarities in mechanical response and differences in the magnitude of the response can be rationally interpreted within a general framework, as proposed by Cotecchia & Chandler. In this framework the magnitude of the clay mechanical properties is related to the stress sensitivity S_σ, which alone identifies the influence of structure on the clay mechanics.

The applicability of the framework to the behaviour of different non-glacial european clays, from soft to stiff, from intact to tectonized and to weathered, has been demonstrated. The framework is shown to be of use to the engineer to interpret the mechanical behaviour of soils in practice. The framework can help forecasting the soil response on the basis of few data, according to the location of the state of the soil being examined in the overall picture. For example, it can help deducing the soil response under shear once the compression behaviour of the soil is known, or, viceversa, can help to deduce the compression and gross yield behaviour of the soil from some of its shear data. The stress sensitivity, that is the structure parameter of the framework, can be implemented in mathematical elasto-plastic constitutive models which simulate the degradation of structure and aim at representing the behaviour of clays of different structural strength.

REFERENCES

Burland, J.B., 1990. On the compressibility and shear strength of natural clays. *Géotechnique* **40**, 329-378.

Cafaro, F., 1998. Influenza dell'alterazione di origine climatica sulle proprietà geotecniche di un'argilla grigio-azzurra pleistocenica. *PhD thesis, Technical University of Bari.*

Cafaro, F., Cherubini, C., & Cotecchia, F., 1999. Use of the scale of fluctuation to describe the geotechnical variability of an italian clay. – *ICASP 8 Conference, Sidney,* 481-486.

Cafaro, F., Cotecchia, F., & Cherubini, C., 2000. Influence of weathering processes on the mechanical behaviour of an italian clay. *GEOENG 2000, Melbourne, GCC 1041. PDF.*

Cafaro, F., & Cotecchia, F., 2001. Structure

degradation and changes in the mechanical behaviour of a stiff clay due to weathering. – *Géotechnique, in press.*

Coop, M.R. & Cotecchia, F., 1995. The compression of sediments at the archaeological site of Sibari. *Proc. 11th ECSMFE*, Copenhagen 8, 19-26.

Coop, M.R., Atkinson J.H. & Taylor, R.N., 1995. Strength, yielding and stiffness of structured and unstructured soils. *Proc. 11th ECSMFE*, Copenhagen 1, 55-62.

Coop, M.R. & Cotecchia, F., 1996. The geotechnical settlements of the archaeological site of Sibari. *Proc. of the Int. Symposium on the Geotechnical Engineering for the Preservation of Monuments and Historic Sites*, Naples, 89-100.

Cotecchia, F., 1996. The effects of structure on the properties of an italian Pleistocene clay. *PhD thesis, London University.*

Cotecchia, F. & Chandler, R.J., 1997. The influence of structure on the pre-failure behaviour of a natural clay. *Géotechnique* 47, 523-544.

Cotecchia, F. & Chandler, R.J., 1998. One-dimensional compression of a natural clay: structural changes and mechanical effects. *2nd International Symposium on Hard Soils and Soft Rocks*, Naples, 103-114.

Cotecchia, F., Coop, M.R. & Chandler, R.J., 1997. "The behaviour of layered clays within a framework for the structure-related behaviour of clays" – *Proc. of the Int. Symposium on the Geotechnical behaviour of soft clays*, Yokohama, Japan, February 1997, 147-164.

Cotecchia, V., Cherubini, C. & Pagliarulo, R., 1994. Geotechnical characteristics of outcropping deposits in the Sibari plain. *Proc. XIII ICSMFE, Delhi, 1*, 245-250.

Cotecchia,V., Dai Pra, G. & Magri, G., 1971. Morfogenesi litorale olocenica tra Capo Spulico e Taranto nella prospettiva protezione costiera. *Geologia Applicata e Idrogeologia*, 6, 65-78.

Fearon, R., 1998. The behaviour of a structurally complex clay from an Italian landslide. *PhD Thesis*, City University.

Graham, J., Crooks, J.H.A. and Lau, S.L.K., 1988. Yield envelopes: identification and geometric properties. *Géotechnique* 38, 125-134.

Hight, D.W., Bond, A.J. and Legge, J.D., 1992. Characterization of the Bothkennar clay: an overview. *Géotechnique* 42, 303-347.

Kavvadas, M., & Amorosi, A., 2000. A constitutive model for structured soils. *Gèotechnique*, 50, No.3, 263-274.

Lambe, T.W. & Whitman, R.V. 1969. *Soil Mechanics.* John Wiley & Sons Inc., New York.

Lerouil, S., 1997. Critical state soil mechanics and the behaviour of real soils. *Proc. Intern. Symposium on Recent Developments in Soil and Pavement Mechanics*, Almeida ed., 41-80.

Lerouil, S., 1998. Contribution to the Round Table: Peculiar aspects of structured soils. *2nd International Symposium on Hard Soils and Soft Rocks*, Naples, in press.

Orombelli, G., & Ravazzi, C., 1996. The Late Glacial and Early Holocene: chronology and paleoclimate. *Il Quaternario. Italian Journal of Quaternary Sciences.* Vol.9, 439-444.

Pagliarulo, R., Cotecchia, F., Coop, M.R. & Cherubini, C (1995) Studio litostratigrafico e geotecnico della piana di Sibari con riferimento all'evoluzione morfologico. *La Citta Fragile in Italia, Proc. 1st Conf., Italian Eng. Geol. Group*, Giardini Naxos, 385-403.

Pagliarulo, R., & Cotecchia, F.,2000 Influenze dell'evoluzione geoambientale e della compressione dei sedimenti sulla scomparsa dell'antica Sybaris. *Proc. GeoBen 2000, "Condizionamenti geologici e geotecnici nella conservazione del patrimonio storico-culturale"*, Moncalieri 8-9 Giugno, 715-722.

Paul, M.A., Peacock, J.D. & Wood, B.F., 1991. The engineering geology of the estuarine deposits at the National Soft Clay Research Site. *Report to the Science and Engineering research Council.*

Picarelli, L., 1998. Properties and behaviour of tectonised clay shales in Italy. *Proceedings of the 2nd International Symposium on Hard Soils-Soft Rocks*, Napoli, Evangelista and Picarelli Eds, 3, in press.

Santaloia, F., Cotecchia, F., & Polemio, M., 2000. Movements in a tectonized soil slope: effects of boundary conditions and rainfalls. *Quarterly Journal of Engineering Geology*, in press.

Schofield, A.N. & Wroth, C.P., 1968. Critical state soil mechanics. *McGraw-Hill Book Co.*, London.

Schmertmann, J. H. (1969). Swell sensitivity. *Géotechnique* 19, 530-533.

Skempton, A.W. and Northey, R.D. (1952). The sensitivity of clays. *Géotechnique* 3, 30-53.

Smith, P.R., Jardine, R.J. and Hight, D.W., 1992. On the yielding of Bothkennar clay. *Géotechnique* 42, 257-274.

Terzaghi, K., 1941. Undisturbed clay samples and undisturbed clays. *J. Boston Soc. Civ. Eng.*, 28, 211-231.

Coastal Geotechnical Engineering in Practice, Nakase & Tsuchida (eds)
© 2002 Swets & Zeitlinger, Lisse, ISBN 90 5809 151 1

Physico-chemical and engineering behavior of Ariake clay and its comparison with other marine clays

A.Sridharan
Institute of Lowland Technology, Saga University, Japan

N.Miura
Department of Civil Engineering, Saga University, Japan

ABSTRACT: Ariake marine clay deposits are widely distributed around Ariake bay, Kyushu, Japan, and the knowledge of their engineering behavior assumes importance because of large number of civil Engineering activity in this area. It has been reported extensively that the principal clay mineral present in the soil is smectite but its physical, physico-chemical and engineering behavior is that of non-swelling clays. The physical, physico-chemical and the engineering behavior of Ariake clay are discussed and comparison is made with other marine clays and that of typical clay minerals. Analysis of clay minerals alone in the clay size fraction can mislead in assessing the physical and engineering behavior of Ariake clay. The behavior is controlled by several parameters like, pore medium chemistry, different types of clay mineral present, and their percentages the non clay mineral portion and the clay particle aggregation due to pore medium chemistry.

1 INTRODUCTION

Marine clay deposits of 15-40 m thick are widely distributed around Ariake Bay, Kyushu, Japan. They exhibit high sensitivity, more than 50 at reduced concentrations like quick clays. Extensive investigations have been carried out on Ariake clay (Egashira & Ohtsubo 1982, Ohtsubo et al. 1982, Torrance 1984, Ohtsubo et al. 1985, Rao et al. 1991 &. 1993, Torrance & Ohtsubo 1995, Miura et al. 1996, Ohtsubo et al. 1996, Li et al. 1998, Yamadera et al. 1999, Ohtsubo et al. 2000, Hong and Tsuchida 2000, to name a few). The parameters affecting the physical, physico-chemical and engineering behavior are many and they are complex. In this note some aspects of Ariake behavior is discussed with particular reference to the physico-chemical parameters affecting them, the role of clay minerals present in these clays and its comparison with other marine clays.

2 PHYSICAL PROPERTIES

Table 1 shows typical properties of Ariake clay. It may be seen that the Ariake clay can have very high liquid limit. Figure 1 shows the plasticity chart representing the Norwegian (Bjerrum 1954), Canadian (Eden & Crawford 1957, Grillot 1979), Ariake clay of Japan (Egashira & Ohtsubo 1982) and Indian (Rao, 1974). The Norwegian clays plot predominantly in the CL region, the Canadian clays in the CH region, the Japanese clays in the MH region and the

Indian clays in the CH region. Most of the Ariake clay plots below the Casagrande A-line (Li et al. 1998). Although in figure 1, it is seen that all marine clays plots around the A line, their physico-chemical and engineering behavior differ greatly. The very high liquid limit of Indian marine clays can be attributed to the presence of considerable amount of smectite mineral. In spite of the fact that the Ariake clays of Japan are smectite rich (as reported in many investigations), they plot in the MH region because of the effects of other clay minerals, non clay fractions and pore medium chemistry probably leading to aggregation (Rao et al. 1993) and behaves like a non swelling soils. The Canadian clays comprising of illite, quartz, feldspar and mica show higher plasticity characteristics by plotting in the CH region. However, plasticity in this case can be attributed to the presence of considerable amount of amorphous sesquioxides (15-25 %) (Sridharan et al. 1989), which would contribute to the higher water holding capacity of the marine clay.

Table 1. Physical properties.

	Ariake clays	usual marine clays of Japan
Specific gravity	2.56- 2.70	2.75
Clay content (%)	25.0 - 78.0	-
Liquid limit (%)	53.0 - 169.5	45.0- 100.0
Plastic limit (%)	28.0 - 59.0	20.0- 40.0
Activity	0.6 - 2.2	-
Natural water content (%)	50.0 - 200.0	-
Liquidity index	> 1.0	< 0.8

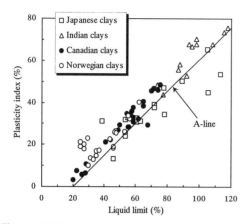

Figure 1. Plasticity index vs. liquid limit for various marine clays.

For marine soils a good relationship exists between liquid limit and plasticity index (Fig. 2). A statistical equation of the form

$$I_P = 0.77 \left(w_L - 17.7\right) \qquad (1)$$

where I_P is plasticity index and w_L is liquid limit, is obtained for 125 marine soils covering different regions including Indian marine clays with a correlation coefficient of 0.97. However, one cannot be misled that their physico-chemical and engineering behavior will be similar.

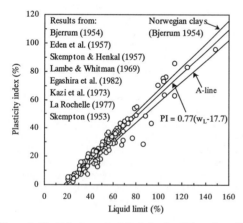

Figure 2. Liquid limit vs. plasticity index for 125 marine soils (Sridharan 1999).

3 MINERALOGY

Many investigators have studied the mineralogy of Ariake clay. For example Egashira & Ohtsubo (1981) reported that the <2μm clay fraction of the Ariake marine mud contains the principal clay mineral as smectite followed by kaolinite, mica and

vermiculite. It is to be noted that the % <2μm is 54%. Egashira & Ohtsubo (1982) mention that the clay fractions of Ariake bay contain considerable smectite, contrary to the mineralogical composition of quick clays of Scandinavia and Eastern Canada. The % clay fraction is varied from 26 to 45%. Ohtsubo et al. (1982) reported that the principal clay mineral in the clay size fraction (<2μm) is smectite accompanied by kaolinite, illite and vermiculite and the percent of clay fraction varied from 26% to 53%. Egashira & Ohtsubo (1982) reported that Ariake Kantaku clay with % clay size of 45% has a smectite content of 39% i.e. the smectite content in the whole soil is 17.55% and another clay with % clay size of 26% has a smectite content of 38% i.e. the smectite content in the whole soil mass is only 9.88%. Ohtsubo et al. (1985) supports the earlier results that smectite is the principal clay mineral in the clay size fraction which could be much less than 50% in the total soil fraction. Although smectite has been identified as the principal clay mineral in the clay size fraction the Ariake bay clay soil behaved much differently than what would normal smectite rich clay soil would behave. This aspect is discussed further.

4 SEDIMENT VOLUME/LIQUID LIMIT BEHAVIOR

4.1 Montmorillonitic and kaolinitic soils

Several investigators have brought out that montmorillonitic and kaolinitic soils behave quit oppositely in their physical and engineering behavior (Sridharan 1991). Figure 3 shows the sediment volumes of dry soil in polar water visa vie in non-polar carbon tetra chloride (Sridharan et al. 1986a).

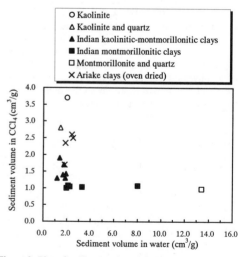

Figure 3. Plot of sediment volume of clays and soils in water versus their sediment volumes in carbon tetrachloride.

The results are disposed in two distinct groups; soil dominated by swelling smectite resulted minimum sediment volume of 1.0 to 1.1 cc/gm in carbon tetra chloride with all the points lying on a plateau, the non swelling kaolinitic soils were positioned well above the plateau and occupied sediment volume greater than 1.1 cc/gm in carbon terra chloride, the values ranging from 1.3 to 3.75 cc/gm. This property of occupying a higher sediment volume in carbon tetra chloride is typical of the non-swelling clays with exposed gibbsite surfaces. They are also randomly oriented in a non-polar solvent. Ariake clay is also shown in the figure indicating that it belongs to non-swelling clays. Table 2 presents the effect of pore medium chemistry on sediment volume/liquid limit of non-swelling (kaolinitic) and swelling (montmorillonitic) soils based on the numerous results available in the literature (Sridharan and Prakash 1999).

Table 2. Sediment Volume/Liquid Limit Behavior.

Increase in	Sediment Volume/Liquid Limit	
	Kaolinitic/ non swelling soil	Montmorillonitic/ swelling soil
Dielectric constant	decrease	increase
Concentration of ions in pore medium	increase	decrease
Hydrated size of ions in pore medium	decrease	increase
Valence of cation	increase	decrease

Egashira & Ohtsubo (1981) reported that the sediment volume of Ariake marine mud increases at the concentration of NaCl increased from 0.02 N to 1.0 N. Similarly the liquid limit of Ariake marine mud got increased significantly from 89.0% to 166.0% as the concentration increased from 0.01 N to 1.0 N of NaCl and 151.0% to 168.0%, when the concentration increased from 0.01 N $CaCl_2$ to 1.0 N of $CaCl_2$. Effect of valence is seen from the change of 0.01 N NaCl to 0.01 N $CaCl_2$ and corresponding change in liquid limit is the increase from 89.0% to 151.0%, which is quite significant. Torrance (1984) also reported significant decrease of liquid limit at low salinity for Meguire natural Ariake clay. This once again confirms that the behavior of Ariake clay is similar to that of non-swelling clays.

Ohtsubo et al. (2000) have compared mineralogy and chemistry of Bangkok clay with Ariake clay. Ariake clay was obtained from the Ariake Bay, western Kyushu, Japan. Both these clays contain smectite as principal clay mineral in their clay size fraction followed by kaolinite and mica. In the Bangkok clay, the sediment volume performed on <2μm clay fraction got reduced for Ca homo-ionization when compared with Na homo-ionized clays. Whereas for Ariake clay no perceptible difference in sediment volume

is noticed between Ca and Na homo-ionized clays. Even the sediment volumes on clay fraction (not the total soil) did not behave similar to that of smectite clay mineral behavior. Similar test carried out on Wyoming montmorillonite showed a significant reduction in their sediment volume for Ca compared with Na homo-ionization.

Results on liquid limit for Ariake clay showed moderate increase upon increase in concentration of Ca cation and increased ≈100% with increase in Na ion concentration. Further, Ca Ariake clay resulted in much higher liquid limit values at lower concentration than Na Ariake clay. For Bangkok clay, increase in concentration of ion did not show increase in liquid limit for Ca Bangkok clay where as Na-Bangkok clay shows moderate increase.

The behavior especially of Ariake clay is similar to that of non-swelling clay type. Recently Sridharan et al. (2000) carried out careful experiments on the Ariake clay obtained from Fukudomi (Fukudomi town, Shimoku, Saga Prefecture), Kubota (Kubota section, Ariake Kantaku, Saga Prefecture), Kawezoe (Nanri, Kawazoe, Saga Prefecture) and Isahaya (Nishiricho, Isahaya, Nagasaki Prefecture), and carried out the X-ray diffraction pattern, the Atterberg limits (both moist and oven dried) and sediment volume test. The X-ray diffraction pattern of the clay size fraction indicated the clay minerals as shown in Table 3.

Table 3. The distribution of clay minerals in Ariake clay in clay size fraction.

Ariake clays	Minerals content (%)			
	Smectite	Kaolinite	Vermiculite	Illite
Kawazoe	26.1	15.4	20.0	38.5
Kubota	22.5	19.5	21.1	36.9
Isahaya	19.4	22.3	19.5	38.8

It may be seen that the percentages of different types of clay minerals; smectite, kaolinite and vermiculite are in the same range except that of illite, to be on the higher side. Their sediment volume both in water and carbon tetra chloride are shown in figure 3, which clearly indicate that they fall on non-swelling region.

Table 4 presents the Atterberg limits of the Ariake clays for both natural moist sample and initially oven-dried condition (Sridharan et al. 2000). The initial condition, moist/oven dried has significant effect on the Atterberg limit especially on the liquid limit. Similar behavior with respect to Cochin marine clay (Indian) has been reported by Jose et al. (1988). A detailed discussion on the influence of drying on the liquid limit behavior of marine clays can be seen in Rao et al. (1989). It has been shown that the presence of calcium and magnesium as the dominant exchangeable ions and of a high pore salt concentration facilitates strong inter particle attraction of van der Waals' and Coulombic bonds. Liquid limits tests were also carried out for Fukudomi clay (initially oven dried) with carbon tetra chloride as fluid which resulted in a value of 90.0% (corrected for specific gravity of carbon tetrachloride), which is higher than what was obtained with water (86.6%), which is typical of the behavior non swelling clays (Sridharan & Rao 1975). Sridharan and Rao (1975) demonstrated that non-swelling soil shows high liquid limit in carbon tetra chloride where as swelling soils shows higher liquid limit values in water.

Table 4. The liquid and plastic limits of Ariake clays.

Ariake clays	Liquid limit (%)		Plastic limit (%)	
	Moist	Oven dried	Moist	Oven dried
Kawazoe	115.9	68.6	45.5	34.1
Kubota	68.0	57.5	25.0	24.4
Isahaya	169.5	90.1	54.4	49.4
Fukudomi	134.4	86.6	45.0	42.4

Figure 4 shows the effect of concentration of Na ions in pore fluid on the liquid limit of Isahaya soil (Koumoto et al. 1998).

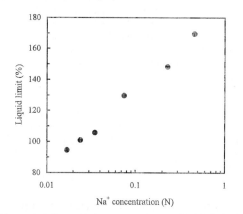

Figure 4. Effect of salt concentration on liquid limit for Isahaya sample.

The liquid limit increased with increase in concentration significantly, typical behavior of a non-swelling soils.

Based on the test results obtained on bentonite equilibrated with the artificial seawater for a period of 1 year and subsequent leaching on ions, Rao et al. (1993) concluded that no mineralogical changes takes place to the bentonite. However, the exchangeable cations positions become prominently populated by divalent magnesium ions as leaching takes place. Results show that in the pore salt concentration range of 42 g/L to 1.1 g/L. The diffuse-double layer thickness is mainly contributed by dissolved sodium ion concentration in the pore water. The thickness of the diffuse double layer governs the Atterberg limits of the bentonite clay specimens. At pore salt concen-

tration below 1.1 g/L the attraction forces mainly contributed by the divalent exchangeable cations and the dissolved magnesium + calcium + potassium ions present in the pore fluid phase out-weigh the diffuse double layer repulsion forces and cause the Atterberg limits of clay specimens to be clearly governed by the mode of particle arrangement (clay fabric), typical of non-swelling clay. The attraction forces cause aggregation of the clay unit layers into domains. Rao et al. (1993) have further compared the results of Ariake clay with those of bentonite, equilibrated with artificial sea water and opined that considerable substitution of Fe^{2+} for Al^{3+} occurs in octahedral sheet which suppresses the clays natural swelling tendency and to behave as a non-swelling clay.

5 COMPRESSIBILITY BEHAVIOR

As has been noticed for sediment volume and liquid limit behaving quite oppositely for kaolinitic and montmorillonitic soils, the compressibility also behaves. Figures 5a & 5b brings out the effect of dielectric constant on the compressibility behavior of kaolinite and montmorillonite (Sridharan & Rao, 1973) similar results have been presented by Olsen and Mesri (1970).It can be seen that there are basically different mechanisms controlling the compressibility behavior viz: Mechanism 1 wherein the compressibility of a clay is primarily controlled by the shearing resistance at the near contact points and volume changes occur by shear displacement and/or sliding between particles, and Mechanism 2. in which compressibility is primarily governed by the long-range electrical repulsive forces (primarily double layer repulsive forces).

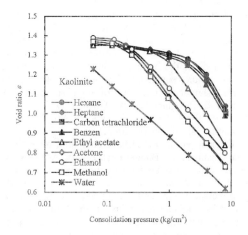

Figure 5a. One-dimensional consolidation curves for kaolinite with different pore fluids.

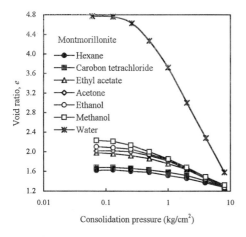

Figure 5b. One-dimensional consolidation curves for mont-morillonite with different pore fluids.

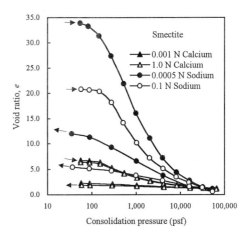

Figure 6b. One-dimensional consolidation curves for smectite.

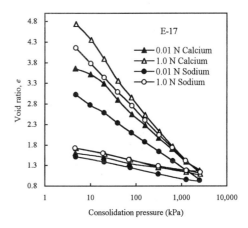

Figure 7. One-dimensional consolidation curves for Ariake clay (sample E-17) in water.

Although both these mechanisms operate simultane-ously, mechanism 1, primarily controls the volume change behavior in non-expanding lattice type clays (non-swelling clays) and mechanism 2 operates in the case of expanding lattice-type clays like montmoril-lonite (Sridharan & Rao 1973).

Figure 6a & 6b (results from Olsen and Mesri, 1970) illustrate the effect of ion valence and ion con-centration on the compressibility behavior of kaolin-ite and montmorillonite clays. The dominating influ-ence of clay mineralogy is quite evident.

Figure 7 illustrates the effect of ion concentration and ion valence on the compressibility behavior of Ariake clay designated as E-17 (Ohtsubo et al. 1985). Soil E-17 is collected from the sea bottom in the reclamation area in Ariake.

As reported by Ohtsubo et al. (1985) the soil con-tains predominantly smectite followed by kaolinite, il-lite and vermiculite. The percent clay size is 54.2% and the liquid limit and plastic limits respectively are 127% and 49%. Comparison of the results of figure 7 with figures 6a & 6b, leads to an obvious conclu-sion that the Ariake clay behaves more of non-swelling kaolinite type material.

Figure 8 shows the compressibility behavior of Ariake clay (initially oven dried) obtained with pore fluid as carbon tetra chloride and water (Sridharan et al. 2000). The conventional oedometer tests were carried out on samples starting with their initial fluid content nearer to the liquid limit. This once again in-dicate the behavior of non-swelling soil type. Com-paring the results of figure 8 with there of figure 5a, clearly indicate that the Ariake clay behaves like a non-swelling soil.

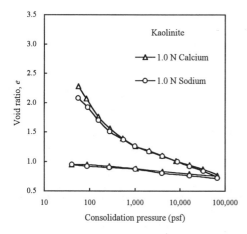

Figure 6a. One-dimensional consolidation curves for kaolinite.

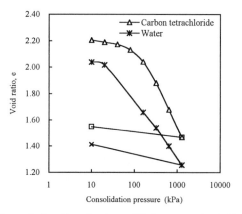

Figure 8. One-dimensional consolidation curves for Ariake clay with different pore fluids (Sridharan et al. 2000).

6 EFFECT OF CHEMICAL TREATMENTS ON THE INDEX PROPERTIES OF REPRESENTATIVE COCHIN AND MANGALORE MARINE CLAYS

Extensive investigations have been carried out by Rao et al. (1991) on Cochin and Mangalore marine clays of Indian west coast. X ray diffraction pattern had revealed that Cochin marine clay contains comparable amount of kaolinite (38%) and smectite (42%) and much less illite (20%). Comparatively, Mangalore marine clay contains kaolinite as the dominant clay (56%) with a lesser but notable amount of smectite (32%) and much less illite (12%). Experiments have been performed by Rao et al. (1991) to assess whether the smectite rich Cochin and Mangalore clays that were deposited in a marine environment show remolded response typical of expanding or fixed lattice type clays. Table 5 presents the index properties of the representative pore salt extracted and calcium treated Cochin and Mangalore clay specimens. Results in Table 5 show that a decrease in pore salinity leads to a visible increase in the liquid and plastic limits of the representative Cochin marine clays, while saturation with divalent calcium ion leads to a small decrease in the liquid limit and notable increase in the plastic limit of the clay. The increase in liquid and plastic limits on pore salt removal and the decrease in liquid limit on calcium saturation is typical of expanding lattice-type clays, where the diffuse double layer forces govern the index properties. In Mangalore marine clay (Table 5), removal of pore salt has negligible influence on the liquid limit, but results in a notable increase in the plastic limit of the clay. Treatment with calcium ions leads to a small increase in the Atterberg limits of the clay (Table 5). The results suggests that Atterberg limits of the Mangalore marine clay, although

containing a notable amount of smectite (32%), are mainly governed by the mode of particle arrangement, typical of fixed-lattice-type clays, which are dominated by strong attraction and weak repulsion forces. The Mangalore marine clay is characterized by a high organic content (~ 13%). The naturally occurring organic matter is also recognized to interact with clay minerals and act as a cementing agent (Ayyar 1971, Rashid & Brown 1975). It is probable that the high organic content in the Mangalore clay binds individual clay platelets into coarser aggregated (clusters and peds) and thereby significantly reduces the available specific surface area for diffuse double layer formation. The consequent marked reduction in diffuse double layer repulsion brought about by the organic matter cementation may allow the attraction forces to dominate over the repulsion forces, typical of fixed lattice clays. Cochin marine clay differs from Mangalore clay in having a much lower organic content (3%); the extent of cementation of clay platelets induced by the organic matter is possibly inadequate in effectively curbing diffuse double layer formation, with the net result that the smectite rich Cochin marine clay exhibits consistency limits response typical of expanding lattice clays.

Table 5 also presents the Atterberg limits and grain size distribution of the pore salt removed, organic matter extracted, sesquioxides (iron oxide, aluminum oxide, and silicon dioxide) extracted Cochin and Mangalore marine clay specimens. After the removal of various cementing agents, the Cochin and Mangalore marine clays were repeatedly washed with distilled water to remove excess chemicals and the decemented specimens in the process were reduced to an electrolyte free condition. The representative Cochin marine clay, however, has a pore salinity of 9.4 g/L. To have similar pore electrolyte conditions for the decemented and untreated marine clay specimens, the Atterberg limits of the decemented marine clays are compared with their respective pore salt extracted specimens. The role of citrate-dithionite extractable sesquioxides in influencing the index properties of a fixed lattice type, kaolinitic soil was investigated by Rao et al. (1988). Results indicated that the accumulated sesquioxides bind individual clay platelets into coarser aggregates (clusters and peds) and contribute to the development of a random soil structure, which facilitates a higher liquid limit because of greater water entrapment in the soil's open structural units. Extraction of the sesquioxides leads to breakdown and dispersion of the fabric units (clusters and peds) and a decrease in the extent of particle flocculation, which results in a reduction in void space for water entrapment and a lower liquid limit of the soil. The naturally occurring organic matter also interacts with clay minerals and acts as a cementing agent (Rashid & Brown 1975). Removal of organic matter

Table 5. Effect of chemical treatment on the index properties of representative Cochin and Mangalore marine clays.

S1 Number	Specimen	Description	Grain size Distribution (%)			Liquid Limit (%)	Plastic Limit (%)	Plasticity Index (%)
			Sand	Silt	clay			
1	Cochin marine	Representative	25	23	52	116	35	81
2	clay	Pore salt extraction	27	20	53	125	49	76
3		Calcium homoionized	25	33	42	111	54	57
4		Sesquioxides extracted	24	16	60	157	61	96
5		Organic matter extracted	23	20	57	129	47	82
6	Mangalore marine	Representative	14	48	38	179	72	107
7	clay	Pore salt extracted	15	45	40	178	82	96
8		Calcium homoionized	49	15	36	184	87	97
9		Sesquioxides extracted	13	40	47	159	58	101
10		Organic matter extracted	5	45	50	150	58	92

should also have effects similar to that of extraction of sesquioxides on the Atterberg limits of fixed-lattice-type clays.

In comparison with expanding–lattice-type clays, removal of cementitious materials such as sesquioxides and organic matter would increase the available surface areas for diffuse double layer formation and enhance the liquid limit values (Rao et al. 1988). Results in Table 5 illustrate that removal of organic matter slightly increased the liquid limit of the smectite rich Cochin marine clay, while the plastic limit is unaffected. Extraction of the sesquioxides, however, leads to a notable increase in the liquid and plastic limits of the Cochin marine clay specimen. The increase in the clay-sized fractions on extraction of organic matter and sesquioxides (Table 5) apparently provides a greater surface area for development of diffuse double layer, resulting in the higher liquid limit values. The trend of the results on removal of cementing agents suggests that the Atterberg limits of the smectite rich Cochin marine clay are governed by electrical diffuse double layer forces. In comparison, the removal of organic matter and sesquioxides from the Mangalore marine clay decreases the liquid and plastic limits, in spite of an increase in the clay-sized content (Table 5). The variations in Atterberg limits of removal of cementing agents are typical of fixed-lattice type clays; wherein the mode of particle arrangement governs the index properties. Form the collation of Atterberg limit results on pore salt removal, ion saturation and cementing agents removal it can be inferred that the index properties of the mineralogically similar Cochin and Mangalore marine clays are governed by different physico-chemical mechanisms. The Atterberg limits of the Cochin marine clay are mainly dependent on the diffuse double layer thickness, typical of expanding lattice-type clays. It logically follows that the strong diffuse double layer formation associated with the Cochin marine clay would cause the osmotic repulsion forces to dominate over the attraction forces. Comparatively, the Atterberg limits of the Mangalore clay are mainly governed by the mode of particle arrangement, similar to the

fixed-lattice type clays. Such a situation apparently arises as the attraction forces dominate over the repulsion forces.

The behavior of Ariake clay can be placed akin to that of Mangalore marine clay. In spite of the fact that Ariake clay can have smectite clay mineral in its clay size fraction its behavior could be seen akin to non-swelling soil behavior like Mangalore marine clays.

7 COMPARISION OF MINERALOGICAL, CHEMICAL AND ENGINEERING PROPERTIES OF COCHIN AND MANGALORE MARIN CLAYS WITH ARIAKE MARINE CLAY

The chemical, mineralogical, and engineering data of Ariake clay are that reported by Ohtsubo et al. (1982) for the marine clay from the Yamaashi profile. In the case of Cochin and Mangalore marine clays, besides the results obtained in this study, the data reported by Rao (1974), Rao and Pranesh (1978), and Jose et al. (1988) have also been included for wider representation of the Indian marine clay properties. Table 6 compares the mineralogical and chemical properties of Cochin and Mangalore clays with those of Ariake clay. X-ray diffraction characterization of Ariake clay from the Yamaashi profile had indicated a smectite presence of 30-40% in the clay size fraction (Egashira & Ohtsubo 1982), which is comparable with that present in Cochin and Mangalore marine clays (42 and 32%, respectively).

The total exchangeable cation content (or cation exchange capacity) of Ariake clay from the Yamaashi profile ranges between 17 and 40 meq/100g (Ohtsubo et al. 1982) and is notably lower than that of Cochin (54.1 meq/100 g) and Mangalore (70.4 meq/100g) marine clays. The pore salinity of Ariake clay from the Yamaashi profile ranges between 0.1 and 1.1 g/L (Ohtsubo et al. 1982), while that of the comparable amounts of dissolved sodium and calcium ions and fewer magnesium and potassium ions. Comparatively, the pore water of Cochin and

Table 6. Chemical and mineralogical characteristics of Ariake, Cochin and Mangalore marine clays.

Soil Description	Pore Salinity (g/L)	Ion Concentration in the Pore water (g/L)					CEC (meq/100g)	Citrate Extractable Sesquioxides (%)	Soil Organic Matter (%)	X-ray Diffraction Minealogy
		Ca	Mg	Na	K	Total				
Ariake marine clay	0.1-1.1[*]	0.92	0.43	1.31	0.36	3.02[*]	17-40+	0.05-10[**]	1-3[***]	Smectite (30-40%)[*], kaolinite mica, vermiculite
Cochin marine clay[****]	9.4	0.67	1.74	5.04	0.15	7.6 5	54.1	3.8	3.0	Smectite (42), kaolinite, illite, feldspar, quartz, mica
Mangalore marine clay[****]	0.6	0.12	0.23	0.16	0.01	0.52	70.4	2.8	12.8	Smectite (32%), kaolinite, illite, feldspar, quartz, mica, hematite, hornblende

[*]Ohtsubo et al. (1982.)
[**]Egashira & Ohtsubo (1982), Ohtsubo et al (1982), Torrance (1984).
[***]Ohtsubo (1983).
[****]Rao et al. (1991).

Mangalore clays is predominantly sodium ions, with fewer magnesium, calcium, and potassium ions (Table 6). The amounts of citrate-dithionite-extractable sesquioxides and organic matter present in Ariake marine clay from the Yamaashi profile are unavailable in literature. Egashira and Ohtsubo (1982) report the presence of about 10% sesquioxides in Ariake marine clay from the Saga Agricultural Experimental Station site; Ohtsubo et al. (1982) & Torrance (1984) report much lower values ranging from 0.05 to 4.5% in Ariake marine clay specimens from the Megurie and Okishin clay profiles. Ohtsubo et al. 1982) also report the presence of 1-3% organic matter in Ariake marine clay from the Megurie and Okishin profiles. Comparatively, Cochin and Mangalore marine clays in the present study (Table 6) were determined to contain about 3-4 % sesquioxides and 3% organic matter, respectively.

7.1 Index properties

Table 7 compares the index properties of Cochin and Mangalore marine clays with that of Ariake clay from the Yamaashi profile (data from Egashira and Ohtsubo (1982) and Ohtsubo et al. (1982)). As indicated earlier, the index properties of Cochin and Mangalore marine clays reported by Rao (1974), Rao & Pranesh (1978), and Jose et al. (1988) are also considered. Data in Table 5 show that Cochin clay possesses a higher range of consistency limits than Ariake clay, while the Atterberg limits data of Rao (1974) and Rao & Pranesh (1978) for Man galore clay appear to be comparable with those of Ariake clay. The natural water content of Ariake clay exceeds the liquid limit water content, while in case of Cochin and Mangalore clays, natural water content is always lower than the liquid limit water content. The net consequence is that Ariake clay exhibits much higher ranges of liquidity index (1.19-2.79) than

Cochin and Mangalore clays (0.3-0.96). Figure 9 plots the consistency limits of Cochin, Mangalore, chart. The figure shows that Ariake clay largely exhibits lower consistency limits (liquid limit between 45 and 80%) than Mangalore and Cochin clays. Ariake clays. Ariake clay generally plots below the A-line in the MH region, while Cochin and Mangalore clays plot close to or above the A-line. Data in Table 6 had shown that Cochin, Mangalore, and Ariake marine clays contain comparable amounts of smectite, despite which Ariake clay exhibits much lower consistency limits than the Indian marine clays. The plasticity characteristics of a clay arise as a consequence of Coulombic and van der Waals attraction forces and diffuse double-layer repulsion force (long-range interparticle forces) associated with the charged clay particle surfaces (Yong & Warkentin 1975, Sridharan et al. 1986a, 1989). The more well developed the long-range forces, the higher is the plasticity exhibited by clay (Cabrera and Smalley 1973). The lower consistency limits of Ariake marine clay in comparison to Cochin and Mangalore clays may then be attributed to a relative deficiency in the long range interparticle forces associated with Ariake clay. Egashira & Ohtsubo (1982) observe that considerable substitution of Fe^{2+} for Al^{3+} occurs in the octahedral layer of smectite in Ariake clay, which acts to suppress the long-range forces normally associated with smectite and, hence, is presumably responsible for the lower plasticity of Ariake marine clay.

7.2 Strength characteristics

Figure 10 plots the undisturbed and remolded undrained strength of Cochin, Mangalore, and Ariake marine clays as a function of their liquidity index values. Once again, for Cochin and Mangalore clays, the strength data obtained by Rao (1974), Rao &

Table 7. Comparison of index properties of Ariake, Cochin and Mangalore marine clays.

Soil Description	Depth of Sampling (m)	Specific Gravity	Grain size Distribution (%)			Water content (%)	Liquid Limit (%)	Plastic Limit (%)	Liquidity Index
			Sand	Silt	clay				
*Ariake marine clay	2-20	-	-	-	-	79-106	53-72	27-40	1.19- 2.79
Cochin marine clay	8-15	2.75**	25**	23	52	90-128*** 113**	108-138*** 116**	43-48*** 35**	0.63-0.89*** 0.96**
Mangalore marine clay	8-19	2.72**	14**	48	38	58-70**** 139**	77-98**** 179**	29-41**** 72**	0.3-0.8***** 0.63**

*Egashira & Ohtsubo et al. (1982.), Ohtsubo et al. (1982).
**Rao (1974), Rao & Pranesh (1978).
***Jose et al. (1988).
****Rao et al. (1991).

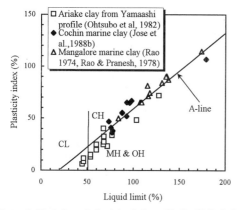

Figure 9. Variation of plasticity index with liquid limit for representative Ariake, Cochin, and Mangalore clays.

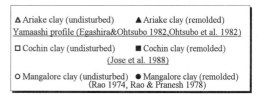

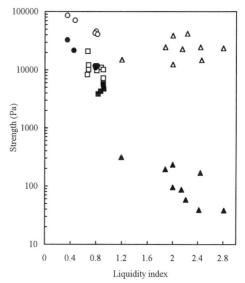

Figure 10. Variations of undisturbed and remolded, undrained strengths with liquidity index for Ariake, Cochin, and Mangalore marine clays.

Pranesh (1978), and Jose et al. (1988) have been included; the undisturbed and remolded strengths of the Ariake clay from Yamaashi profile correspond to those reported by Egashira and Ohtsubo (1982) and Ohtsubo et al. (1982). As noted in Table 7, data in figure 11 shows that Cochin and Mangalore marine clays exhibit a liquidity index less than 1, while Ariake clay exhibits a liquidity index ranging from 1.19 to 2.79. Another notable feature is that Ariake clay possessing much higher liquidity indices has undisturbed undrained strengths comparable those of Cochin and Mangalore clays. Upon remolding, Cochin and Mangalore clays show a 1.28 to 4.4-fold decrease in strength; comparatively, Ariake clay exhibits a much larger decrease of 48 to 714 fold in undrained strength. Because the undisturbed strengths of Cochin, Mangalore, and Ariake marine clays are comparable, the high sensitivity of Ariake clay evidently results from its low, remolded, undrained strengths. The remolded, undrained strength of a clay depends on the liquidity index (Skempton & Northey 1952, Mitchell & Houston 1969) and, hence, in effect, on the water content and consistency limits of the clay, An increase in water content would decrease the remolded, undrained strength, while an increase in plasticity characteristics would increase the remolded, undrained strength

211

(Cabrera & Smalley 1973, Torrance 1975). Clay with high plasticity has a large remolded, undrained strength owing to the preponderance of long-range interparticle forces, which gives it a viscous nature.

The natural water content of Cochin and Mangalore clays is low (in comparison to their liquid limit water content) and their consistency limits are high (Fig. 9). The combined effects of low natural water contents and high plasticity lead to low liquidity index (<1) and high remolded, undrained strength values, responsible for the low sensitivity of the marine clays.

8 CONCLUDING REMARKS

Understanding the engineering behavior of marine soils has become important because of increased off shore and near shore geotechnical activities. Extensive studies have been carried out with Ariake clay soil to understand its physical, physico-chemical and engineering properties. It has been widely reported that the Ariake clay contains smectite as principal clay mineral. Most of the studies try to relate the physical, physico-chemical and engineering behavior of Ariake clay to the principal clay mineral viz smectite. While smectite could be the principal clay mineral in the clay size fraction, in many reported instances other factors like pore medium chemistry, percent clay size fraction in the total soil, the different clay minerals present and their percentages can control the overall behavior. The electrical attraction and repulsion (diffuse double layer repulsion) play a dominant role in controlling the engineering behavior. In marine soils the effect of pore medium chemistry consisting of the type of exchangeable ions, their concentration and valence, and organic matter is highly complex and hence requires in-depth study to explain the overall engineering behavior. Some of the above factors with respect of Ariake clay have been discussed and it has been brought out that although the clay may contain smectite as principle clay mineral in the clay size fraction, because of the pore medium chemistry, high percentage of non-clay soil, and presence of other clay minerals in the clay size fraction makes it to behave as a non-swelling soil. Sediment volume per unit dry weight of total soil (not the clay size fraction) in water and in carbon tetrachloride can form a simple test to identify the soil from swelling and to non-swelling type.

9 ACKNOWLEDGEMENT

Authors thank Mr. Ahmed El-Shafei, a doctoral student, for his help in getting the paper prepared.

REFERENCES

Ayyar, T.S.R. 1971. Role of bonds in organic clays. *J. of Institution of Engineers (India) Soil Mechanics and Foundation Engineering Group*. 52:1-7.

Bjerrum, L. 1954. Geotechnical properties of Norwegian marine clays. *Geotechnique*. 9(4): 49-69.

Cabrera, J.G. & Smalley, I.J. 1973. Quick clays as products of glacial action: a new approach to their nature, geology, distribution and geotechnical properties. *Engineering Geology*. 7:115-133.

Eden, W. J. & Crawford, C.B. 1957. Geotechnical properties of Leda clay in Ottawa area. *Proc. 4th Int. Conf. On Soil Mechanics and Foundation Engineering, London*. 1:22-27.

Egashira, K. &Ohtsubo, M. 1981. Low swelling smectite in a recent marine mud of Ariake bay. *Soil Sci. Plant Nutr.* 27(2):205-280.

Egashira, K. &Ohtsubo, M. 1982. Smectite in marine quick clays of Japan. *Clays and Clay Minerals*. 30:275-280.Gillot, J.E. 1979. Fabric, composition and properties of sensitive soils from Canada, Alaska and Norway. *Engineering Geology*.14: 149-172.

Hong, Z. & Tsuchida, T. 1999. On Compression characteristics of Ariake clays. *Canadian geotechnical J.* 36:807-814.

Jose, B.T., Sridharan, A. & Abraham, B.M. 1988a. Physical properties of Cochin marine clays. *Indian Geotechnical J.* 18(3): 226-244.

Jose, B.T., Sridharan, A. & Abraham, B.M. 1988b. A study of geotechnical properties of Cochin marine clays. *Marine Geotechnology*. 7: 189-209.

Koumoto, T., El-Shafei, A. & Ohtsubo, M. 1998. Effect of salinity on consistency limits and remolded strength of Ariake clay. *Proc. Int. Symp. on Lowland Technology, Institute of Lowland Technology, Saga University*: 75-82.

Li, X, Hyashi, S., Nakaoka, H. & Yamauchi 1998. Statistical characteristics of geotechnical properties of marine and non marine clays in Saga plain, Japan. *Proc. Int. Symp. on Lowland Technology, Institute of Lowland Technology, Saga University*: 75-82.

Mitchell, J.K. & Houston, W.N. 1969. Causes of clay sensitivity. *ASCE, J. of Soil Mechanics and Foundations Division*. 95: 845-871.

Miura, N., Akamine, T. & Shimoyama, S. 1996. Study on depositional environment of Ariake clay formation and its sensitivity. *J. Jpn. Soc. Geotech. Engrg*. 541(III-35): 119-131 (in Japanese).

Ohtsubo, M., Takayama, M. & Egashira, K. 1982. Marine quick clays from Ariake Bay area, Japan. *Soils and Foundations*. 22(4): 71-80.

Ohtsubo, M., Egashira, K. & Takayama, M. 1985. Properties of a low swelling marine clay of interest in soil engineering. *Canadian Geotechnical J.* 222: 241-245.

Ohtsubo, M., Egashira, K. & Takayama, M. 1996. Mineralogy and chemistry, and their correlations with the geotechnical properties of marine clays in Ariake bay, Japan: Comparison of quick and nonquick clay sediments. *Marine Georesources and Geotechnology*. 14: 263-282.

Ohtsubo, M., Egashira, K., Koumoto,T. & Bergado, D.T. 2000. Mineralogy and chemistry, and their correlation with the geotechnical index properties of Bangkok clay: comparison with Ariake clay. *Soils and Foundations*. 40(1):11-21.

Oslen, R.E. & Mesri, G. 1970. Mechanisms controlling compressibility of clays. *Proc. ASCE, J of Soil Mechanics and Foundations*. 96(SM6):1863-1878.

Rao, B.M. 1974. Geotechnical investigation of the marine deposits in the Mangalore harbour site. *Indian Geotechnical J.* 4:78-92.

Rao, B.M. & Pranesh, M.R. 1978. Marine clays along west coast of India. *Proc. of Indian Geotechnical Conf., New Delhi*.1: 404-409.

Rao, S.M., Sridharan, A. & Chandrakaran, S. 1988. the role of iron oxide in tropical soil properties. *Proc. II Int. Conf. on Geomechanics in Tropical Soils, 1. Singapore*: 43-48.

Rao, S.M., Sridharan, A. & Chandrakaran, S. 1989. Influence of drying on the liquid limit behavior of a marine clay. *Geotechnique*. 39(4): 715-719.

Rao, S.M., Sridharan, A. & Chandrakaran, S. 1991. Engineering behaviour of uplifted smectite-rich Cochin and Mangalore marine clays. *Marine Geotechnology*. 9(4): 243-259.

Rao, S.M., Sridharan, A. & Chandrakarn, S. 1993. Consistency Behavior of Bentonites exposed to sea water. *Marine Georesources and Geotechnology*. 11: 213-227.

Rashid, M.A. & Brown, J.D. 1975. Influence of marine organic compounds on the engineering properties of a remolded sediment. *Engineering Geology*. 9:141-154.

Skempton, A.W. & Northey, R.D. 1952. The sensitivity of clay. *Geotechnique*. 9:198-203.

Smalley, I.J, Ross, C.W. & Whitton, J.S. 1980. Clays from New Zealand support for the inactive particle theory of soil sensitivity, *Nature*. 288:576-577.

Sridharan, A., Rao, S.M. & Murthy, N.S. 1986a. Liquid limit of montmorillonite soils. *ASTM, geotechnical Testing J*. 9(3): 156-159.

Sridharan, A., Rao, S.M. & Murthy, N.S. 1986b. A rapid method to identify clay type in soils by free swell technique. *ASTM, geotechnical Testing J*. 9(4): 198-203.

Sridharan, A., Rao, S.M. & Chandrakaran, S. 1989. Analysis of index properties of marine clays. *Proc. III National Conf. on Dock & Harbour Engineering, Suratkal*: 156-159.

Sridharan, A. 1991. Engineering Behavior of Fine Grained Soils,-A Fundamental Approach. *Indian Geotechnical J*. 21(1):1-136.

Sridharan, A. 1999. Engineering behavour of Marine clays. *Keynote Lecture 3. Geo Shore, Int. Conf. on off Shore and Near Shore Geotechnical Engineering, Dec.2-3, Bombay*. Oxford & IBH Publishing Co. PVT.LTD., New Delhi: 49-64.

Sridharan, A. & Prakash 1999. Influence of clay mineralogy and pore medium chemistry on clay sediment formation. *Canadian Geotechnical J*. 36(5):961-966.

Sridharan, A , Miura, N. & El-Shafei, A. 2000. Geotechnical behavior of Ariake clay and its comparison to typical clay minerals. *Proc. Int. Symp. on Lowland Technology, Institute of Lowland Technology, Saga University*.

Torrance, J.K. 1975. On the role of chemistry in the development and behvior of sensitive marine clays of Canada and Scandinavia. *Canadian Geotechnical J*. 12: 326-335.

Torrance, J.K. 1984. A comparison of marine clays from Ariake Bay, Japan and the South Nation River landslide site, Canada. *Soils and Foundations*. 24: 75-81.

Torrance, J.K. & Ohtsubo, M. 1995. Ariake Bay quick clays: a comparison with general model. *Soils and Foundatons*. 35(1): 11-19.

Yong, R. N. & Warkentin, B.P. 1975. *Soil Properties and Behavior*. New York: Elsevier Scientific.

Coastal Geotechnical Engineering in Practice, Nakase & Tsuchida (eds)
© 2002 Swets & Zeitlinger, Lisse, ISBN 90 5809 151 1

Problems related to the evaluation for the soil characteristics of Gadukdo clay – Case studies

S.S.Kim
Han Yang University, Kyonggi, Korea

C.S.Jang
Chun-il GEO Consultant Co., Ltd, Seoul, Korea

ABSTRACT: In designing and construction for soft ground, accurate estimation of the consolidation behavoir is very important part in geotechnical engineering, and it is often influenced by a much of factors such as soil sampling technique, testing methods and analyzing techniques. There have been different opinions for the estimation of consolidation state around the mouth of the Nak-Dong river in Korea. In this paper, those opinions are presented by the case studies, and by the comparison of each opinion and also the possible errors that can be occurred when the estimation of consolidation state is investigated and reconsideration of geotechnical properties is preformed.

1 INTRODUCTION

In recent years, the Korean Government has launched a new port construction project called 'Pusan New Port' at the west of Pusan city, as shown in Figure 1. The mouth of the Nak-Dong River is located at the east of the site.

The geotechnical properties in this site are revealed by a lot of data from the site investigations that has been performed since early 1970's, especially during the last 6 years. In this site, however, because some parts of soil layer have more than 70m depth of layer as show in Figure 3, many different opinions have been presented about the possible existence of artesian pressure as well as determination of OCR, sampling technique, and mechanical properties.

In this paper, the reconsideration of geotechnical characteristics such as deposit conditions, drainage conditions, sampling disturbance, viscosity effect, and methods for the determination methods for soil parameters comparing to the properties of the nearby site based on Gaduk-Do, is carried out to provide the fundamental reference for better solutions of various geotechnical problems.

2 GEOTECHNICAL PROPERTIES

In order to investigate the subsoil conditions of the project site, 129 exploratory boreholes were driven to the bedrock formation. Field tests such as field vane shear test, cone penetration test and standard penetration test were carried out. Undisturbed soil samples for the laboratory tests were also collected according to a systematic layout scheme. For undisturbed sampling thin-walled piston samplers were used.

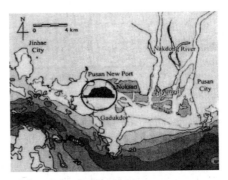

Figure 1. Location map of Pusan area

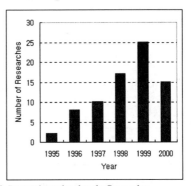

Figure 2. Researches related to the Pusan clay (Chung et al., 2000)

The Sea-coast line near Gadukdo has been formed by the continuous up-lift of sea-water (Kim et al., 1997). Soil properties in the base of this site are composed of granite rock accompanied by volcano action since Cretaceous period.

The subsoil layers shown in Figure 3 describes very simple stratography categorized into two groups, marine sediment and bedrock formation, depending on the origins and properties. Marine sediments are largely classified into two layers, the upper fine-grained soil layer and the lower coarse-grained soil layer. Occasionally coarse-grained soil layer appears within the fine-grained soil layer. However, it is relatively thin and its continuity is not revealed from the field investigation. The soft clay layer, which is about 35m to 40m thick with N-value less than 8, covers 80% of the whole area. Some boreholes hit the thin layer of residual soil of which extent is quite limited compared to the whole area of the project site. Most of the fine-grained soil layers are underlain directly by bedrock formation classified as granodiorite. Typical subsoil profile and its broad stratography are shown in Figure 4, and summarized in Table 1, respectively.

Kim et al. (1997) estimated that this site may have different soil properties comparing to the other site, such as, the deltaic zone in the mouth of the Nak-Dong river, because there are no upper sand layers.

Table 1. Subsoil Characteristics of the Project Site

Layer	USCS	Thickness	N value
Marine Deposit (Fine Grained)	CL/CH	0 - 63m	0 ~ 30
Marine Deposit (Coarse rained)	SM/SW/GP/ GP-GM	0 - 28.5m	1 ~ 50
Residual Soils	SM	0.7 - 14m	50
Bedrock	Granodiorite		

Figure 5. Basic geotechnical properties of the soil profile (S. R. Kim, 1999)

2.1 Geotechnical conditions of the site

In a recent study by Locat and Tanaka(1999), mineralogy of the Pusan clay was investigated by means of X-ray diffraction and EDAS system coupled to the scanning electron microscope for samples collected from different depths from 5.4m to 21.4m (Chung et al. 2000).

The mineralogy seems to be quite consistent for the whole depth investigated and was identified as mainly consisting of illite with significant amounts of chlorite and vermiculite, including small amount of kaolinite, quartz and horblende. Kim et al.(1998) compared the minerals included in Gadukdo clay with other site and presented its results as in Table 2.

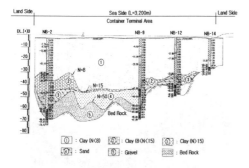

Figure 3. Typical subsoil properties of the site

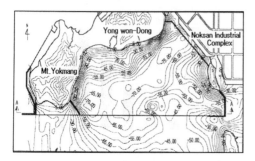

Figure 4. Close-up view and bedrock contour (S. R. Kim, 1999)

2.2 Mechanical properties of Gaduk-Do clays

Figure 6 shows the relation between liquid limit and natural water content. If natural water contents are greater than the liquid limits, soil shows a liquefaction behavior due to overburden pressure. From the above result, it is estimated that there's a little possibility of liquefaction when water content lose their originality below the liquid limit.

Figure 8 shows that clay classified as a normality active clay because the activity expressed as $A=I_p/(<2\mu m)$ has the value of 0.75~1.25 and average the value of 0.99.

216

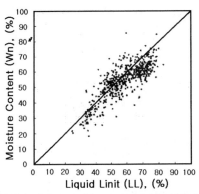

Figure 6. Relation between moisture content and liquid limit
(S. R. Kim, 1999)

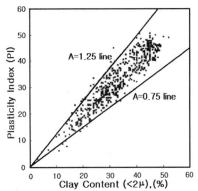

Figure 8. Relation between plasticity and clay fraction
(S. R. Kim, 1999)

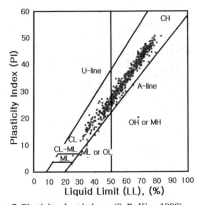

Figure 7. Plasticity chart indexes (S. R. Kim, 1999)

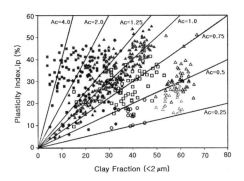

Figure 9. Activity of Pusan area with clay content
(Chung et al., 2000)

Table 2. Comparison of X-ray analysis (Kim et al., 1998)

Location		Clay Minerals						Non-Clay Minerals		
		Kaoliite	Illite	Montmorillonite	Halloysite	Chlorite	Amphibole group	Muscovite	Feldspar	Quartz
Inchon	Yongjongdo	O	O	O	X	O	O	O	O	O
	Shihwa	O	O	O	X	O	O	O	O	O
Asan Bay	Asan	O	O	O	X	O	X	O	O	O
	Sokmun	O	O	O	X	O	O	O	O	O
Kum River Estuary	Teachon	O	O	X	X	O	O	O	O	O
	Kunsan	O	O	O	X	O	O	O	O	O
	Kisan	O	O	X	X	O	X	O	O	O
Yongsan River Estuary	Muan	O	O	X	X	O	O	O	O	O
Posong Bay	Posong	O	O	X	X	O	X	O	O	O
Kwangyang Bay	Yosu	O	O	X	X	O	X	O	O	O
Nakdong River Estuary	Noksan	O	O	X	X	O	O	O	O	O
	Yangsan·Mulkum	O	O	X	X	O	O	O	O	O
	Kadokdo	*O*	*O*	*X*	*X*	*O*	*O*	*O*	*O*	*O*
The East Coast	Tonhae	O	O	X	X	O	X	O	O	O

As an example, Figure 9 shows that the activity of clay from different site near Pusan area can be classified into several groups according to its properties.

Pusan clays have the activity of 0.25~4 which Kimhae clay are distinguished (which have the activity of 0.25~0.75 in lower part) from Noksan clay(which have the activity of 1.25~6.0 in higher part) and this activity decreases as it becomes distant from sea-shore.

Figure 10(b) presents the variation of C_c with depth. C_c values seem to have the constant range regardless of depth with its large variations. C_c values are quite sensitive in void ratio(e) and liquid limit(LL) when these are compared to those of other's results shown in the same Figures. The relationships of C_c with LL and e are similar to those of Ariake clay. However, if data are screened according to each zoning scheme, clear trends are noticed. These trends demonstrate that the project area should not be considered as uniform soil but characterized locally based on the variations of properties. The water contents and clay contents have close relationship each other and show the same trend. It may be thought that the compressibility of clay is affected by the clay contents as well.

The C_v values and coefficient of permeability from the consolidation tests are plotted in Figure 10(c) and Figure 10(d). C_v values tend to be constant with depth while the permeability of clay decreases linearly with depth. The compression index, C_c varies from 0.4 to 1.1 with depth, and it has a quite good correlation with Liquid Limit (LL) and void ratio(e) as seen in Figure 11 and 12. Such relations can be formulated as, $C_c = 0.013(LL-14)$, $C_c = 0.94e-0.75$, respectively.

It is generally known that the $e - log\ P$ curve obtained from consolidation tests is influenced by sample disturbance. If samples are disturbed during sampling and preparing test, P_c and C_c values are often underestimated. It is interesting to examine Figure 14 in which the results of disturbed samples and undisturbed samples are shown separately. It clearly demonstrates that the sample disturbance significantly affect the P_c values.

The coefficient of secondary consolidation, C_a should be evaluated for the prediction of the residual settlement. Table 3 summarizes the C_c and C_a values obtained from oedometer tests with the different load increment ratio(LIR). C_a is about 3.5% of C_c values which agrees to the results reported by other researchers.

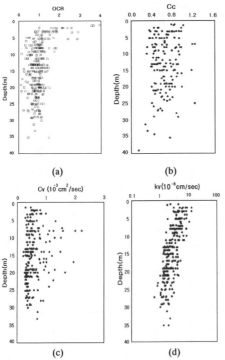

(a)   (b)

(c)   (d)

Figure 10. Variations of OCR, C_c, C_v and k_v with depth : (a)Variation of OCR with depth, (b)Variation of C_c with depth , (c) Coefficient of consolidation with depth, (d) Coefficient of permeability with depth (S. R. Kim, 1999)

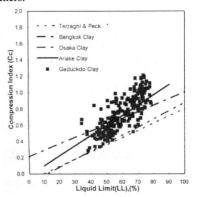

Figure 11. Relation between C_c and LL (S. R. Kim, 1999)

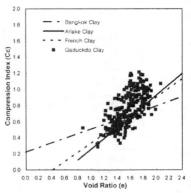

Figure 12. Relation between C_c and e (S. R. Kim, 1999)

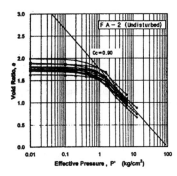

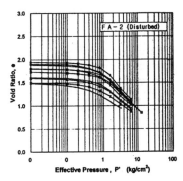

Figure 13. *e - log P* plot from oedometer test
(S. R. Kim, 1999)

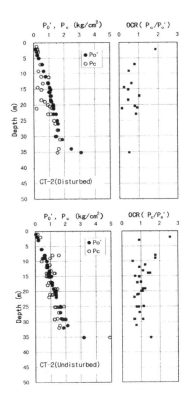

Figure 14. Variation of P'_o, P_c and OCR with depth
(S. R. Kim, 1999)

Table 3. Coefficient of secondary consolidation
(S. R. Kim, 1999)

Borehole	Depth (m)	Load Increment Ratio(LIR) = 0.5			Load Increment Ratio(LIR) = 1.0		
		C_c	C_α	C_α/C_c	C_c	C_α	C_α/C_c
NB-33	6.0	0.46	0.016	0.035	0.43	0.016	0.038
	31.0	0.63	0.013	0.031	0.62	0.013	0.021
NB-95	5.0	0.55	0.019	0.035	0.59	0.019	0.032
	15.0	0.63	0.021	0.033	0.69	0.021	0.032
	25.0	0.99	0.043	0.043	0.79	0.043	0.055
Average				0.033			0.035

Firstly, the variation of permeability with void ratio are surveyed from the result of the oedometer test (ILcon test) and the constant rate of strain test (CRS test).

Secondly, the application of Gadukdo clays are examined from the void ratio - permeability relation

based on the other countries, and after then the permeability change index (C_k), C and n of Gadukdo clays are determined.

Figure 15 shows a correlation between the consolidation curve and the void ratio(e) - premeability(k) relation.

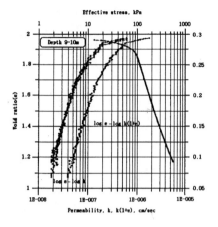

Figure. 15 Relation between the void ratio and the permeability of Gadukdo clay determined from the CRS test

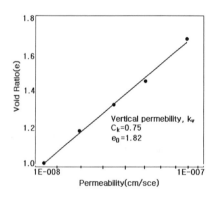

(a) Result form the oedometer test

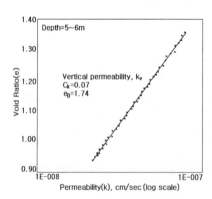

(b) Result from the CRS consolidation test

Figure. 16. Index of permeability examined from Gadukdo clay

Figure 16 shows the variation of permeability with the reduction of void ratio from the result of oedometer and CRS test, using the specimens sampling from the depth of 5~6m. The permeability ranges, in general, between 0.64 and 0.987 with depths.

Figure 17 shows the experimental relationship, $C_k=0.5e$, between the index of permeability and the initial void ratio. Relation between permeability and initial void ratio are in the narrow range, as it is mentioned above.

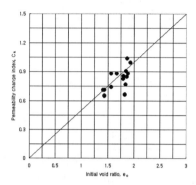

Figure 17. Relation between permeability and initial void ratio

Gadukdo clay generally shows, as shown in Figure 18, the linear $log\ e$ - $log\ k(1+e)$ curve within the general strain ratio. The value of n and C with depth were determined as $n=3.74 \sim 4.26$, $C=2.08 \times 10^{-8} \sim 4.38 \times 10^{-8}$ cm/sec.

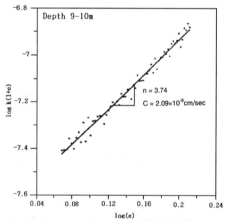

Figure 18. $log e$-$log k$ $(1+e)$ relation of Gadukdo clay

3 CASE STUDIES

Two different opinions, underconsolidation and normally consolidation, in the estimation of consolidation states of clay layer in the Nak-Dong River are discussed in this section.

3.1 Viewpoint concerning with underconsolidation

3.1.1 Yang-San Clay
Most of the data points indicating liquid limit for sedimentary soils are below 80% of natural water contents as shown in Figure 19(b).

OCRs are mostly below 1.0 except ground surface. Therefore, in this area, sedimentary layers are in the state of underconsolidation.

3.1.2 Myung-Gi Clay
Figure 20 shows that liquid index is less than 1.0 and drainage strength increases with depth. Moreover, this region is also in the underconsolidation state as indicated by the reduction of OCR with depth as shown.

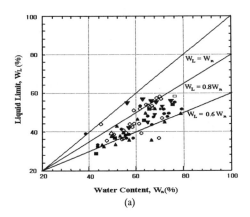

(a)

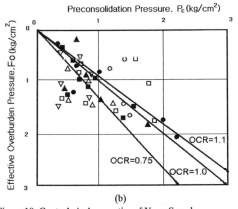

(b)

Figure 19. Geotechnical properties of Yang-San clay ;
(a) Relation between preconsolidation pressure and Effective overburden pressure, (b) Relation between water content and liquid limt.

3.1.3 Shin-Ho Clay
The underconsolidation state is evidenced by the data points of liquid index being almost 1.0 as shown in Figure 21. It also can be seen that the undrained shear strengths are constant with depth and OCRs are below 1.0.

3.1.4 Nok-San Clay
In Figure 22, the sedimentary properties of this region are assumed to be underconsolidated due to natural water contents being larger than liquid limits and OCRs being under 1.0 throughout the layers.

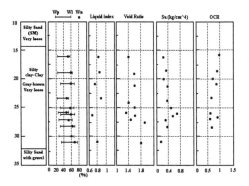

Figure 20. Geotechnical properties of Myng-Gi clay

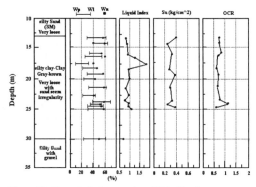

Figure 21. Geotechnical properties of Shin-Ho clay

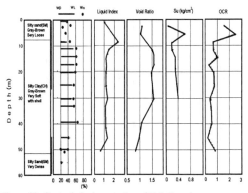

Figure 22. Geotechnical properties of Nok-San clay

221

3.2 Geotechnical problems for underconsolidated ground

Underconsolidated ground can cause a lot of geotechnical problems, such as large consolidation settlement causing land subsidence, and economical problems related to the bearing capacity under heavy weight construction.

Especially, in Kim-Hae Airport, very high air pressure due to the existence of air bubble caused many geotechnical and social problems. In addition, the reduction of artesian pressure due to the excavation and soft ground treatment induced the acceleration of ground settlements in Yang-San region.

3.3 Viewpoint concerning with normally consolidation

Figure 23 shows the distribution of OCR with depth for the Gaduk-Do clay. OCR values show a different tendency according to the test site. OCR values of the east and the west breakwater range from 0.5 to 1.8. Consolidation state of this region cannot be estimated easily because they include both of the state, underconsolidation and normally consolidation.

It is reported that the Pusan Clays can be classified as normally consolidated clays for all reclamation areas investigated. In some locations where the clay layer is on the top of the soil profile, its upper part may have been undergone some overconsolidation. It is not excluded that at some sites, especially near the coastline or in the Nakdong river mouth area where clays are younger clay, in the lower part of the clay layer there might be some underconsolidation due to the recent sedimentation process as well as the possible effect of artesian pressure exsistence.

It is also reported that the Pusan Clay is a normally consolidated deposit that is very sensitive to sampling. On the other hand, it is clear from the result that the upper layer has many of the attributes of "cemented" profile namely, little change in water content and strength, although CPT values show a linear increase even in this section of the profile.

One possible explanation for this behavior would be that the upper layer being composed of a well structured clay with evidences of bridging and potential brittle behavior. From the geotechnical data, the depth at which the bonding strength disappears at about the depth of 15m which coresponds also to the grain size profile.

Field tests illustrate that the undrained shear strength increases with depth as shown in Figure 24. All the results indicate non-existence of stiff crust. Shear strength increment ratio, (S_u/P_o), varies from 0.22 to 0.25.

OCRs for a deltaic clay in the mouth of Nak-Dong river is obtained from the result of field measurements, and the range is between 0.95 and 1.20 (see Figure 25b), although they were estimated from the values of 0.4 to 0.7(Figure 25) by the conventional consolidation test for the clay. From the dissipation test, it was found that the excess pore pressure hardly existed in the clay deposit, therefore the soil was not in the under-consolidation condition.

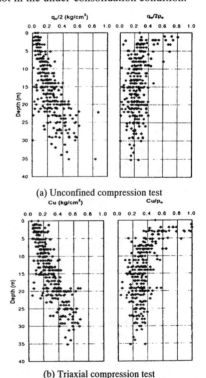

(a) Unconfined compression test

(b) Triaxial compression test

Figure 24. Normalized undrained shear strength distribution with depth of Gaduk-Do clay from field test results

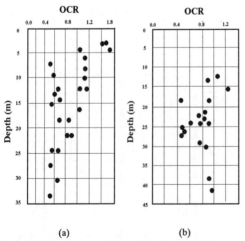

(a)	(b)

Figure 23. Distribution of OCR with depth in Gaduk-Do :
(a) East breakwater, (b) West breakwater

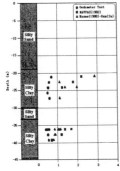

Figure 25. OCRs from the result of conventional consolidation test.

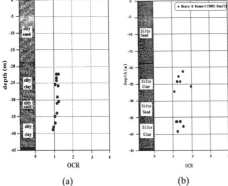

(a)   (b)

Figure 26. OCRs form the analysis of field measutemnet ;

    (a) estimated result by in-situ pore pressure,

    (b) determination by stand penetration test(SPT).

4 FACTOR INFLUENCING ON THE BEHAVIOUR OF CLAYS

4.1 The possibility of confined groundwater layer

One of the main topic in surveying the properties of clay near Pusan is that how can analyze the consolidation settlement considering the existence of confined groundwater layer.

The accurate estimation of in-situ preconsolidation pressure and the variation of confined groundwater are important in consolidation analysis because there are some differences in consolidation settlement according to the confined pressure within the confined groundwater layer and its variation.

Figure 28 and 29 show the result of settlement analysis assuming the stresses in a ground as Figure 27, it shows a good agreements with in-situ values when considering the existence of confined groundwater layer. Table 4, which also show the effect of the confined groundwater layer on the consolidation

settlement summarized by Chung et al.(1999). They suggested that new analysis methods are necessary, because the correct prediction considering of the effect of sample disturbance did not agree with the measured value. For example, the predicted settlements considering the under-consolidated stage, overestimate the consolidation settlements by 240% comparing to the in-situ settlements but it gives good agreements when the normally consolidated stage is considered.

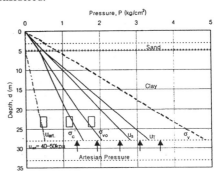

Figure 27. Stresses in a ground with confined groundwater layers (Kim et al., 1999)

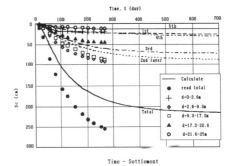

(a) Neglecting the confined groundwater layers

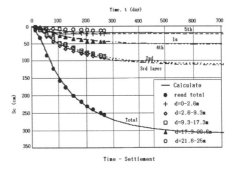

(b) Considering the confined groundwater layers

Figure 28. Comparison between the results of consolidation analysis and measured values

The finite element analysis results for Yangsan clay, considering the confined groundwater layer suggested by Jin et al.(1998), in Figure 29 and 30, also predict the larger settlements than measurement. They especially suggested that the right and the left side of the embankment show different behavior with each other along the stress path. When the confined groundwater layer exits in the centre of the embankment, the failure locus are smaller than the initial condition due to the reduction of effective stress, but does not reach the critical state in spite of the occurrence of hardening from normally consolidated state. When the confined groundwater layer exists in both side of the embankment, the behavior of embankment varied from initial normally consolidated stage, through hardening during construction the embankment. When recover the confined pressure, i.e., the hydrostatic pressure condition, the behavior of embankment showed a constant elastic stage with.

Table 4. Comparison of consolidation settlement of delta in the lower site of Nakdong river

Calculated (cm)			Field	
Immediate	Consolidation	Total	Settlement	T_{90}
4.57	280.36	284.93	118.87	1100

Re-calculated by Terzaghi equation.				
by C_c (cm)	by C_{c1} (cm)	by C_{c2} (cm)	Skempton (cm)	NAVFAC (cm)
116.23	138.54	194.90	81.63	195.13

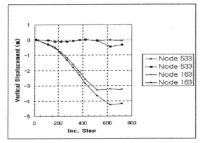

(a) Comparison of vertical displacement

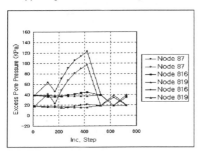

(b) Variation of excess pore pressure

Figure 29. Comparison of effects of confined groundwater layers in finite element analysis

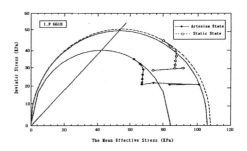

(a) Beneath centre of embankment

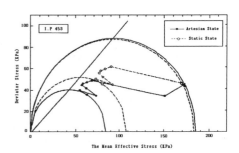

(b) Left and right side of embankment

Figure 30. Variation of stress path in finite element analysis

Table 5. Variation of measured pore pressure (unit t/m^3)

Boring NO.	Depth (m)	N Value	Hydrostatic pressure (a)	Measured pore pressure (b)	Residual Pore pressure (b-a)	error (%)
NB-2	16.9	3	17.4	17.0	0.4	2.4
	27.2	7	28.0	28.4	0.4	1.4
	45.4	16	46.8	47.1	0.3	0.6
NB-77	21.3	3	21.9	23.5	(1.6)	6.8
	26.6	2	27.4	27.8	0.4	1.4
	40.3	1.3	41.5	41.9	0.4	1.0
ANB-11	42.2	36	43.5	44.4	0.9	2.0

In Gadukdo clay, to be different from so called deltaic clay which is in the lower part of Nakdong river, the confined groundwater layer can be ignored because they are assumed to be in the hydrostatic pressure state with residual pore pressure of 0.04 to 0.16kg/cm²(4~16kPa).

4.2 Overconsolidation ratio

Chung et al.(1999) suggested the OCR using SPT, CPT and CPTU tests with the depth of clay layer for the purpose of comparison to the results of oedometer test for the Nakdong clay, as it is presented in Figure 31.

The results showed that the OCR of clay analyzed using the pore pressure occurring from the result of embankment is considered to be a normally consolidated state with OCR of 0.95~1.2 and the dissipation test could support this result. Again, it could be known that the laboratory test results underestimate the OCR due to the disturbance occurred when sampling, and that the OCR could be well estimated by the in-situ values using cone penetration test.

In Gadukdo clay, more elaborate studies are necessary because the OCR using oedometer test show the feature of under-consolidated clay with OCR of 0.6~0.8 by the depth.

As it is shown in Table 6, by the OCR values obtained by in-situ vane tests of Mayne & Mitchell type and Larsson type, the soil is considered as normally consolidated clay with the average OCR over 1.0.

When the stress path by the K_o triaxial test as shown in previous Figure 30, it is considered to be a normally consolidated state because the failure envelope passed through the origin, and also it can be decided as a normally consolidated clay from the pore pressure measurements presented in Table 6.

For the purpose of estimation for overconsolidation ratio(OCR) of extraordinarily sensitive clay in the lower part of Nakdong river, the analysis and dissipation test was performed and compared with the OCR determined by laboratory and in-situ soil test. The results from this study can be summarized as below:

1) The OCR of clay analyzed using pore pressure occurred during embankment was about 0.95~1.20, and the deltaic clay was almost in the normally consolidated state. This result was different from the past assumption of under-consolidated state using oedometer test result, and this was supported by the dissipation test.

2) In determination of the OCR using the results of laboratory and in-situ test, the methods suggested by Mayne(1991) and Cao(1996) using piezocone test and Mayne & Kemper(1998) using cone penetration test showed more agreement with in-situ measured values.

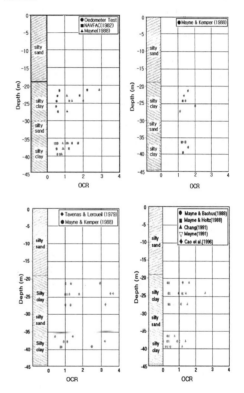

Figure 31. OCR decided by in-situ experimental formulation (S. G. Chung, 1999)

Table 6. OCR due to the results of in-situ vane test

Site No.	Mayne & Mitchell type		Larsson type	
	Range	Average	Range	Average
I	0.71 ~ 3.18	1.63	0.66 ~ 2.95	1.51
II	1.06 ~ 2.62	1.41	0.97 ~ 2.39	1.28
III	0.69 ~ 4.05	1.45	0.62 ~ 3.83	1.37
IV	0.85 ~ 2.79	1.43	0.79 ~ 2.57	1.31
V	0.81 ~ 3.99	1.46	0.75 ~ 3.67	1.35
VI	0.89 ~ 4.00	1.61	0.83 ~ 3.73	1.51

4.3 Effects during sampling

4.3.1 Soil sampling techniques

The block sampler of large diameter consists of three main components, i.e., (i) An inner PVC tube, having a 30cm inside diameter, 7 to 10cm thickness and a 600 angle of cutting edge. The inner tube is fixed to a cross-shaped plate that is connected to a rod. The area ratio of this PVC tube is less than 10%, but if a thin wall steel tube would be used instead of PVC it may be smaller. As this sampling tube has two splitting hakes which are tied together by a steel ring, it would be easy to get the undisturbed sample by separating two halves apart, reducing in this way the sample disturbance; (ii) An outer casing, which is composed of four shoes and open windows(holes). The four shoes at the tip are used for opening during penetration, cutting the sample and closing to prevent the inner sample from falling down. The outer casing therefore helps to get a complete sample recovery as well as to prevent the sample from distortion; (iii) A jetting apparatus, which is attached at the tip of the outer casing. The water jetting through pipe nozzle would help the outer casing penetrate easily and deeper than the inner tube.

4.3.2 Soil disturbance

The effects of soil disturbance due to different sampling techniques on a conventional consolidation test results are shown in Figure 32 and CIU triaxial test results are shown in Figure 33. In both Figure, for a Yangsan clay sample collected at a depth of 9m, a common trend was observed, and namely, larger sampler diameter less soil disturbance. However, judged from that specimens sampled from different depths do not show such a tendency, sampling techniques as well as soil sample transportation and storage would certainly have a very strong influence on soil disturbance, and consequently on the results of geotechnical testing. In recent work, the effects of soil sample disturbance on reducing the values of unconfined compressing strength, preconsolidation pressure of the Pusan clays are reported.

4.4 Strength properties

4.4.1 Field vane shear test

As shown in Figure 34, strength distribution from the triaxial compression test(Figure 29(b)) is more scattered than those from the field vane shear test. The field vane shear test, therefore, seems to produce more reliable results for soft and thick clay deposit.

Figure 35 contains plots of the OCR, unconfined compressive strength, axial strain at failure and pore pressure measured by the piezometers in the project site respectively.

Figures 35(c) and (d) show that the lower strengths are measured at relatively larger strain($\varepsilon_f >$ 6%), and it consequently results in lower OCR value which can be interpreted as an underconsolidated state. However, the strength points from all the samples failed at the strain level less than 6% fall on the straight line. Furthermore, pore pressure measured by the piezometers is nearly the same as hydrostatic pressure as shown in (b). These are the evidences that the clay deposit in the project site may not be in the underconsolidated state but be in the normally consolidated state. It is also interesting, however, to point out here that ε_f values are relatively big compared to that of cemented clay or aged clay. Therefore, the entire clay deposit can be classified as normally consolidated young sensitive clay.

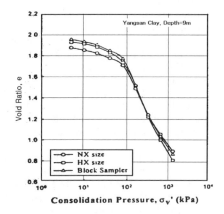

Figure 32. Effects of soil sampling techniques on conventional consolidation test results

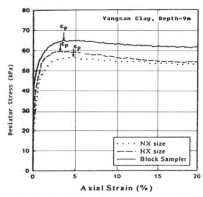

Figure 33. Effects of soil sampling techniques on CIU triaxial test results

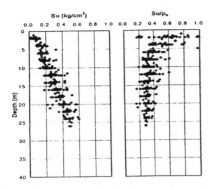

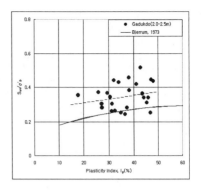

Figure 34. Field vane shear test (S. R. Kim, 1999)

Figure 37. Variation of S_{uv}/σ'_p with plasticity index (Chung et al., 2000)

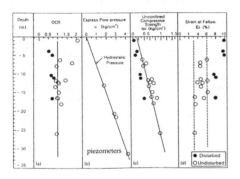

Figure 35. Variation of OCR, q_u, ε_f, and u with depth (S. R. Kim, 1999)

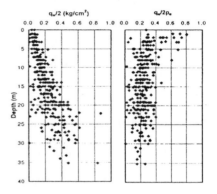

(a) Unconfined compression test

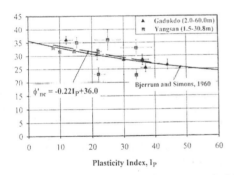

Figure 36. Friction angle in the NC range vs. plasticity index(Chung et al., 2000)

Figure 36 shows the relation between the friction angle and the plasticity index in the normally consolidated zone. The friction angle tends to decrease as the plasticity index increases when the friction angles are in the range of $25°~36°$.

The relation between clays in this site can be expressed as $\Phi'_{nc}=-0.22I_p+36$, and this expression shows a good agreement with those of Bjerrum and Simon(1960).

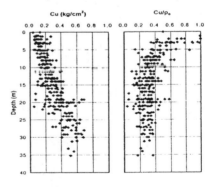

(b) Triaxial compression test

Figure 38. Shear strength profiles with depth from different tests (S. R. Kim, 1999)

Figure 37 shows the relation between the ratio of the shear strength to load(S_{uv}/σ'_p) and plasticity index expressed as

$$S_{uv}/\sigma'_p = 0.27+0.0021I_p$$

This expression seems to be larger value of 0.1 comparing to the result of Bjerrum(1973) with the same plasticity index.

Field tests illustrate that the undrained shear strength increases with depth. All the results indicate non-existence of stiff top crust. Shear strength increment ratio, (S_u/P_o) varies from 0.22 to 0.25. Figure 38 implies the sample disturbance inevitably caused by sampling and testing process.

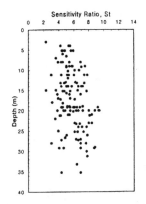

Figure 39. Sensitivity ratio with depth (S. R. Kim, 1999)

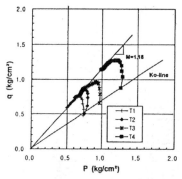

Figure 40. (q, p) plot from CK$_0$U test (S. R. Kim, 1999)

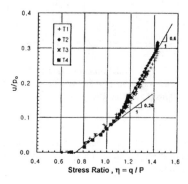

Figure 41. Relation between u/P_0 and η (S. R. Kim, 1999)

The possibility of sample disturbance is also supported by Figure 39, which contains the sensitivity ratio of all samples with depth. It varies from 3 to 8 and tends to increase with depth. It also averages to about 5, which belongs to the sensitive clay category suggested by Skempton(1952) and Bjerrum(1954). It is convinced that this sensitivity of clay certainly plays significant role in the estimation of the preconsolidation pressure probably resulting in OCR value less than unity.

Only few CK$_0$U tests were performed and (q, p), $(u/P_0, \eta)$ relationships are plotted in Figures 40 and 42, respectively. Samples were initially K_0-consolidated under the in-situ effective stresses and then axially compressed with undrained condition. K_o values of the samples are about 0.5 and the relationship between u/P_0 and stress ratio $\eta = q/p$ is bilinear which is typically seen from the K_o normally consolidated clay samples. This behavior also indicates the existence of strain independent nature of the definition of two pore pressure parameters. The failure envelope in Figure 40 also passes through the origin. The angle of internal friction obtained these tests is about 30°. These test results are another evidences that the clay deposit can possibly be normally consolidated state

5 CONCLUDING REMARKS

Recently, there have been arguments whether Gaduk-Do clays have similar geotechnical properties as Pusan clays. Some findings investigated and examined in this study are summarized as follows ;

1) This paper presents some different opinions about the geotechnical properties of the Gaduk-Do clay. Previous studies presented that the sedimentary ground in the mouth of the Nak-Dong River is in the underconsolidated state. However, this study suggests that experimental results can be affected by some factors. The Gaduck-Do clay is soft sensitive clay with high compressibility; they are, thus, easily affected by soil disturbance during soil sampling. Apparent soil disturbance, however, is caused by sampling procedure.

2) Because the Gaduck-Do clay layer has the thickness larger than 70m, the physical properties are varied with depth, and there are various trends according to the location. The variation of the water content with depth can reflect the possible underconsolidated state. However, it is revealed from detailed analysis that the clay content can be affected by deposition characteristics or surrounding environmental conditions.

3) A large number of preconsolidation pressures obtained from the oedometer tests are smaller

than in-situ effective overburden pressure, and this can be an evidence of under consolidation. It is, however, concluded from the careful analyses of available test data, that the sample disturbance which can easily take place in such a sensitive clay, could be the main reason for the underestimation of P_c values. This conclusion is also supported by the tendency that the strains at failure of the samples with OCR less than unity, are considerably large, and that any notable field excess pore pressure was not measured. However, undisturbed samples mostly show relatively larger strain at failure compared to those of cemented or aged clay. Therefore, the entire clay deposit is seemed to be a young clay body under the normally consolidated state.

4) As the geotechnical properties of this sedimentary ground in the mouth of the Nak-Dong River can be presented by different ways with location, new testing methods and analytical techniques considering different properties, are required.

REFERENCES

Bjerrum and L. Simon (1960), The Effects of Time on the Shear Strength of Soft Marine Clay, Proc. Brussels Conf. On Earth Pressure Problems, Vo. 1, pp. 148-158

Cao L. F., M. F. Chang and C. I. (1996), Cavity expansion in modified Cam clay and its application to the interpretation of piezocone tests, Geotechnical Research Report NTU/GT 9603, Nanyang Technoligical University, Singapore

Cho K. Y., Lee D. M. & Chung S. G. (1997), Engineering properties of alluvial clays around estuary of Nakdong river, Research Works, Research Institute of Construction Technology and Planning, College of Engineering, Dong-A University, Vol.21, No.1, pp.47-55 (in Korean)

Chung H. I., Jin H. S., Lee Y. S., Jin K. N. and Lee J. S. (2000). The behavior of soft ground with artesian pressure under embankment, ATC-7 Workshop, Feb.24, Dong-A University, pp.38-47 (in Korean)

Chung S. G., Giao P. H., Kim G. J. and Yoon D. D. (2000). A review of hydraulic characteristics of Yangsan soft clay with reference to settlement analysis, ATC-7 Workshop, Feb.24 Dong-A University, pp.78-91.

Chung S. G., Je H. K. and Jin H. S.(1999). Consolidation analysis for partially penetrating vertical drains in thick soft clay ground, Research institute of construction technology and planning, College of engineering, Dong-A University, Vol.23, No.1, pp.81-88(in Korean)

Chung S. G., Jin H. S. and Baek S. H.(1999). Effect of consolidation behaviour due to penetrate depth of vertical drains soft clay ground under artesian pressure, Research Works, Research Institute of Construction Technology and Planning, College of Engineering, Dong-A University, Vol.22, No.2, pp.77-84(in Korean)

Chung S. G.(1999). Engineering properties and consolidation characteristics of Kimhae estuarine clayey soil, Thick Deltaic Deposit, ATC-7 Workshop, Special Publication at the 11th Asian Regional Conference on soil mechanics and geotechnical engineering, Seoul, Korea, pp.93-108(in Korean)

Chung S. G., Eun S. M., Baek S. H. and Lee D. M.(1999). PDSS analysis on partially penetrated band drains in soft clay ground, KGS Spring '99 National Conference, Korean Geotechnical Society, Seoul, pp.365-372(in Korean)

Chung S. G., Lee N. K., Cho K. Y. and Kim M. K.(1997). Behavior of soft clay due to removal of preloading and recompression, Research Works, Research Institute of Construction Technology and Planning, College of Engineering, Dong-A University, Vol.21, No.2, pp.111-128(in Korean)

Chung S. G.(1988). An experimental study on constant rate of strain consolidation for the Nakdong river clays, Research Report, Research Institute of Korean Resource Development, Dong-A University, Vol.12, No.1, pp.11-25(in Korean)

Chung S. G.(1988). Constant rate of strain consolidation test on Nakdong river estuarine clay, Research Report, Institute of Korean Resource Development, College of Engineering, Dong-A University, Vol.12, No.1, pp.11-25(in Korean)

Giao P. H. and Chung S. G.(2000) Some theoretical and practical considerations on the use of dewatering to accelerate consolidation of Pusan clay, Invitation Lecture on Soft Clay Engineering, March 30, Dong-A University.

Giao P. H., Chung S. G. and Yoon D. D.(2000). Settlement analysis of Yangsan soft clay, ATC-7 Workshop, Feb.24, Dong-A University, pp.18-37

Han Y. C.(2000). A study on stratigraphical history for soft clay deposit in Yangsan/Mulgeum site, ATC-7 Workshop, Feb.24, Dong-A University, pp.11-17

Jin H. S., Cho K. Y., Yoo G. Y. and Chung S. G.(1998). Finite element analysis for embankment on soft ground under artesian pressure, KGS Spring '98 National Conference, Korean Geotechnical Society, Seoul, pp.127-134(in Korean)

Kim D. H., Lim H. D., Kim J. W. and Lee W. J.(2000). Evaluation of permeability characteristics of Kimhae clay by laboratory tests, KGS Spring 2000 Naitonal Conference, Korean Geotechnical Society, pp.647-654(in Korean)

Kim S. K.(1999). Consolidation analysis of vertical drain considering artesian pressure, '99 Seminar, Technical Committee on Soft Ground Treatment, Korean Geotechnical Society, Feb. Seoul. Pp.62-69(in Korean)

Kim S. K.(1999). Relevance of foundation design to engineering characteristics of Kimhae clay, Proceeding of Sang-Kyu, Kim Symposium on Geotechnical Engineering, April 17, Seoul, pp.205-230(in Korean)

Kim S. K., Lim H. D. and Moon S. K.(1998). Clay minerals and their distribution in the soft ground deposited along the coastline, Journal of the Korean Geotechnical Society, Vol.14, No.6, pp.73-80(in Koran)

Kim S. K. and Koh S. Y.(1997). Under consolidation characteristics of deposits in the mouth of the Nakdong river and its neighboring coast, KGS Fall '97 National Conference, Korean Geotechnical Society, Seoul, pp.3-18(in Korean)

Kim S. R.(1999). Some factors affecting the ground improvement design for Pusan new port project, Thick Deltaic Deposits, ATC-7 Workshop, Special Publication at the 11th Asian Regional Conference on soil mechanics and geotechnical engineering, Seoul, Korea, pp.65-91(in Korean)

Mayne P. W. and Bachus, R. C. (1988), Profiling OCR in clays by piezocone soundings, Proc. 1st Int'l. Symp. On Penetration Testing, Orlando, pp. 857-864

Mayne P. W. and Kemper J. B., Jr. (1988), Profiling OCR in stiff clays by CPT and SPT, ASTM, Geotechnical Testing Journal, 11(2), pp. 139-147

Mayne P. W. (1991), Determination of OCR in clays by piezocone tests using cavity expansion and critical state concepts, Soils and Foundations, 31(2), pp. 65-76

Tanaka H., Mishima O., Tanaka M., Park S. Z. & Jeong G. H. (1999), Consolidation characteristics of Nakdong river clay deposit, Proceeding of '99 Dredging and Geoenvironmental Conference, Seoul, Korea, pp. 3-14(in Korean)

Coastal Geotechnical Engineering in Practice, Nakase & Tsuchida (eds)
© 2002 Swets & Zeitlinger, Lisse, ISBN 90 5809 151 1

Characteristic features of soft ground engineering in Bangladesh

A.Siddique, A.M.M.Safiullah & M.A.Ansary
Department of Civil Engineering, Bangladesh University of Engineering & Technology, Dhaka, Bangladesh

ABSTRACT: In this paper, the current practice of soft ground improvement in Bangladesh has been presented. Problems encountered with soft soils and typical foundations adopted for soft soils in this country are addressed. "Cut and replacement" technique of foundation construction has been effectively used for low rise buildings in Bangladesh. This technique essentially reduces total and differential settlement built over soft clay. Current methods of strengthening the soft sub-soil in Bangladesh includes preloading and vertical drainage in conjunction with preloading. The vertical drains normally consists of sand wicks and prefabricated vertical drains. Several case studies of buildings, roads and embankments have been presented in this paper to illustrate the important features of soil improvement methods in soft clay deposits.

1 INTRODUCTION

Soft soil deposits are widespread, and they present very special problems of engineering design and construction. Foundation failures in soft clay are comparatively common, high surface loading in the form of embankments and shallow foundations inevitably results in large settlements which must be accommodated for in design, and which invariably necessitate long-term maintenance of engineered facilities. The construction of buildings, roads, embankments, bridges, canals, harbours and railways in soft soils has always been associated with stability problems and settlements.

In this paper a brief introduction of the geology and land formation of Bangladesh is presented. Problems associated with soft soils of Bangladesh are addressed in the paper. Types of foundations used on soft soils of Bangladesh and typical soil improvement techniques adopted in Bangladesh are also presented in this paper. Finally, salient features of soft ground engineering in Bangladesh have been illustrated by presenting case histories of a number of completed and on-going construction projects, including buildings, roads and embankments.

2 LAND FORMATION AND PHYSIOGRAPHIC CONDITION OF BANGLADESH

Bangladesh is the largest delta of the world receiving sediments from three rivers, namely, the Ganges (called the Padma in Bangladesh), the Brahmaputra (called the Jamuna in the lower stretch) and the Meghna, and numerous tributaries. These rivers (about 230 with a total length of 2200 km) form roughly 800 km of perennial waterways. The major river systems originate from India in the Himalayan Range and the Shillong Massif and enter Bangladesh from the west, north and northeast and flow into the Bay of Bengal. Bangladesh is characterized by an almost smooth plain of very low relief with broad ridges and extensive shallow basins. This plain landscape covers more than 80% of the natural area with an altitude varying between 3 and 15 m above mean sea level. The land slopes gently from the north towards the Bay of Bengal at a rate of 1 m in 20 km (Alam et al., 1990). The entire Bangladesh plain is one physiographic province, bounded to the north by the Shillong Massif and to the west by the Rajmahal Hills (India), and by the Tippera Hills to the east and the Chittagong Hills to the southeast.

According to the study of Morgan and McIntire (1959), Bangladesh is filled with sediments of Tertiary to Quaternary age. Bangladesh is divided into two major tectonic units, the Precambrian Platform in the northwest and the Bengal Foredeep in the southeast, separated by the Calcutta-Mymensingh Hinge Line. The thickness of the sedimentary cover on the basement is 180 m in the west, increasing to the southeast, with a thickness of over 18000 m in the eastern part of the country. The plains of Bangladesh are affected by settlement due to consolidation of the sediments, and by tectonic movements. Bangladesh is situated in one of the most active tectonic regions of the world, where three major plates

(Indian Plate, Tibet and Burmese Subplates) collide and thrust over each other. As a consequence, severe seismic activity occurs in the northern and eastern part of Bangladesh, causing major earthquakes. Over 200 earthquakes have been recorded during 1833-1971 with magnitudes of 5.0 to 8.5 on the scale of Richter (Committee of Experts 1979). The geology of the Quaternary and tectonic activities of the Bengal Basin described in detail by Bakhtine (1966), Alam (1971) and Haque (1982), and in a recent Master Plan Organization Report (Harza Engineering et al. 1987).

Based on surfacial geology Bangladesh may be split up into a number of landform patterns that have evolved under relatively uniform geological and climatic conditions with similar topographic features. From topographic mapping and data from Soil Survey Project of Bangladesh, Bangladesh Transport Survey (1974) identified 21 Land systems in Bangladesh. These Land Systems may be grouped into six Soil Units as shown in Figure 1 (after Hunt 1976). General characteristics of these Soil Units are summarized in Table 1. Alluvial Flood Plain deposits are characterized by finer materials (silt and clay) at the surface underlain by coarser materials (fine to medium sand). The Raised Alluvial Terrace Deposits or Pleistocene deposits comprise of a relatively homogeneous clay known as Madhupur day (in the east) or Barind clay (in the west). The thickness of this clay layer varies but is underlain by sandy soil. The two soil units, (i) alluvial flood plain deposits and (ii) estuarine and tidal flood plain deposits are called recent deposits. The raised alluvial terrace deposits and the recent deposits constitute more than 80 percent of the land surface of Bangladesh. Comparison in Standard Penetration N values show that the Terrace deposits are more compact at relatively shallow depths while there is a considerable variation in N values in the Estuarine and Tidal Flood Plain deposits. As far as grain size and mineral contents are concerned, Pleistocene (Terrace) sediments are almost identical with those of recent flood plain. Despite similarity in grain size field differentiation between Terrace and Recent deposits is simple. Recent sediments are typically dark, loosely compacted, and have a high water content and variable but may contain appreciable quantities of organic material. Terrace deposits, on the other hand, are well oxidized and typically are reddish, brown or tan, and are mottled. They commonly contain ferruginous or calcareous nodules. Water content is lower, resulting in firmer, more compact material. Organic material in Terrace sediments is confined to the surface soil profile.

Soft soils in Bangladesh are mainly available in the alluvial flood plain deposits, depression deposits and estuarine and tidal plain deposits. The soft soil, mainly comprise underconsolidated to normally consolidated clays and silts often containing organic

materials. Geotechnical aspects of Bangladeshi soils were reported by Safiullah (1991) and Serajuddin (1998), while Mollah (1993) reported the geotechnical conditions of the deltaic alluvial plains of Bangladesh.

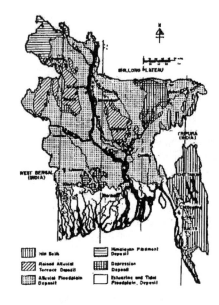

Figure 1. Simplified soil units of Bangladesh (after Hunt 1976)

Table 1. Simplified soil units of Bangladesh (modified after Hunt 1976)

Soil Unit	General Characteristics
Hill Soils	Variable soil types which are function of underlying geology. Frequently sandy clays and clays grade into disintrigated rock at shallow depths.
Raised Alluvial Terrace deposits	Comprises a relatively homogeneous clay known as Madhupur clay and Barind clay. (liquid limit = 30 to 80, plasticity index = 12 to 50. Variable depth, underlain by fine to medium uniformly graded sand.
Himalayan Piedmont deposits	Mainly sandy silt in higher areas and silty clays in basin areas but often overlying fine sands at shallow depth.
Alluvial Flood Plain deposits	Locally variable but in general silts and silty clays. Silt sizes predominent. (liquid limit = 20 to 50, plasticity index = 4 to 30). Fine sands abounds at depths and dose to rivers. Contains mica.
Depression deposits	In the south alternating organic/clay deposits overlying clay at depth. Elsewhere, predominently silty clays and clays. (liquid limit = 30 to 40, plasticity index = 10 to 16).
Estuarine and Tidal Flood Plain deposits	Generally silt and silty clays. Acid sulphate soils found near coast. Organic soils close to surface in some places. Widely varies in consistency and water content.

3 PROBLEMS WITH SOFT SOILS IN BANGLADESH

The principal foundation problems in Bangladesh are related to the low shearing resistance of the underlying soil. Soils with low shear strength are not strong enough to support the most common structures with conventional shallow foundation systems and, therefore, pose a serious foundation problem for the entire region. In practice, various types of shallow foundations such as pad, strip and compensating types are used for light structures with pressures ranging from 25 to 40 kPa at depths between 1.5 m and 3.0 m.

Settlement is another major foundation problem in Bangladesh related to the loose and compressible nature of the subsoil. Excessive settlement is observed with many structures even with portal frames and boundary walls. The most extreme settlement can be seen in rural roads and also in the major roads connecting the districts. Several segments of these roads are built on 1.5 m to 3.0 m high embankments where settlements up to 40 cm were recorded (Mollah 1993). It is expected that these settlements are the result of consolidation of both fill material and compressible soft soil occurring near the surface. Numerous settlements on a large number of road segments have made the bituminous running surface uneven, causing severe cracking. Embankments, requiring extensive filling work, are constructed throughout the plains of Bangladesh for flood protection, irrigation and the development of a road network. In general, the natural state of the local soils is not suitable for embankment construction and maintenance. The most common problem with embankment construction is the generally soft nature of the top soil as well as the foundation soil.

The position of ground water table in Bangladesh lies between 1 and 5 m. During the rainy season, the water table is higher than 1.0 m and the land is often flooded. Construction activities, therefore, often involve excavations of considerable depth underwater. These excavations lead to unstable situations because the work is made either in highly permeable loose sandy soils or in soft plastic clays.

Liquefaction is the phenomenon by which saturated soils, essentially loose, are temporarily transformed into a liquefied state. In the process, the soil undergoes transient loss of shear strength which commonly allows ground displacement or ground failure. Liquefaction characteristics of a soil depend on several factors, among others, position of groundwater table, grain size distribution, soil density, ground accelerations, sedimentation history, age of sedimentation, thickness of the deposit, location of drainage, magnitude and nature of superimposed loads (Seed & Idriss 1967, Casagrande 1976, Seed 1979). In Bangladesh, soils are typically young, comprising sandy material and occurring in a saturated condition. In general, the soil has a density less than the critical density. Therefore the particular characteristics of the soil in the top 15 m to 20 m together with the seismic history of the region render these soils sensitive to liquefaction. Extensive field and laboratory tests conducted during the feasibility study for the Jamuna Bridge gave the same conclusions (Heiznen, 1988).

4 FOUNDATION ON SOFT SOILS IN BANGLADESH

With rapid expansion of urbanisation, major cities of Bangladesh are facing foundation problems on soft grounds. Structural foundations on soft soils in this country are limited to use of raft foundations, piled footing and well supported footings. The principle of floating foundations (raft foundations) has been frequently used in reducing settlements, specially in soft clays. Wells, sunk by manual digging, are normally used up to depths of 5 m to 8 m. These wells, made of brick masonry, are 1.2 m to 1.6 m in diameters. Piled foundations using timber piles and reinforced concrete piles are popular. The loads created by major civil engineering structures are often transferred through reinforced piles on the sand stratum which has a high bearing value. Precast concrete piles are commonly used in areas where there is soft clay. Because of low labour costs, large diameter cast-in-place (bored) piles are normally used in soft clays to support heavy concentrated loads. Timber piles have been extensively used for a long time to support buildings in soft soils. For smaller loads, timber piles with a diameter of 120 mm to 150 mm and a length of 8 m to 10 m are economical and, therefore, often used. These are used in large groups assigning a load of about 50 kN/pile, mainly governed by the structural capacity. However, the durability of these piles is only guaranteed when the piles remain constantly in the groundwater. The process of drying and wetting increases the possibility of attack by insects and enhance weather action thus rapidly shortening the life of timber piles. Figures 2 and 3 present photographs showing installation of timber and precast piles, respectively. The "Cut and Replacement" technique has been used recently in a number of building construction projects. In these constructions, the top soils, often containing soft organic layers are excavated and replaced by river sands. Spread footings or mat foundations are then constructed. These are used to increase the allowable bearing pressure and achieve uniform stressing of underlying soft layers, thereby minimizing the differential and total settlements. Very recently some use of sand columns, the so called sand piles, has been reported. Sand piles ranging in diameter from 150 mm to 300 mm has been used.

Figure 2. Installation of timber piles in progress

5 TYPICAL STRENGTHENING METHODS USED FOR SOFT SOILS IN BANGLADESH

In the process of designing and building structures on soft soils, or when foundation distress or failure is encounterd, the engineer needs to consider whether the soil properties can be economically improved. Methods and applications of ground improvement techniques in Bangladesh have been summarised by Ansary and Doulah (1993). Typical methods used to increase the bearing capacity of soft soils of Bangladesh include preloading (or precompression) and vertical drainage in conjunction with precompression. Loading berms, or pressure berms, are being used as counter weights to increase the stability of road and railway embankments in Bangladesh. Fabric-reinforced soil, where woven or non-woven fibre materials are placed in road and railway embankments to improve their overall stability. Vertical drainage methods using three types of drains, namely, sand drain, wick drain and prefabricated vertical drain are being used. Due to lack of mechanical skills and equipment, other ground improvement techniques like dynamic compaction, admixture stabilization, stone columns, electroosmosis and lime piles and columns are yet to gain popularity in Bangladesh.

5.1 Preloading

Preloading (or precompression) has been used in a number of projects, specially road and railway embankments. The technique of improving soil properties by precompression consists in surcharging the ground with uniformly distributed surface loads prior to construction of the intended structure. The surcharge results in one or more of the three effects, namely, (i) primary consolidation settlement; (ii) secondary compression settlement, and (iii) increased undrained shear strength of the soil. In preloading technique, staged construction is mainly employed as a means of gradually increasing the shear strength of soft clay layer which would otherwise be inadequate to carry the intended structure without failure. Precompression surcharges are usually effected by means of earth fills.

5.2 Sand drains and sand wicks

Sand drains are used for many years to increase the consolidation rate and shear strength of primarily soft clay. They range in diameter from 180 mm to 450 mm and they have installed to great depths. The classical sand drains used in augured holes, however, have the disadvantage of becoming clogged and discontinuous during consolidation process Sand-wicks have also been used in Bangladesh. One of the vertical drains that have been successfully used in Bangladesh is the sand wicks. Sand wicks are of ready-

Figure 3. Installation of precast concrete piles in progress

made small diameter sand drains which are contained in long canvas bags are also used in Bangladesh. The most common diameter is about 65 mm. This type of vertical drain is economical since labour cost are low, since the making and filling of the canvas bags is generally done by hand, and installation in soft clay are accomplished entirely by hand labour.

5.3 Vertical drains with preloading

When a layer of soft clay is very thick or the permeability is exceptionally low, the preloading technique is likely to be inefficient when used alone because an inordinately long period of time will be needed to bring about significant compressions. In these circumstances, radial improvements in the preloading time can be effected by installation of vertical drains to shorten the drainage path under which the clay consolidates. This has become a commonly employed device with soft clays in Bangladesh. The benefits of vertical drains used in conjunction with preloading are related almost entirely to acceleration of the primary consolidation of the clay, since they bring about the rapid dissipation of excess pore pressures. In Bangladesh, sand wicks and prefabricated vertical drains (PVDs) are usually used in conjunction with preloading.

6 CASE HISTORIES OF "CUT AND RE-PLACEMENT" FOUNDATION TECHNIQUE

6.1 Khulna university buildings

A typical example of "cut and replacement" foundation technique is the foundation construction of Khulna University of Bangladesh. Khulna University is located on the south western part of Bangladesh in a lowlying area. Six boreholes driven to depths between 30 m and 49 m were made to assess the soil condition at this 39 acre land. A typical soil bore log is shown in Figure 4. From the boings four distinct layers can be recognized. The top layer about 1 to 4m thick consisted of grey soft clay. Below this layer is found a very soft dark gray and black organic soil with thickness of about 3.5m. The third layer consisted of soft clay with silt and some organic matter up to a depth of 18 m to 21 m resting on a dense sand layer. It is the upper three layers up to depths of 18 to 21 m that is the problematic soil that had in places standard penetration values as low as zero. Within upper 6 m, soil is very soft and has in places water content as high as 400 percent. Presence of organic matter was also evident within this zone. The percent of organic matter in the different boreholes in various depth varied from 3% to 50%. Void ratio close to 3 in some undisturbed samples were also noted. Undrained shear strength of undis-

turbed samples in the soft organic clay layer varied from 2 kPa to 25 kPa.

Two four storied structures, a students residential hostel (Boy's Hostel) and an Academic Building were built in the above location. Foundation for the buildings consisted of shallow continuous footing (raft foundation) over fine sand fill placed after removal of about 4m of soft ground at the surface and the peat layer. Because of the low level of the surrounding area the sand fill was extended to an additional 1.4 m above the surrounding ground level. The filling sand having fineness modulus of 2.2 and 1.2 was mixed at a ratio 1:1 and compacted properly by sheep foot roller. During compaction the optimum water content and the layer of sand which was approximately 230 mm for each compaction was ensured to attain the maximum dry density around 16.5 kN/m³. Mat depth of 305 mm to 457 mm was cast over mixed compacted sand filling. From mat up to plinth, fine sand (fineness modulus = 0.8) was used as filling materials. Details of the foundation system is shown in Figure 5.

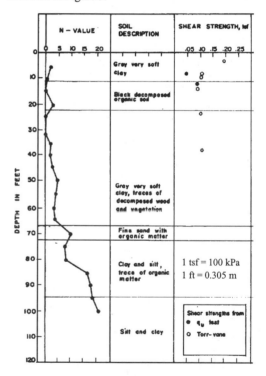

Figure 4. Typical borelog at Khulna university site

The construction of mat foundation of the Academic Building commenced in November 1992 and was completed in February 1994. Razzaque & Alamgir (1999) investigated the long-term settlement of the Academic Building. To record the settlement, some settlement plates were installed in the bottom layer of sand filling as shown in Figure 5. Figure 6 shows an average of the recordings obtained from all the settlement plates. 760 mm settlement occurred during the last six years based on the last recording on 16th March, 1999. The result showed that most of the settlement, which was around 508 mm, occurred during the first 1.5 years. The rate of settlement decreases as the elapsed time increases. The settlement remained constant during the last four recording. This study revealed that the observed settlement is more than 3 times than that of the predicted. Due to the replacement of peat layer by compacted sand cushion, the load on the subsoil below increased near about 3 times. The compacted sand cushion with mat foundation consolidated the underlying soft fine grained soil layer leading to downward movement. The measurements obtained from different settlement plates and plinth indicate that almost uniform settlement occurred. Settlement readings were also taken for 221 days from November 3, 1994 to June 11, 1995 at eight locations after the completion of the Boy's Hostel. Figure 7 shows settlement records at 8 locations. Although settlement was still continuing the maximum settlement was found to be of the order of 92 mm. Settlement of these points have been relatively uniform. The maximum differential settlement limited me maximum angular distortion to a value of 7.7×10^{-4} which is very much within the allowable limit for most framed structures. Photographs of the Academic Building and Boy's Hostel are shown Figures 8 and 9, respectively.

The above demonstrate that "cut and replacement" method has successfully reduced total and differential settlement of low rise buildings built over a soft clay. Here the sand filling acted as stiffening layer distributing the building load uniformly over the soft clay.

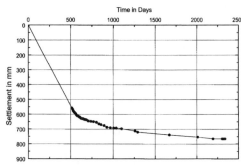

Figure 6. Time-settlement observation of academic building at Khulna university (after Razzaque & Alamgir 1999)

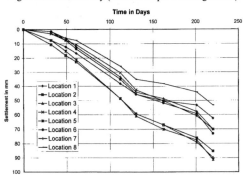

Figure 7. Time-settlement observation of boy's hostel at Khulna university (after Razzaque & Alamgir 1999)

Figure 8. Academic building of Khulna university

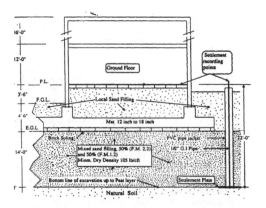

Figure 5. Foundation system of the 4-storied academic building at Khulna university (after Razzaque & Alamgir 1999)

Figure 9. Boy's hostel of Khulna university

6.2 Khulna medical college buildings

Kabir et al. (2000) reported on the design and constructional aspects of geotextile included granular mattress foundation of Khulna Medical College (KMC) of Bangladesh using "cut and replacement" technique. The granular mattresses were constructed up to 2 m thick. The philosophical and analytical bases of the project came from previous research (Kabir et al. 1992). The functions of the granular mattress for the KMC building foundations may be described as the following. (i) increase in bearing capacity, (ii) distribution of stress over a large area and (iii) minimize total and differential settlement. The geotextile layer acted as separator, filter, containment layer and reinforcement. As separator the geotextile prevented mixing of clean aggregate layer with the clay underneath. This function was crucial during the vibro compaction stage but also quite dominant during the full loading of the footings.

The subsoil in the area is composed of recent soft soils with interlayers of decomposed and partially decomposed vegetative organic matters originating from subsidence of mangrove forest. The DPL (Dynamic Probing Light) tests were conducted to ascertain the depth of soft top layer more precisely than that revealed by SPT.

Non existence of bearing layer at moderate depths poses problems for construction of buildings between 4 and 6 stories, which are the more desired types in this area. Buildings above 6 stories are founded on piles whose length often exceed 30 m.. Granular mattress foundations were envisaged, designed and constructed for KMC buildings for cost and time savings, ease of construction by available local technical skill with the prospect of easy adaptability and sound load carrying capacity, settlement and distortion. Initially four buildings, including an Academic Building , Nurses Training Centre, Girls Hostel and Boys Hostel were constructed. The Academic Building has a framed structure with wings 3 to 5 storied in height. The other three buildings are load bearing wall types with height between 4 and 5 stories. Photograph of the Girls Hostel buildings is presented in Figure 10.

Figure 10. Girl's hostel building of Khulna medical college (after Kabir et al. 2000)

The granular mattresses were constructed by laying a nonwoven needle punched geotextile on top of the soft soil at the base of excavation, 1.8 m to 2 m in depth. A graded mix of crushed brick aggregates and coarse sand was placed on the top of the geotextile. The aggregate layer consisted of 2 parts crushed brick aggregates and 1 part coarse sand. The crushed brick aggregate consisted of 25 mm down graded aggregate conforming to the ASTM grading for concrete aggregates. The coarse sand consisted of river sand from the north east of the country, called Sylhet Sand, having fineness modulus ≥ 2.5. The aggregates were vibro compacted, in layers, to desired density, by using twin drum vibratory rollers. A layer of locally available fine river sand was placed by vibro compaction on top of the coarse aggregate layer. The sand fill layer consisted local river sand having fineness modulus greater than 1.0. Fines passing number 200 sieve was limited to 5% for fineness modulus up to 1.5 and 10% for fineness modulus greater than 1.5. The densities of the materials were monitored by Transport Research Laboratory dynamic penetrometer tests. A geotextile separator and filter layer was placed at the bottom of the excavation on the soft clay layer. A nonwoven needle punched geotextile was used. The weight, grab tensile strength and permeability was greater than or equal to 200 g/m^2, 750 Newton and 1 x 10^{-3} m/s, respectively. Typical cross sections for the mattress for load bearing wall footing and continuous column footing for the framed structure are presented in Figures 11 and 12, respectively. Continuous inverted Tee beam type footings were used for load bearing walls and continuous tapered footings were used for column foundations of framed structures. A beam on elastic foundation program based on finite element analysis developed by Hulse and Mosley (1986) was used to analyze both the types of footings. Settlement and distortion under working load dictated the proportioning and design of the footings. The deflection of the wall and column footings are presented in Figures 13 and 14, respectively.

Four cases of foundations were considered to provide a comparative representation of cases with and without mattress foundation as well as those under drained and drained conditions. The results in Figure 13 is for an inverted Tee beam 1.45 m wide, 33.2m long having 250 mm wide web and 300 mm thick flange. The results show that provision of the mattress will reduce the settlement to less than 1/3. This is also the order of values for differential settlement for this case. Figure 14 shows a typical strip footing for carrying column loads. The footing is 5.3 wide, 410 mm deep and 27.3 m long. It can be seen that provision of mattress foundation more than halved the total settlement as well as the differential settlement of the footing.

The construction of the buildings was completed in 1997. Settlements at gauge points were recorded time to time, but not on a regular basis. The order of the values of maximum settlement agreed well with the predicted values.

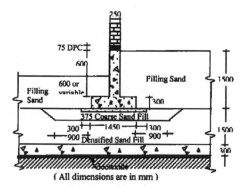

(All dimensions are in mm)

Figure 11. Details of wall footing (after Kabir et al. 2000)

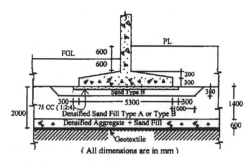

(All dimensions are in mm)

Figure 12. Details of column footing (after Kabir et al. 2000)

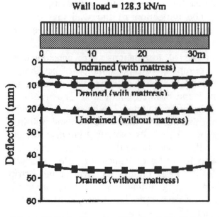

Figure 13. Deflection of wall footing (after Kabir et al. 2000)

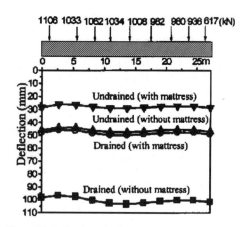

Figure 14. Deflection of column footing (after Kabir et al. 2000)

7 CASE HISTORY OF PRELOADING

Ali (1993) reported a case study on soil improvement method used for a factory building in Pabna. Pabna is situated in the south-west region of Bangladesh. The structure of the factory consists of two and three storey frame type of structure. The factory consists of three blocks, Block-A, Block-B and Block-C. Adjacent to Block-B is located the Ice Flaking Unit which is a five storey high framed building. A total of eight boreholes were drilled each up to depth of 13 m to determine the soil characteristics at the project site. The soil at the top clay layers has low undrained shear strength (22 kPa to 25 kPa) and bulk density and high void ratio, water content and compression index values. The soil improvement technique consisted of replacement of the fill and top sand layer by compacted sand layer and preloading of the whole foundation area. Settlements of foundations due to building load before preloading were estimated and it was found that the predicted settlement of foundations varied from 105 mm to 165 mm.

The factory covers an area of approximately 1045 m². Preloading the entire area at a time required huge amount of loading materials. Therefore, it was decided to carry out preloading operations in three steps. First the entire area covered by Block-B and half of the corridor block were loaded. After completion of preloading operation in Block-B, materials were shifted to Block-C and the remaining half of the corridor block. Finally, materials were shifted to block-A for preloading. In this way material requirements for preloading were significantly reduced and the materials purchased for the construction of the factory could be used for preloading. Moreover, since the construction of different blocks could be carried out independently, the time required for completion of the project was not significantly affected. The Ice Flaking unit was preloaded in the last stage. For all preloading operation loads were placed beyond the exterior column lines by about 2.5 m for effective consolidation of the soil in these areas. For all loaded areas bricks as well as crushed stones and sands contained in gunny bags were used to apply preload. An average stress of about 58 kPa could be produced in all loaded areas by applying preload in this manner. Preloading was performed in four increments. As these materials could be handled easily, application and removal of loads for preloading was quite fast.

Settlement during preloading of blocks B, C, A of the main building and also of the ice flaking unit, were measured by installing settlement markers. For all the loaded areas, the time of settlement was found to be very high as the settlement occurred instantaneously with the application of load. Maximum settlement of around 25 mm was recorded after 45 days of loading. When the settlement had stabilised after the application of the final load, the loads were removed. The area was then cleaned and the foundations were constructed. It was found that after the completion of preloading N-value from SPT increased in the upper layers. The construction of the factory building was completed in 1992.

8 CASE HISTORIES OF PRELOADING WITH SAND WICKS

8.1 Goran Land Project (Phase - I), Dhaka

Dastidar (1989) reported on ground treatment and foundation at Goran, Dhaka. Goran land project is located in the low lying areas of the eastern part of Dhaka city. Filling up of the phase I of the project was started in 1985 by hydraulic fill. The hydraulic fill material was obtained by cutter suction draggers from low lying areas and transported by pipelines. The hydraulic fill materials consists of silts and clays at the upper levels and silty sands at the lower level.

Towards the end of 1986 when the land at Goran Project was raised to within the few feet of the proposed finished level a sub-soil investigation of the site indicated the ground conditions to be poor. From the borehole logs, it appeared that there had been a wide variations in the properties of the recent fill and also at the upper part of the virgin ground. In the upper 6m of the virgin ground two types of soils were noted, namely, (i) firm dark grey peaty clay (0.6 m to 1 m thick) overlying 4.5 m to 6 m soft bluish grey clay and (ii) firm to stiff mottled brownish grey clay. The undrained shear strength of the soft clay samples were less than 30 kPa with high compression index of 0.6. From the subsoil conditions, at the end of the filling in early 1988, the following problems were anticipated.

(i) The soil up to 4.5 m depth is incapable of supporting more than 20 kPa load.
(ii) The virgin ground below the recent fill will settle 150 mm to 300 mm under the weight of the fill.
(iii) Under the load of a 5-storied building the ground may settle 0.6 m to 1.2 m.
(iv) Piles will be more than 15 m long.

After considering different options of ground improvement it was found that in order to strengthen the weak sub-soil preloading with sand wicks would be most effective. In this project sandwicks having 64 mm diameter with 1.22 m centers were used up to a depth of 6 m to 12 m. For preloading, soil heaps and brick stacks were used. Settlement was recorded at 12 locations three to five times every week. It was noted that more than 90% consolidation took place within 4 to 5 weeks of application of load at each stage. This was in fairly good agreement with the theoretical estimate. Piezometers installed at 3 m and 7.5 m depth showed the excess pore pressure steadily reduced with time. Standard penetration tests before and after preloading showed substantial improvement in the ground conditions. Tests on samples recovered from boreholes after preloading showed an average value of undrained shear strength of 45 kPa.

8.2 Embankments at the Dhaka export promotion zone area in Savar, Dhaka

Construction work for embankments in Savar, Dhaka within the South and Western end of the Bangladesh Export Processing zone area was started in January, 1994. During the construction period the earth fill at two areas subsided in April, 1994. The construction work was stopped and an investigation was carried out to assess the causes of subsidence.

Figure 15 shows embankment section and soil condition at a typical section. It can be observed that below the embankment, a very soft to soft clay layer exist. Thickness of this layer varies in places from 1 m to 7 m depending on location. The N-value of this layer is as low as zero at some locations. It was therefore concluded that low shear strength of this clay layer contributed to the failure. The undrained shear strength and extent of this soft layer is the main concern in the embankment construction and its stability. Table 2 shows typical geotechnical properties of a soft organic sample collected from 4 m depth. Stability analysis fill height of 7m and undrained shear strength of the foundation soil of 15 kPa provided a factor of safety of less than 1.0. It was therefore necessary to increase the shear strength of the foundation soil (soft clay layer) by consolidation and preloading before 7 m of embankment fill is placed.

Table 2. Properties of a soft organic clay sample

Properties	Values
Natural water content	346 %
Liquid limit	220
Plastic limit	57
Plasticity Index	163
Initial void ratio, e_0	4.52
Compression index, C_c	2.1
Coefficient of consolidation, c_v in cm^2/s	1 to 11 x 10^{-4}

Figure 16. Steps for construction of embankments

In order to accelerate the consolidation of the soft clay, 65 mm diameter sand wicks encased in jute cloth at a triangular spacing of 1.5 m centre to centre was recommended. Installation equipment for this type of cores are available with local contractors. This can be manufactured locally as well. The following steps were recommended for the reconstruction of the fill:

Step 1: Removal of excess soil
Step 2: Installation of sand wicks
Step 3: Placement of 300 mm thick coarse sand (enclosed in jute fabric of adequate strength)
Step 4: Placement of 1m fill on jute fabric with compaction of soil at 0.3 m interval
Step 5: Construction of the embankment up to the required level would be carried out, after a waiting period of 9 to 10 months for allowing consolidation.
Figure 16 shows steps for construction of the embankment using sand wicks.

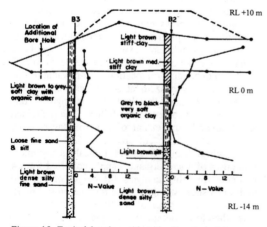

Figure 15. Typical borelog with embankment section at Dhaka EPZ site, Savar

9 CASE HISTORIES OF PRELOADING WITH PREFABRICATED VERTICAL DRAINS IN ROAD AND RAILWAY PROJECTS

9.1 Soil improvement study for Jamuna bridge access roads project

The subsoil of part of the alignment namely the Kaliakoir Bypass Road (3 km stretch) in Jamuna Bridge Access Roads Project, consists of very soft clay deposits with peats. The proposed bypass runs through 7 sections of soft ground (soft organic clays). The total length of the soft ground sections in the 3 km stretch bypass is approximately 1.4 km. The total thickness of the soft clay layers varies from 4 m to 13 m. Figure 17 shows a photograph of a part of the site of the Jamuna access road project. Figure 18 presents borelogs along a typical section showing the soil profiles. Comprehensive laboratory tests were conducted to determine the strength and compressibility characteristics of soft organic clays. Results indicated values of compression index, initial void ratio and natural moisture content of soft clay samples were as high as 4.5, 7.2 and 500%, respectively. The undrained shear strength of the soft samples varied from 2 kPa to 25 kPa.

From stability analysis of embankment on unimproved ground, it was found that the embankment (inclusive of the extra fill required to compensate for consolidation settlement) that is constructed to its proposed design level on the unimproved soft clay areas has a factor of safety of less than 1.0. Table 3 shows the calculated factor of safety for various heights of embankment and limit height of the first stage embankment determined based on the minimum factor of safety 1.2. The limit height of the embankment on the unimproved soft ground varies from 5.5 m to 8 m. It was found that the limit height is less than the proposed height of embankment and staged construction is required for all the soft ground sections except the section CH 2570 to 3000.

Figure 17. Jamuna bridge access road project site at Kaliakoir

soft ground, some sort of soil improvement was therefore required to construct the proposed embankment and to limit the post construction settlement to an acceptable level. The soil improvement proposals for the soft ground sections is shown in Table 4, depending on the time required for the expected consolidation settlement. Preloading with or without vertical drains appeared to be the most economical method of the soil improvement for this project. Installation of vertical drains for the sections where the time for nearly completing the consolidation settlement is in an order of 5 to 10 years or more has been recommended. For the sections where the time required for completing most of the consolidation settlement is in an order of 0.5 to 2 years, preloading without vertical drain has been recommended.

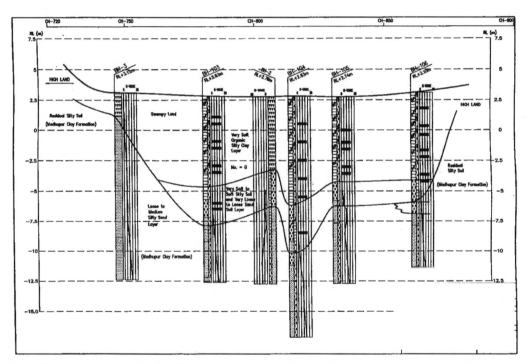

Figure 18. Typical soil cross profile along the soft ground alignment at Jamuna bridge access road project site at Kaliakoir (after Kiso-Jiban 1999)

Based on the soil profiles, the maximun length of the drainage path for the soft ground is in a range of 1 m to 6.5 m. The estimated time required to achieve a degree of consolidation of 90% is 0.3 to 2 years if the length of the drainage path is less than 2m for sections CH 520 to 550, CH 1865 to 2361 and CH 2570 to 3000. The estimated time for the 90% consolidation is more than 5 years for sections CH 415 to 490, CH 740 to 880, CH 1015 to 1145 and CH 1570 to 1660, where the length of the drainage path is greater than 4.5 m. In these seven sections of the

The use of vertical drains can shorten the time required for consolidation settlement. If the vertical drains are used with a spacing of 1.0 m. The time required to achieve the 90% consolidation can be shortened to 6 months regardless the actual thickness of the soft clay layer. The above estimate is based on a coefficient of consolidation for vertical drainage, c_v, of 3×10^{-4} cm^2/sec and coefficient of consolidation for horizontal drainage, c_h, of 6×10^{-4} cm^2/sec. The predicted magnitude of final settlement for the proposed embankment varies from 0.5 to 3.2 m de-

Table 3 Factor of safety of embankment on unimproved ground and limit height of embankment

Section	Factor of safety of embankment (H = Height)			Limit height of first stage filling (m)	Required thickness of fill to be placed (m)
	H=5m	H=6m	H=7m		
CH 415 to 490	1.30	1.10	-	5.5	9.8
CH 520 to 550	1.41	1.25	1.15	6.5	7.3
CH 740 to 880	1.30	1.10	-	5.5	11.0
CH 1015 to 1145	-	1.45	1.24	7.0	12.2
CH 1570 to 1660	1.35	1.13	-	5.5	8.9
CH 1865 to 2361	1.48	1.28	1.13	6.5	7.7
CH 2570 to 3000	-	-	1.36	8.0	6.9

Table 4 Proposed methods of construction for soft ground areas in Jamuna bridge access roads project

Section	Expected settlement (m)	Total thickness of soft clay (m)	Time required for 90% consolidation with no vertical drain (Years)	Method of soil improvement
CH 415 to 490	2.0	9	5 to 10	Preloading with vertical drain
CH 520 to 550	0.5	4	1.5 to 2	Preloading
CH 740 to 880	3.2	12	10 to 15	Preloading with vertical drain
CH 1015 to 1145	3.0	11.5	10 to 15	Preloading with vertical drain
CH 1570 to 1660	2.9	12.7	10 to 20	Preloading with vertical drain
CH 1865 to 2361	1.7	9.5	1 to 2	Preloading
CH 2570 to 3000	1.0	3.7	0.3 to 0.5	Preloading

pending on the ground conditions and thickness of the fill placed.

Soil improvement using vertical drains would be carried out together with multi-stage (6 steps) construction of the proposed embankment on soft clay ground. The objective of installing the vertical drain is to speed-up the rate of consolidation of the soft clay soil and thereby increasing the rate of strength gain in the soft clay layer. Before installing the vertical drain, construction of a 1.5 m thick sand mat at the ground surface to assist the dissipation of excess pore water pressure in the soft clay layer. Vertical

drains shall be installed at a spacing of 1m in a square pattern in the sections CH 1015 to 1145 and CH 1570 to 1660. The vertical drain shall be installed to the top of the Madhupur Clay layer or the top of the very loose to loose sandy soil layer immediately above the Madhupur Clay layer. Based on the above vertical drain spacing, the preload or surcharge fill shall be maintained for at least 6 months (during the rainy season) to allow the soft clay to achieve 90% of consolidation. It is recommended that a 0.5m high surcharge be applied to cater for traffic loading under service condition. It is recommended to construct counter berm at both sides of the embankment for the sections CH 415 to 490, CH 740 to 880 and CH 1015 to 1145. The counter berm is needed to improve the stability of the embankment during the critical construction period when the positive effect of the soil improvement measures has not been fully achieved. The proposed width of the counter berm is 5 m to 30 m. A part of the counter berm will be removed later, anytime after the maintenance period of the second stage filling of 6 months. The counter berm for the section CH 1015 to 1145 can be removed to original embankment shape after the maintenance period. It is also recommended that two geo-synthetic layers, a layer of geo-textile and a layer of geogrid, be placed in the first stage fill. The first geo-textile layer shall be laid before placing the sand mat. This layer of geotextile will act as a separation layer to prevent loss of sand fill into the soft clayey soil. It is proposed to lay the layer of geo-grid at approximately 3 m above the ground level. This layer of geogrid will function as a reinforcement layer to prevent slope failure from occurring within the fill during the construction of the embankment. The proposed sequence for the multistaged construction of the embankment is illustrated in Figure 19.

To monitor stability and consolidation settlement of the embankment during the construction period lateral displacement pegs and settlement plates were recommended respectively. A row of 4 lateral displacement pegs shall be installed on the adjacent ground at each side of the embankment. The spacing between the pegs is maintained at 2.5 m and 10 m. The length of each peg shall be 3 m and installed to a depth of 2.7 m below the ground surface. Two rows of lateral displacement peg (one row at each side of the embankment) shall be installed along one cross-section of the embankment. Proposed interval between instrumented section shall every 50 m along the road alignment for embankment that is constructed on area with soft clay soil only. A row of 5 settlement plates shall be installed along the section of the embankment. One plate shall be installed at the center of the embankment and the remaining four at both side of the crest. One row of settlement plates shall be installed along one cross-section of the embankment. Proposed interval between instrumented section shall every 25m along the road alignment for embankment that is constructed on area with soft clay soil only. The proposed layout of the settlement plates and lateral displacement pegs are schematically shown in Figures 20 and 21, respectively.

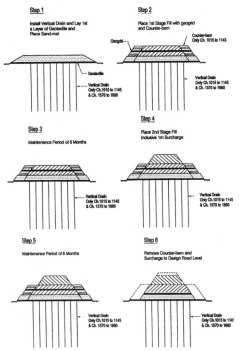

Figure 19. Construction sequence of embankment at Jamuna bridge access road project site at Kaliakoir (after Kiso-Jiban 1999)

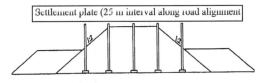

Figure 20. Schematic layout of settlement plates in embankment at Jamuna bridge access road project site at Kaliakoir (after Kiso-Jiban 1999)

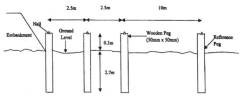

Figure 21. Schematic layout of lateral displacement peg in embankment at Jamuna bridge access road project site at Kaliakoir (after Kiso-Jiban 1999)

9.2 Dhaka Integrated Flood Protection Project

In the wake of the floods of the 1987 and 1988, the Government of Bangladesh established a Committee for Flood Control and Drainage of Greater Dhaka in October, 1988. The primary objective of the Committee was the preparation of a flood control plan for the Greater Dhaka area. In January 1989, the Committee submitted a detailed plan for phased investments in flood protection and drainage for Dhaka and other cities, which was approved by the Government of Bangladesh in March, 1989. In view of the high priority assigned to the Dhaka flood protection project, the Government of Bangladesh immediately initiated Phase I of the recommended works on a crash program basis using their own resources. These works, which were designed to provide protection for about 137 sq. km in the highly urbanized western part of the city of Dhaka, included, among others, the construction of: (i) a 29.2 km of embankment along the west side of the city, and (ii) an 8.5 km of reinforced concrete wall bordering the densely populated south-western side of the city. Figure 22 shows the location of Dhaka Integrated Flood Protection Embankment while Figure 23 shows a photograph of the embankment.

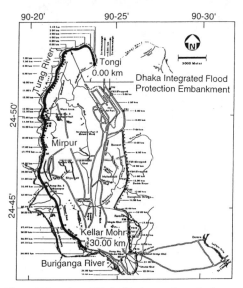

Figure 22 Location of Dhaka flood protection embankment

Ansary et al. (1998) reported the results of recent soil investigations carried out at the Dhaka Integrated Flood Protection Embankment site. The subgrade soil profile beneath the western embankment was found to be fairly consistent along the embankment alignment. The subgrade soils consisted of an upper 1m to 30 m layer of soft clayey silt with high plasticity or non-plastic clay with silt. This layer was underlain by medium dense silty sand or sand at

243

depth. The clayey silt or silt layer encountered below the embankment varied in thickness from about 1m to 30m. Undrained shear strength varied between 25 to 50 kPa. Initial void ratio, compression index and coeficient of consolidation (c_v) varied from 0.78 to 0.89, 0.06 to 0.09 and 0.001 to 0.008 cm^2/s, respectively. The upper layer of clayey silt to silt soils were interbedded with very soft, high plasticity organic clays or silts at several locations along the embankment alignment. These soil layers were typically less than 3m in thickness. The organic clay and silt layers were of high plasticity, very weak and highly compressible. The upper layer of clayey silt and soils are underlain at depth by silty sand and sand layer.

Figure 23. A section of Dhaka flood protection embankment

A damage survey was performed by Louis Berger International Inc. and Geosyntec Consultants in May 1991 and October 1991. Among other things they concluded that parts of the embankment, totaling about 4.7 km may be subjected to sudden failure resulting from inadequate subgrade shear strength. These areas were classified as Class I areas requiring immediate remedial action. In an additional 3.1 km of the embankment deep foundation failure was not likely. These areas were classified as Class II and required short-term remedial action. In remediation of Class I areas use of synthetic prefabricated vertical drain and high strength geosynthetics were recommended. Only synthetic prefabricated vertical drains were recommended for Class II areas. The wick drains consisted of a continuous polypropylene drainage core wrapped in a needlepunched non-woven geotextile. The wick drains in the Class I area to be installed to an average depth of 23m. It was found that if the vertical drains were spaced at about 1.5 m on center, 90 percent consolidation would be achieved in about 10 months. The critical mode of failure changed from deep circle to a shallow circle at an average degree of consolidation of about 33 percent, which would be achieved in about 3.5 months. In Class II areas requiring subgrade improvement, vertical drains would be installed through the existing embankment. The wick drains

would be extended either to the top of the sand layer or to the top of the medium stiff clayey silt layer. After the embankment reaches about 90 percent consolidation under the existing load, the embankment will be extended to the final elevations. Class II areas requiring monitoring and inspection have been remediated by constructing toe berms, flattening the slopes and reconstructing the embankment. The previous remedial actions in these areas have increased the factors of safety and reduced the probability of failure. While the existing factor of safety might still be low, it was anticipated that an acceptable factor of safety of 1.2 would be achieved in these areas over time. In addition, failure of these sections of the embankment, if it did occur, would not likely be catastrophic. Therefore, it was recommended that a monitoring and inspection program be developed.

Under the backdrop of very expensive remedial measures suggested by the bridging period consultants, a pilot project proposal for use of jute fibre drains (JFD) was prepared by the Civil Engineering Department of the Bangladesh University of Engineering and Technology (BUET), Dhaka in 1993. The BUET took the initiative for a number of reasons. These are described in the following:

(i)   Developing a very cost effective solution for the problem of Dhaka embankment.

(ii) Development of a technology which will bring substantial cost saving in soil improvement (land development) technology in Bangladesh. Cost of construction of embankment on soft soils and hydraulic structures like gates, pump houses, barrages, etc. on soft soil may be substantially minimized.

(iv) Bring confidence in soil construction technology, where there is lack in confidence in the field of hardware development and their proper use. As part of this project vertical drain installation technology will be developed.

(v)   Development of analytical and testing capability of the BUET in the area of vertical drains, which will remain available for future projects involving vertical drains.

It was also decided that a preliminary trial work of sample jute drain production by Bangladesh Jute Mills Corporation (BJMC) and trial installation of a number of drains will be conducted by Mechanical Engineering Directorate of the Bangladesh Water Development Board. This will be done under the supervision of the BUET, before the commencement of the real pilot project. As part of trial production exercise the BJMC produced a number of samples of drains using jute rope core and jute fabric sleeve, replicating samples brought from Singapore. A series of laboratory tests were performed on the drain materials and JFDs. These results have been reported by Kabir et al. (1994). The trial installations were conducted in May, 1994 at the Dhaka Embankment. A total of seven JFDs were installed up to a depth of

5 m under the supervision of BUET Box type mandrels having a cross-section of 125 mm x 25 mm and wedge shaped steel carrier shoes were used. A 20 ton tire mounted crane with a hydraulically operated vibratory hammer was used satisfactorily to perform the job. Figure 24 shows the complete jute drain made of Plain Hycess jute fabric filter and coconut coir strands. It consists of two layers of jute burlap (Plain Hycess Jute fabric) wrapped around four of coconut coir strands, held together by three continuous longitudinal stitches. Its width is about 100 mm, and thickness varies from 10 to 14 mm. It weighs about 525 g/m. Average Grab tensile strength of jute drains was 5290 kN and average elongation at break was about 12 percent.

It was anticipated that JFDs with coconut wire would be installed in the Class I and Class II areas. The purpose of the jute drains was to provide vertical drainage, thus reducing the length of the drainage path. Since the undrained shear strength of the soil is closely related to the degree of consolidation of the soil, providing vertical drainage increases the shear strength of the soil with time. Increase in undrained shear strength of the sub-soil of embankment already occurred due to preloading for the last decade. Compared with the previous investigation (Technoconsult 1994), factor of safety of the embankment section also increased as reported by Siddique et al. (1998). It was therefore concluded that jute drain or other wick drains may not be necessary for enhancing vertical drainage in order to achieve sufficient shear strength of the embankment sub-soil.

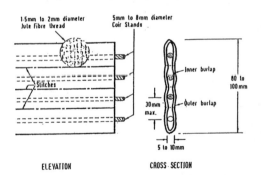

Figure 24. Complete jute drain made of jute fabric filter and coconut coir strands

9.3 South-west road network development project

This project consists of Gopalgonj Bypass (6 km long) and Noapara-Mollarhat road section (15 km). The project road alignment traverses several stretches of peat soft ground which fall mostly on the swampy area. Thin layers of problematic peat (0.3 to 1.5 m deep) exist at several areas near to surface, which are a part of soft sub-soils. Thickness of soft sub-soils along Gopalgonj Bypass and Noapara-Mollarhat road was found vary between 5 m and 16

m. Moisture content of the peaty soft found was as high as 375%. Primary consolidation settlements and consolidation time of the peaty soft ground caused by loads transmitted with the proposed embankment heights were estimated. The estimated settlements showed maximum 85 cm depending on embankment heights and thickness of peaty soft ground. It was found that in all road sections 70% consolidation settlements will reach within two years while 90% of consolidation will take maximum four years.

It was anticipated that the proposed height of the embankment will be flattened and cause critical excessive settlements due to peat soils. To overcome this problem two stages of filling operation has been recommended. Figure 25 shows detailed embankment construction sequence in peaty soils.

In case of very soft ground, installation of PVDs has been recommended to accelerate consolidation settlement in order to countermeasure instability of high embankments on very soft ground. Close monitoring of settlement and movements of filling and surrounding soils using settlement plates has also been recommended. Detailed embankment construction procedure using PVDs is shown in Figure 26.

9.4 Jamuna bridge railway link project

The embankment for the Jamuna Bridge Railway Link Project at Kaliakoir, Dhaka runs through 8 zones of soft ground. The total length of the soft ground zones in the Jamuna Bridge Railway Link Project is 4.05 km. The maximum height of the embankment to be constructed is 6 m. Some sections of the embankment have been designed with lateral berm. The soft ground consists of very soft organic clay. The thickness of the soft organic clay layers varied from 3.5 m to 14 m in the 8 zones. The N-value from SPT in these soft layers varied from 0 to 3. The values of compression index, initial void ratio in the compressible layers varied from 0.4 to 0.65 and 1.2 to 1.6, respectively. Primary consolidation settlements and consolidation time of the organic soft ground caused by loads transmitted with the proposed embankment heights were estimated. The estimated primary settlements for the 8 zones varied between 33 cm and 148 cm depending on the thickness of the compressible layer. It was found that without installation of any drain 80% consolidation settlements will reach within 5 to 70 years, depending on the thickness of the compressible layer. The coefficient of consolidation for vertical drainage and radial drainage were taken as 2.5×10^{-4} cm^2/s and 5.0×10^{-4} cm^2/s, respectively.

Due to bad soil quality and poor mechanical properties, it has been recommended to construct the embankment in 2 stages. The critical height of the first stage construction should provide a factor of safety of at least 1.3 by short time. The vertical drain spacing for the 8 soft ground zones has been calculated to obtain al least 80% consolidation in four months. This time has been considered to be sufficient to improve enough supporting ground and construction of the next stage. The recommended vertical drain spacing for square pattern and equilateral triangular pattern varied from 1 m to 2 m and 1.07 m to 2.15 m, respectively. Figure 27 presents a photograph showing the site with installed PVDs.

Close monitoring of settlement and pore pressure has been recommended. Settlement will be measured using settlement gauges placed at the surface of natural ground. Pore water pressure wiil be measured with gauge drilled in the compressible layers at various levels. These apparatus are essential to check the soil behaviour and to prevent failure in case of stage construction. Each measurement profile will be composed of 2 settlement gauges and to water pressure gauges, in the centreline and drilled at 2.5 m and 5.0 m depth.

complex, most of the country can broadly be divided into older deposits (raised alluvium or terrace deposits) and recent deposits (on flood plain and estuaries). In both the formations percentage of silt sized materials predominate.

The principal foundation problems in Bangladesh are related to the low resistance of the underlying soil to shearing resistance. Settlement is another major foundation problem in Bangladesh related to the loose and compressible nature of the soil. Major cities of Bangladesh are facing foundation problems on soft grounds. Structural foundations on soft soils in this country are limited to use of raft foundations, piled footing and well supported footings. Piled foundations using timber piles and reinforced concrete piles are popular. Because of low labour costs, large diameter cast-in-place (bored) piles are normally used in soft clays to support heavy concentrated loads. For smaller loads, timber piles are economical and, therefore, often used. The "cut and replacement" technique has been used recently in a

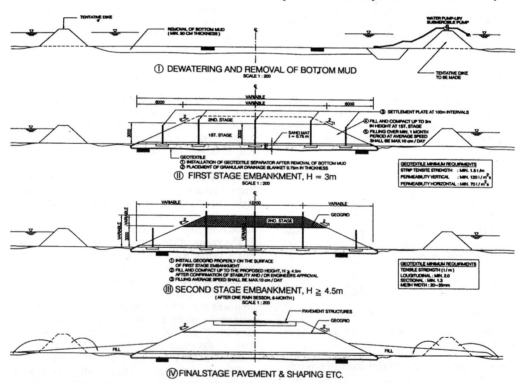

Figure 25. Detailed embankment construction sequence in peaty soils for south-west road project at Gopalgonj

10 CONCLUSIONS

Bangladesh is formed by three major rivers whose sedimentation is controlled by climate and hydrologic conditions, fluctuation of sea levels and tectonic activities. Although the soil formations are

number of building construction projects. In these constructions, the top soft soils, often organic layers, are excavated and replaced by river sands. Spread footings or mat foundations are then constructed. "Cut and replacement" method increases the allowable bearing pressure and achieve uniform stressing of underlying soft layers, thereby successfully re-

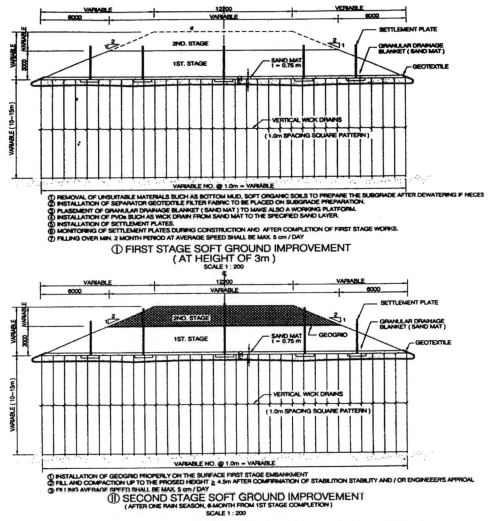

Figure 26. Detailed embankment construction procedure using PVDs for south-west road project at Gopalgonj

Figure 27. A section of Jamuna bridge railway link project site with installed PVDs

duced total and differential settlement of low rise buildings built over a soft clay. Typical methods used to increase the bearing capacity of soft soils of Bangladesh include preloading (or precompression) and vertical drainage in conjunction with precompression. Different vertical drainage methods using sand drains, sand wicks and prefabricated vertical drains (PVDs) are usually used. Case histories of a number of completed and on-going construction projects, including buildings, roads and embankments, have been presented to demonstrate the salient features of soft ground engineering in Bangladesh. It appeared that vertical drains in conjunction with preloading could be an effective solution for the improvement of soft clay deposits of Bangladesh.

ACKNOWLEDGEMENT

The authors gratefully acknowledge the support of Prof. M. Humayun Kabir and Mr. Md. Jahangir Alam of the Civil Engineering Department, Bangladesh University of Engineering and Technology, Dhaka for providing necessaary technical information on the foundation for Khulna medical college for presentation in this paper. The authors express their deep gratitude to Mr. Zafar Sadeq Chowdhury, Senior Pavement and Materials Engineer, Mott MacDonald, Dhaka for providing the necessary data on the soil improvement study for Jamuna Bridge Access Roads Project for presenting in this paper. The authors are also grateful to Mr. M.A. Aziz, Director, Bangladesh Consultants Limited for his sincere support in providing the relevant necessary technical information on the soil improvement for South-West Road Network Development Project and Jamuna Bridge Railway Link Project.

REFERENCES

Alam, M., 1971. Tectonic classification of Bengal Basin. *Bull. of the Geol. Soc. of America*, Vol. 83.

Alam, M.K., Hassan, A.K..M.S., Khan, M.R. and Whitney, J.W., 1990. Geological map of Bangladesh. Geological Survey of Bangladesh, Dhaka.

Ali, M.H. 1993. Soil improvement bu preloading – A case study. *Proc., First Bangladesh – Japan Joint Geotechnical Seminar on Ground Omprovement*, 59-65.

Ansary, M.A. & Daulah, I.U. 1993. Ground improvement – State of the art report and application prospect in Bangladesh. *Proc., First Bangladesh – Japan Joint Geotechnical Seminar on Ground Omprovement*, 1-25.

Ansary, M.A., Siddique, A., Al-Hussaini, T.M., Kabir, M.H. & Islam, S. 1998. Soil investigations at a Dhaka Integrated Flood Protection Project site. *Proceedings of the 13th South East Asian Geotechnical Conference, Taipei, Taiwan, Republic of China*, 13-18.

Bangladesh Transport Survey 1974. The soils of Bangladesh. *A report by the Economist Intelligence Unit Ltd. in association with Scott Wilson Kirkpatrick & Partners*. Submitted to the Government of the People's Republic of Bangladesh.

Bakhtine, M.I., 1966. Major tectonic in Pakistan, Part 2, The Eastern Province. Science. Ind., 4(2).

Casagrande, A., 1976. Liquefaction and Cyclic Deformation of Sand-Critical Review. Harvard Soil Mech. Ser., 88. Harvard Univ. Press, Cambrige, Mass., USA.

Committee of Experts, 1979. Seismic zoning of Bangladesh and outline of a Code for earthquake resistant design of structure. Geol. Surv. Bangladesh. Final Report. Comm. Experts on Earthquake Hazard Minimisation, Dhaka.

Dastidar, A.G. 1989. Report on ground treatment and foundations at Goran Phase I, Eastern Housing Limited

Haque, M., 1982. Tectonic set-up of Bangladesh and its relation to hydrocarbon accumulation, Phase-1. Center for Policy Research, Dhaka Univ., Univ. Field Staff Int. (UFSI), USA.

Harza Engineering Co. Int., Sir M. MacDonald and Partners Ltd. (UK), Meta Systems Inc. (USA) and Engineering and Planning Consultants Ltd. (Bangladesh), 1987. Geology of Bangladesh. Master Plan Organization (MPO), Minist. Irri-gation, Water Development and Flood Control, Government of Bangladesh, MPO Tech. Rep. 4, Dhaka.

Heijnen, W.J., 1988. Liquefaction study for the Jamuna bridge: Field and laboratory testing. Delft Geotechnics Rep. BO-291600/143, Dhaka.

Hulse, R & Mosley, W.H. 1986. *Reinforced concrete design by computer*. Macmillan Education Ltd., UK, 288p.

Hunt, T. 1976. Some geotechnical aspects of road construction in Bangladesh. Geotechnical Engineering 7(1):1-33.

Kabir, M.H., Abedin, M.Z., Siddique, A. & Akhtaruzzaman, M. 1992. Foundations for soft soils in Bangladesh. *Proc., Institute of Lowland Technology Seminar on Problems of Lowland Development, Saga University, Saga, Japan*, 225-230.

Kabir, M.H., Ahmed, K.U. and Hossain, S.S. 1994. Mechanical properties of some jute fabrics and fibre drains. *Proc. 5th Int. Conf. on Geotextiles, Geomembranes, Singapore.*

Kabir, M.H., Alam, M.J., Hamid, A.M. & Akhtaruzzaman, A.K.M. 2000. Foundations on soft soils for Khulna medical college buildings in Bangladesh. Accepted for publication in the proceedings of the Int. Conf. on Geotech. Engrg. to be held during 19 –24 November, 2000 in Melbourne, Australia.

Kiso-Jiban Consultants Co. Ltd. 1999. Report on soil improvement study for Jamuna bridge access roads project, Bangladesh. Submitted to Japan Overseas Consultants.

Mollah, M.A. 1993. Geotechnical conditions of the deltaic alluvial plains of Bangladesh and associated problems. *Engineering Geology*. 36: 125-140

Morgan, J.P. & McIntire, W.G., 1959. Quaternary geology of the Bengal Basin, East Pakistan and India. *Bull. of the Geol. Soc. of America*, 70: 319-342.

Razzaque, M.A. & Alamgir, M. 1999. Long-term settlement observation of a building in a peat deposits of Bangladesh. *Proc., Civil and Environmental Engineering Conf. New Frontiers & Challenges, Bangkok, Thailand, Vol. 2 (part I) : Geotechnical & Geo-environmental Engineering*, 85-94.

Safiullah, A.M.M. 1991. Some geotechnical aspects of Bangladeshi soils. PanelistÔs Report, Theme 2, *9th Asian Regional Conference on Soil Mechanics and Foundation Engineering, Bangkok, Thailand*. Vol. 2, 5-13.

Seed, H.B. & Idriss, I.M., 1967. Analysis of soil liquefaction-Nigata earthquake. *J. Soil Mech. Found. Engrg. Div., ASCE*. 93(3): 83-108.

Seed, H.B., 1979. Soil liquefaction and Cyclic mobility evaluation for level ground during earthquake. *J. Gcotech. Engrg. Div. ASCE*.105(2): 201-255

Serajuddin, M. (1998). Some geotechnical studies on Bangladesh soil: A summary of papers between 1957-96. *Journal of Civil Engineering, The Institution of Civil Engineers, Bangladesh*. 26(2): 101-128.

Siddique, A., Ansary, M.A., Al-Hussuaini, T.M., Kabir, M.H. & Islam, S. 1998. Stability analysis and remedial design of damaged sections of Dhaka Flood Protection Embankment. *Proc., 13th South East Asian Geotechnical Conference, Taipei, Taiwan, Republic of China*, 649-654.

Technoconsult Int. Ltd. & Louis Berger Int. Inc. 1994. Report on remedial design for repair of western embankment Tongi to Mirpur, BWDB, GOB, Dhaka Integrated Flood Protection Project, November, 1994

Late paper for Volume 1

Measurement of initial stress condition of saturated soft grounds of Caspian Sea coasts using newly developed device

A.Z.Zhuspbekov
Karaganda Metallurgical Institute, Temirtau, Kazakhstan

A.C.Zhakulin
Karaganda National Technical University, Karaganda, Kazakhstan

H.Z.Kuanshaliev & A.S.Kuanshalieva
JSC 'Intergazstroi', Aksai, West-Kazakhstan area, Kazakhstan

ABSTRACT: Authors have developed a method of estimating pressure induced under natural ground. A stress measuring device having gauges of stress and pore water pressure has been designed. The results of measurement of stresses in saturated natural soil deposit of the Caspian coast are analyzed. The estimation of an initial stress for saturated soil is given.

1 INTRODUCTION

Soil element in saturated ground is under influence of an own weight. In the geomechanics the self weight of saturated natural ground in a rest condition is defined as the stress coming from self weight of the ground as shown below.

$$\sigma_\chi = \gamma_i' h_i;$$

$$\sigma_y = \sigma_z = \xi_0 \sigma_\chi \tag{1}$$

$$\tau_{xy} = \tau_{yz} = \sigma_{xz} = 0,$$

where $\gamma_i' = (\gamma_s - \gamma_w)/(1-n)$ - unit weight of a ground, in view of weighing action of water,

$\xi_0 = \dfrac{\mu_0}{(1-\mu_0)}$ - factor of lateral stress of a ground in

a condition of rest, μ_0 - Poisson's ratio;

Thus, saturated ground foundation which is subjected to self weight, requires its estimation, which serves as a source of the necessary information.

The estimation of initial stress deformation condition of saturated ground can be carried out in any point of a profile of a ground by considering the external influence and measurement of its response as stress from this influence.

2 TECHNIQUE OF AN ESTIMATION OF INITIAL STRESS-DEFORMATION CONDITION

2.1 *Essence of a method of research*

Installation of the device having measuring probes, in a natural saturated ground is generally done by a method of cave-in. The process of cave-in consists of shearing apart a ground and replacing it with measuring probe. As a result of it, at the contact "point of ground" there is stress concentration at contract point on cave-in, integral of which reflects natural stress working in the given point, and additional stress caused by external influence due to tracture of the ground. Thus, total stress developed at contact-point of devices can be expressed as:

$$P_w^m(\sigma^m) = P_w(\sigma^{nat}) + P_w^{tot}(\sigma^{tot}); \tag{2}$$

At cave-in of a saturated natural ground and for given thickness of a plate, deformation and the reaction measured is constant, which depends on the ground type and its mode of relaxation.

At the initial moment of such tests, the reaction in ground is maximum, and after a period of time pore water pressure, and total stress decrease. The rate of relaxation of total stress and pore water pressure reduction is determined by properties of a ground. Thus relaxation of only additional total stress (σ_{tot}) and pore water pressure (P_w), occurs due to external influence of shear deformation of a ground. The natural stress (σ^{nat}) and pore water pressure (P_w^{nat}) remain constant in the given point. The device is kept at the given depth until the stable stress (σ_m) and pore water pressure (P_m) are obtained. Thus depending on physical properties of a ground and additional the ultimate stress relaxation is $\sigma_{tot} > \sigma_{nat}$. Pore pressure in consolidated ground at the given depths is:

$$P_w^m = P_w^{nat} = \gamma_w h_i \; ; \tag{3}$$

And in not consolidated saturated ground, it may expressed as:

$$P_w^m > \gamma_w h_i \; ; \tag{4}$$

where γ_w - unit weight of water.

A method to examine release of pore water pressure in a saturated natural ground thus is proposed using cave-in device having measuring probe. A cave-in device causes shearing of a saturated ground. Dissipation of an additional stress and pore water pressure are measured at the interface of soil and the device.

2.2 Formulation of model

The estimation of an initial stress in a saturated ground can be done from:

$$\sigma_{ef} = \sigma_m - P_w^m \; ; \tag{5}$$

Natural stress on saturated ground due to self weight of soil:

$$\sigma_{nat} = \gamma' h \; ; \tag{6}$$

Initial pore water pressure ( ) is:

$$\beta = \frac{P_{w,t_0}^m}{\sigma_{m,t_0}} , \tag{7}$$

where P_{w,t_0}^m - pore pressure at t=0, σ_{m,t_0} - stress at t_0.
Hydrostatic stress at depth h is:

$$P_z = \gamma_w h \; . \tag{8}$$

To examine the nature of relaxation of pore water pressure and the stress a new parameter is defined:
Value of attenuation factor K_3 - depends on

probing speed, pore water pressure and stress:

$$K_3 = \frac{P_{w,t}^m (\sigma_{m,t})}{P_{w,t}^m (\sigma_{m,t})} \tag{9}$$

where $P_{w,t}^m$ and σ_{m,t_0} - pore pressure and stress at time t, P_{w,t_0}^m and σ_{m,t_0} pore pressure and stress at time t_0.

The relaxation (K_p) is:

$$K_p = \frac{P_z(\sigma_{nat})}{P_{w,t_0}^m} \tag{10}$$

Figure 1 shows a typical characteristics of relaxation of stress and pore water pressure with time in a saturated ground.

2.3 Design feature of the device

To measure the initial stress - deformation condition of saturated ground, device having gauges for stress and pore water pressure measurement is developed. The device has pointed thin plate of thickness 12 mm, width 120 mm, length 220 mm and length of the projected part 50 mm. Tip angle is 160^0.

The device has three jacks: two for measurement of stress gauges and pore water pressure, third jack is on opposite side, which is used for connection of wires to the measuring equipment. Electrical gauges have diameters of 35 mm, thickness of 6 mm and pore water pressure gauge has a porous stone.

3. ANALYSIS OF RESULTS OF RESEARCH

By examining the nature of relaxation of stress and pore water pressure on the given depths of saturated ground foundation of natural addition the following diagrams are constructed:
    -relaxation of pore water pressure and stress with time; (Figure 1)
    -distribution of pore water pressure and stress with depth. (Figure 2)
    In figure 1 the relations for relaxation of pore water pressure and stress from t=0 to 72000 seconds at depths of 5, 10, 15 and 20 meters are summarized. From the relaxation diagrams it

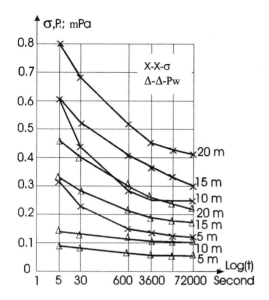

Figure 1. Relaxation of pore water pressure and stress with time

follows, that the process of cave-in of the device entails occurrence of excess pore water pressure and additional stress, which with time starts decreasing. Also, excess pore water pressure and additional stress increases with depth of cave-in is naturally increased. Most intense relaxation process occurs in the first 30 minutes after inserting the measuring device.

The measurement was made at a depth of 5 meters from the ground surface, where soil consists of saturated silt. Initial excess pore pressure has made – 0.051 MPa and in 20 hours, it reduced to – 0.050 MPa. Thus the initial excess pore water pressure is 90% higher than the final value. Relaxation of additional stress is different in nature, than pore water pressure. Value of initial stress is 2.5 times larger than the final. Factor of attenuation of stress is 0.39, and pore water pressure is 0.625. Following this, measurement was made at a depth of 10 meters, where soil is saturated sand. Its initial excess pore was 0.13 MPa, in 30 minutes 0.1 MPa, and even after 20 hours it did not change from 0.1 MPa. The character of relaxation of additional stress has sharp features, having high initial values of 0.6 Mpa, which within one hour completely reaches the final measured value. Factors of attenuation of stress is 0.40 and pore water pressure is 0.76. Similarly, initial pore water pressure factors at depth of 5 meters is 0.29 and at 10 meters it is 0.21.

Subsequent measurements were carried out at

depths of 15 and 20 meters, where was saturated. At t=0 the excess pore pressure at depth of 15 meters is 0.32 MPa and at 20 meters is 0.45 MPa, in 30 minutes which reduced to 0.20 MPa and 0.28 MPa, and in 20 hours it decreased to 0.16 MPa and 0.22 MPa. The value of initial excess pore pressure is 2 times higher than final at both depths. Relaxation of additional stress has similar character. Factors of attenuation pore water pressure is 0.49 at these depths and stress is 0.48 – 0.50. Initial pore water pressure factor changes within the limits of 0.52 – 0.54.

Distribution of pore water pressure and stress with depth in saturated natural deposit (Republic of Kazakhstan)

The size of initial additional stress and pore water pressure depends also on kind and density of soil. For soil, initial excess pore pressure and stress exceeds 2 - 3 times final value, whereas for additional stress it is 1 - 2 times the final value. Also follows, that the characters relaxation of stress and pore water pressure are similar. Also the characters of relaxation of stress and pore water pressure are similar. For sandy soil character of stress and pore water varies depending on its nature and intensity.

Values of initial pore water pressure for sand is 0.20, for silt is 0.29 and for loam is 0.52 – 0.54. For sandy soil pore pressure is less, which is reasonable.

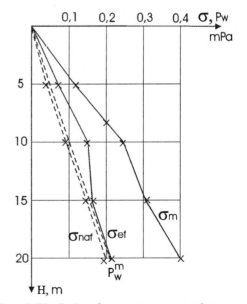

Figure 2. Distribution of pore water pressure and stress with depth.

In figure 2 the distributions of pore water pressure and stress with depth after 20 hours of inserting the measuring device in saturated ground are shown. From the analysis of the diagrams it follows that, the process of pore water pressure relaxation was completely finished, and their values at given depths correspond to hydrostatic pressure. In sandy soil excess pore pressure becomes equal to hydrostatic stress of water after relaxation; in loam soil there is a small divergence of 0.03 MPa.

Despite the insignificant divergences in excess pore water pressure after relaxation, it is possible to conclude that pore pressure reduces to hydrostatic stress of water at a given depth. It is also clear that saturated soils are completely stable and have strong structures of contact between firm particles. Therefore measured stress could not reach theoretical values, as the initial structures of contact between firm particles are not exactly restored to original position.

4 CONCLUSIONS

1. A technique to estimate initial stress condition for saturated ground of natural deposition is developed.
2. The reduction of excess pore water pressure and stress depends mainly on kind of a ground. In sandy soils, the process relaxation with time is very intense and a stable value reaches within an hour, and its further fall with time are insignificant. For loam soil relaxation in pore water pressure is less intense and it also takes much longer time, than in sandy soil.
3. The attenuation factors confirm the above assumption and accordingly for loam: pore water pressure – 0.48/0.50; and stress – 0.49, and for sand: pore water pressure – 0.76 and stress – 0.40 and for soil: pore water pressure – 0.625 and stress – 0.39. The values of attenuation factors suggest loam have a degree of creep, when sandy soil do not have that property.
4. From this research, the quantitative valus of pore water pressure factor for various kinds of soil are established, which characterizes structure of saturated ground.

5 REFERENCES

1. Campanella R.G., Gillespie D., Robertson P.K. 1982 Pore pressures during cone penetration testing. Proceedings of ESOPT-2, Amsterdam, vol.2, p.570-512

2. Jamilkowski M., Ladd C., Germaine J. New developments in field and laboratory testing of soils. Proceedings of the Eleventh International Conference on Soil Mechanics and Foundation Engineering, San-Francisco,-1985. v. 1-p57-154

Coastal Geotechnical Engineering in Practice, Nakase & Tsuchida (eds)
© 2002 Swets & Zeitlinger, Lisse, ISBN 90 5809 151 1

IS-Yokohama 2000 list of participants

ABE	Hiroshi	Kajima Corp.	Japan
ABE	Kazuo	The 1st District Port Construction Bureau, MOT	Japan
ABOSHI	Hisao	Fukken Co., Ltd.	Japan
ADACHI	Kakuichiro	Shibaura Institute of Technology	Japan
AKAGI	Hirokazu	Waseda Univ.	Japan
AKAI	Koichi	Geo-Research Institute	Japan
ARISU	Seigi	Saeki Kensetsu Kogyo, Co., Ltd.	Japan
ARITA	Naoko	Kowa Co., Ltd.	Japan
ASADA	Hideyuki	Toa Corp.	Japan
ASANUMA	Takeo	Toa Corp.	Japan
ASOGAWA	Manabu	Kubota Corp.	Japan
BANG	Sangchul	South Dakota School of Mines	USA
BERGADO	Dennes T.	Asian Institute of Technology	Thailand
BHANDARY	Netra Prakash	Ehime Univ.	Japan
Bo	Myint Win	SPECS Consultants Pte. Ltd.,	Singapore
BONE	Ludi	Pt. Indonesia Nihon Seima	Indonesia
CHEN	Jing-Wen	National Cheng Kung University	Taiwan
CHEN	Guangqi	Kyushu Univ.	Japan
CHERUBINI	Claudio	Politecnico Di Bali	Italy
CHIEN	Lien-Kwei	National Taiwan Ocean University	Taiwan
CHOA	Victor	Nanyang Technological Univ.	Singapore
CHUNG	Sung Gyo	Dong-A Univ.	Korea
COTECCHIA	Federica	Technical Univ. of Bari	Italy
EGUCHI	Shin-ya	Penta-Ocean Construction Co., Ltd.	Japan
ELA	Noliel C.	310 Construction Specialist Corp.	Philippines
ESCUETA	Denis Noel	Tayabas Western Academy	Philippines
FUJII	Teruhisa	Fukken Co., Ltd.	Japan
FUJII	Hiroaki	Okayama Univ.	Japan
FUJIO	Yoshiya	Toyo Const. Co., Ltd.	Japan
FUJITA	Yuji	Toyo Const. Co., Ltd.	Japan
FUJIWARA	Teruyuki	Geo-Research Institute	Japan
FUKUE	Masaharu	Tokai Univ.	Japan
FUKUHARA	Kazuaki	Chuden Engineering Consultants Co., Ltd.	Japan
FUKUOKA	Hideki	Mitsui Harbor and Urban Construction Co., Ltd.	Japan

FURUNO	Takehide	Saeki Kensetsu Kogyo, Co., Ltd.	Japan
GOTOH	Katsushi	Ohmoto Gumi Co., Ltd.	Japan
HAGIWARA	Toshiyuki	Nishimatsu Construction Co., Ltd.	Japan
HAMABE	Masao	Service Center of Port Engineering	Japan
HAMADA	Eiji	Kiso-Jiban Consultants Co., Ltd.	Japan
HANZAWA	Hideo	Toa Corp.	Japan
HARADA	Kenji	Fudo Construction	Japan
HASE	Kazuya	Hokkaido Development Bureau	Japan
HASHIDA	Hideyuki	Nippon Kaiko Co., Ltd.	Japan
HASHIMOTO	Tamotsu	Kawasaki Geological Engineering Co., Ltd.	Japan
HAYANO	Kimitoshi	Institute of Industrial Science, Univ. of Tokyo	Japan
HAYASHI	Kentaro	Penta-Ocean Construction Co., Ltd.	Japan
HAYASHI	Yasuhiro	Kumamoto Univ.	Japan
HAYASHI	Masahiro	Wakachiku Construction Co., Ltd.	Japan
HIGASHIKAWA	Kiyoshi	Saeki Kensetsu Kogyo, Co., Ltd.	Japan
HIGHT	Devit	Geotechnical Consulting Group	U.K.
HIGUCHI	Toru	Univ. of Durham	U.K.
HIGUCHI	Youhei	Penta-Ocean Construction Co., Ltd.	Japan
HIKIYASHIKI	Hideto	Port and Harbour Research Institute	Japan
HIRANO	Miyukichi	Rinkai Construction Co., Ltd.	Japan
HIRAO	Takayuki	Chuden Engineering Consultants Co., Ltd.	Japan
HIRATA	Masafumi	Toyo Const. Co., Ltd.	Japan
HONDA	Tuyoshi	Univ. of Tokyo	Japan
HONG	Zhenshun	Port and Harbour Research Institute	Japan
HONSYO	Takanobu	Toyo Const. Co., Ltd.	Japan
HORIUCHI	Sumio	Shimizu Corp.	Japan
HORPIBULSUK	Suksun	Saga Univ.	Japan
HYDE	Adrian F.	University of Sheffield	U.K.
HYODO	Masayuki	Yamaguchi Univ.	Japan
IBARAKI	Makoto	Waseda Univ.	Japan
ICHIHARA	Michizo	Nitto Daito Construction Co., Ltd.	Japan
IGARASHI	Masaru	Dia Consultants Co., Ltd.	Japan
IIZUKA	Atsushi	Kobe Univ.	Japan
IKENOUE	Shigehiro	Japan Industrial Land Development Co., Ltd.	Japan
IMAI	Makoto	Coastal Development Institute of Technology	Japan
IMAI	Goro	Yokohama National Univ.	Japan
IMANISHI	Hajime	Geo-Research Institute	Japan
INOUE	Koji	Coastal Development Institute of Technology	Japan
INOUE	Toshiyuki	Fukken Co., Ltd.	Japan
ISHIHARA	Kenji	Science Univ. of Tokyo	Japan
ISHIHARA	Masatoyo	Port and Harbour Research Institute	Japan
ISHII	Ayumu	Coastal Development Institute of Technology	Japan
JANG	Chan-Soo	Chunil Geo-Consultants Co., Ltd.	Korea
JARDINE	Richard	Imperial College of Science, Technology and Medicine	U.K.
KAKEHASHI	Takaharu	Japan Industrial Land Development Co., Ltd.	Japan
KAMATA	Ryuji	Toa Corp.	Japan
KANATANI	Mamoru	Central Research Institute of Electric Power Industry	Japan

KANG	Min-Soo	Port and Harbour Research Institute	Japan
KANNO	Yuichi	Fukken Co., Ltd.	Japan
KASAMA	Kiyonobu	Kyushu Univ.	Japan
KATAGIRI	Masaaki	Nikken Sekkei Co., Ltd.	Japan
KATSUMI	Takeshi	Ritsumeikan Univ.	Japan
KATSUUMI	Tsutomu	Coastal Development Institute of Technology	Japan
KAWABE	Tomoyuki	Saeki Kensetsu Kogyo, Co., Ltd.	Japan
KAWAMURA	Satoshi	Toa Corp.	Japan
KAZAMA	Motoki	Tohoku Univ.	Japan
KIKUCHI	Yoshiaki	Port and Harbour Research Institute	Japan
KIM	Sang-Kyu	Dong-Pusan College	Korea
KIM	Soo Sam	Chung-Ang Univ.	Korea
KIM	Gyu Jong	Dong-A Univ.	Korea
KIMURA	Makoto	Kyoto Univ.	Japan
KIMURA	Mitsutoshi	WAVE	Japan
KISHIDA	Takao	Toa Corp.	Japan
KISHIDA	Hideaki	Science Univ. of Tokyo	Japan
KITAHARA	Shigeshi	Hokkaido Development Bureau	Japan
KITAMURA	Ryosuke	Kagoshima Univ.	Japan
KITAYAMA	Naoyoshi	Fukken Co., Ltd.	Japan
KITAZAWA	Sosuke	Ministry of Transport	Japan
KITAZUME	Masaki	Port and Harbour Research Institute	Japan
KOBAYASHI	Nobuo	Nippon Slag Association	Japan
KOBAYASHI	Shun-ichi	Kyoto Univ.	Japan
KOBAYASHI	Masaki	Coastal Development Institute of Technology	Japan
KODA	Masayuki	Railway Technical Research Institute	Japan
KOHATA	Yukihiro	Muroran Institute of Technology	Japan
KOKEGUCHI	Kiyoshi	The 3th District Port Construction Bureau, MOT	Japan
KOKUSYO	Takaji	Chuo Univ.	Japan
KOTAKE	Nozomu	Toyo Const. Co., Ltd.	Japan
KU	Chih-Sheng	National Cheng Kung University	Taiwan
KUDO	Hidetaka	Kubota Construction Co., Ltd.	Japan
KURIMOTO	Taku	Obayashi Corp.	Japan
KUROSAKA	Masahiro	Honma Corp.	Japan
KUSAKABE	Osamu	Tokyo Institute of Technology	Japan
KUWANO	Jiro	Tokyo Institute of Technology	Japan
KUWANO	Reiko	Univ. of Tokyo	Japan
KUZUGAMI	Koji	Service Center of Port Engineering	Japan
LE	Vinh Ba	Yokohama National Univ.	Japan
LEE	Der-Her	National Cheng Kung University	Taiwan
LEE	Ki Ho	Tohoku Univ.	Japan
LEE	Nam Ki	Dong-A Univ.	Korea
LEROUEIL	Serge	Laval University	Canada
LEUNG	C.F.	National Univ. of Singapore	Singapore
LIN	Wang	Chuo Kaihatsu Company	Japan
LOCAT	Jacques	Laval University	Canada

LOHANI	Tara Nidhi	Yokohama National Univ.	Japan
LUNNE	Tom	Norwegian Geotechnical Institute	Norway
M.D. ZAYEDUR	Rahman	Natioal Univ. of Singapore	Singapore
MAEGAWA	Futoshi	Newjec Inc.	Japan
MATSUDA	Hiroshi	Yamaguchi Univ.	Japan
MATSUMOTO	Kouki	Hazama Corp.	Japan
MATSUMOTO	Yukihisa	Wakachiku Construction Co., Ltd.	Japan
MATSUMOTO	Hideo	The 2nd District Port Construction Bureau, MOT	Japan
MATSUOKA	Gaku	EPDC	Japan
MIKASA	Masato		Japan
MIKI	Takayuki	Penta-Ocean Construction Co., Ltd.	Japan
MIMURA	Mamoru	Kyoto Univ.	Japan
MISHIMA	Osamu	Port and Harbour Research Institute	Japan
MITACHI	Toshiyuki	Hokkaido Univ.	Japan
MITARAI	Yoshio	Toa Corp.	Japan
MITSUKURI	Koji	The 4th District Port Construction Bureau, MOT	Japan
Miyachi	Yousuke	The 3th District Port Construction Bureau, MOT	Japan
MIYAJI	Yutaka	Shiogama Port&Airport Construction Office, MOT	Japan
MIYAJIMA	Shogo	Port and Harbour Research Institute	Japan
MIYAZAKI	Yoshihiko	Kanmon Kowan Kensetsu Co., Ltd.	Japan
MIYAZAKI	Keiichi	Nishimatsu Construction Co., Ltd.	Japan
MIYOSHI	Toshiyasu	Penta-Ocean Construction Co., Ltd.	Japan
MIYOSHI	Akihiro	Fudo Construction	Japan
MIZUGUCHI	Hiroshi	Hokkaido Electric Power Co., Ltd.	Japan
MIZUKAMI	Junichi	Port and Harbour Research Institute	Japan
MIZUNO	Kenta	Wakachiku Construction Co., Ltd.	Japan
MIZUNO	Hiroshi	Kokusai Kogyo Co., Ltd.	Japan
MIZUTANI	Takaaki	Port and Harbour Research Institute	Japan
MIZUTANI	Susumu	Pacific Consultants Co., Ltd.	Japan
MOCHIDA	Fumihiro	Oyo Corp.	Japan
MOCHIZUKI	Akitoshi	The Univ. of Tokushima	Japan
MORI	Nobuo	Hokkaido Development Bureau	Japan
MORI	Syozi	Fukuoka City Hall	Japan
MORIKAWA	Yoshiyuki	Port and Harbour Research Institute	Japan
MORIWAKI	Takeo	Hiroshima Univ.	Japan
MUKUNOKI	Toshifumi	Kumamoto Univ.	Japan
MURAKAMI	Tadashi	Amano Corp.	Japan
MURAKAMI	Satoshi	Ibaraki Univ.	Japan
MURAMOTO	Tetsuji	Coastal Development Institute of Technology	Japan
MURATA	Osamu	Railway Technical Research Institute	Japan
NAGASAKA	Yuji	SLS Corp.	Japan
NAGAYAMA	Hideki	Toyo Const. Co., Ltd.	Japan
NAKAHAMA	Akito	WAVE	Japan
NAKAJIMA	Mitsuo	The City of Yokohama	Japan
NAKAJIMA	Hidekatsu	Kiso-Jiban Consultants Co., Ltd.	Japan
NAKAMURA	Shuichi	Penta-Ocean Construction Co., Ltd.	Japan

NAKANO	Masaki	Nagoya Univ.	Japan
NAKANO	Toshihiko	Kansai International Co., Ltd.	Japan
NAKASE	Akio	Nikken Sekkei Co., Ltd.	Japan
NAKASE	Hitoshi	Tokyo Electric Power Service Co., Ltd.	Japan
NAKAZONO	Yoshiharu	Kanmon Kowan Kensetsu Co., Ltd.	Japan
NGUYEN	Nam Hong	Univ. of Tokyo	Japan
NISHIE	Shunsaku	Chuo Kaihatsu Company	Japan
NISHIMURA	Tomoyoshi	Ashikaga Institute of Technology	Japan
NISHIMURA	Masato	Nikken Sekkei Co., Ltd.	Japan
NISHIMURA	Shinichi	Okayama Univ.	Japan
NITAO	Hiroshi	Fudo Construction	Japan
NODA	Setsuo	Coastal Development Institute of Technology	Japan
NOJI	Masahiro	Sumitomo Metal Industries, Ltd.	Japan
NORIYASU	Naoto	Chuden Engineering Consultants Co., Ltd.	Japan
NOZU	Mitsuo	Fudo Construction	Japan
OCHIAI	Hidetoshi	Kyushu Univ.	Japan
OGATA	Tsuneo	Ohmoto Gumi Co., Ltd.	Japan
OGINO	Toshihiro	Akita Univ.	Japan
OGURO	Shoji	Toyo Const. Co., Ltd.	Japan
OHISHI	Kanta	Nikken Sekkei Co., Ltd.	Japan
OHMAKI	Seiki	National Research Institute of Fisheries Engineering	Japan
OHMUKAI	Naoki	Oyo Corp.	Japan
OHTA	Hideki	Tokyo Institute of Technology	Japan
OIKAWA	Hiroshi	Akita Univ.	Japan
OJIMA	Takashi	The City of Yokohama	Japan
OKA	Fusao	Kyoto Univ.	Japan
OKADA	Junji	Nakabori Soil Corner Co., Ltd.	Japan
OKI	Takeshi	Kawasaki Steel Corp.	Japan
OKUBO	Yasuhiro	Penta-Ocean Construction Co., Ltd.	Japan
OKUMURA	Tatsuro	DRAM Engineering Inc.	Japan
ONO	Mutsuo	Hazama Corp.	Japan
OTANI	Jun	Kumamoto Univ.	Japan
OWADA	Makoto	Port and Harbour Research Institute	Japan
OYADOMARI	Masataka	Okinawa General Bureau	Japan
Pipatpongsa	Thirapong	Tokyo Institute of Technology	Japan
PORBAHA	Ali	New Technology & Research	USA
RITO	Fusao	Oyo Corp.	Japan
SAEKI	Kimiyasu	National Research Institute of Fisheries Enginnering	Japan
SAKAI	Shinsuke	Hanshin Consultants Co., Ltd.	Japan
SANDANBATA	Isamu	Hazama Corp.	Japan
SANO	Ikuo	Osaka Sangyo Univ.	Japan
SATO	Katsuhisa	Obayashi Road Corp.	Japan
SATO	Kenichi	Fukuoka Univ.	Japan
SATO	Yuichiro	Wakachiku Construction Co., Ltd.	Japan
SATO	Mitisuke	Toyo Const. Co., Ltd.	Japan
SATOU	Yoshinori	Honma Corp.	Japan

259

SEKIGUCHI	Hideo	Kyoto Univ.	Japan
SEKIGUCHI	Koji	NKK Corp.	Japan
SENG	Lip Lee	National Univ. of Singapore	Singapore
SHIBATA	Azuma	Kowa Co., Ltd.	Japan
SHIBUYA	Satoru	Hokkaido Univ.	Japan
SHIGEMATSU	Hiroaki	Gifu Univ.	Japan
SHIMIZU	Yutaka	Fukken Co., Ltd.	Japan
SHIMOBE	Satoru	Junior College, Nihon Univ.	Japan
SHINSHA	Hiroshi	Penta-Ocean Construction Co., Ltd.	Japan
SHIRAI	Akira	Kubota Construction Co., Ltd.	Japan
Shiwakoti	Dinesh R.	Port and Harbour Research Institute	Japan
SHOGAKI	Takaharu	Natioal Defense Academy	Japan
SIDDIQUE	Abu	Bangladesh University of Engineering & Technology	Bangladesh
Skúlason	Jón	Almenna verkfæðistofan hf.	Iceland
SRIDHARAN	Asri	Saga University	India
SUETSUGU	Daisuke	National Defense Academy	Japan
SUGAI	Masazumi	Maeda Corp.	Japan
SUGAMURA	Toshihiko	Nitto Daito Construction Co., Ltd.	Japan
SUMIKURA	Yuichiro	Kyoto Univ.	Japan
SUTOH	Yoshiji	Fukken Co., Ltd.	Japan
SUWA	Seiji	Geo-Research Institute	Japan
SUZUKI	Koji	Toa Corp.	Japan
TAGUCHI	Hirofumi	Toa Corp.	Japan
TAJIKA	Koji	Nippon Koei Co., Ltd.	Japan
TAKAHASHI	Hiroyuki	Kowa Co., Ltd.	Japan
TAKAHASHI	Kunio	Port and Harbour Research Institute	Japan
TAKAHASHI	Shinichi	Obayashi Corp.	Japan
TAKAHASHI	Akihiro	Tokyo Institute of Technology	Japan
TAKAYUKI	Shinoi	Nippon Kaiko Co., Ltd.	Japan
TAKEMURA	Jiro	Tokyo Institute of Technology	Japan
TAKI	Masakazu	Fukken Co., Ltd.	Japan
TAN	Thiam Soon	National Univ. of Singapore	Singapore
TANAKA	Yosuke	Yokohama National Univ.	Japan
TANAKA	Hiroyuki	Port and Harbour Research Institute	Japan
TANAKA	Masanori	Port and Harbour Research Institute	Japan
TANAKA	Akihito	Dia Consultants Co., Ltd.	Japan
TANG	Yi Xin	Kanmon Kowan Kensetsu Co., Ltd.	Japan
TANI	Kazuo	Yokohama National Univ.	Japan
TATSUOKA	Fumio	Univ. of Tokyo	Japan
TATSUTA	Masataka	Nippon Steel Corp.	Japan
TEMMA	Minoru	Hokkaido Univ.	Japan
TEPARAKSA	Wanchai	Chulalongkorn Univ.	Thailand
TERADA	Kunio	Takenaka Corp.	Japan
TERASHI	Masaaki	Nikken Sekkei Co., Ltd.	Japan
THERAMAST	Nattavut	Nagoya Univ.	Japan
TOHDA	Jun	Osaka City Univ.	Japan

TOMITA	RYUZO	Koa Kaihatsu Co., Ltd.	Japan
TOYAMA	Yukio	Toa Corp.	Japan
TSUCHIDA	Takashi	Port and Harbour Research Institute	Japan
TSUCHIDA	Hajime	Nippon Steel Corp.	Japan
TSUJI	Kiyoshi	Japan Port Consultants, Ltd.	Japan
TSUKAMOTO	Yoshimichi	Science Univ. of Tokyo	Japan
TSURUGASAKI	Kazuhiro	Toyo Const. Co., Ltd.	Japan
UEHARA	Hosei	Uehara Geot. Research Institute	Japan
UMEHARA	Yasufumi	Yachiyo Enginnering Co., Ltd.	Japan
UMEZAKI	Takeo	Shinshu Univ.	Japan
UTSUNOMIYA	Tatsushi	Nippon Kaiko Co., Ltd.	Japan
UWABE	Tatsuo	Service Center of Port Engineering	Japan
WADA	Kozo	Coastal Development Institute of Technology	Japan
WAKITANI	Yoshiaki	Okayama Univ.	Japan
WAKO	Tatsuo	Japan Port Consultants, Ltd.	Japan
WATABE	Yoichi	Port and Harbour Research Institute	Japan
WATANABE	Yasuo	East Japan Railway Company	Japan
WATANABE	Masao	Toa Corp.	Japan
WATANABE	Shingo	Yamaguchi Univ.	Japan
YAGI	Norio	Ehime Univ.	Japan
YAMADA	Yasuhiro	Kensetsu-Kiso-Engineering Co., Ltd.	Japan
YAMADA	Koichi	Penta-Ocean Construction Co., Ltd.	Japan
YAMADA	Kazuhiro	Fukken Co., Ltd.	Japan
YAMADA	Keisuke	Wakachiku Construction Co., Ltd.	Japan
YAMADA	Takashi	Fudo Construction	Japan
YAMAGAMI	Takuo	The Univ. of Tokushima	Japan
YAMAGUCHI	Atsushi	Yokohama National	Japan
YAMAMOTO	Yoichi	Mitsui Construction Co., Ltd.	Japan
YAMAMOTO	Nobutaka	Tokyo Institute of Technology	Japan
YAMAMOTO	Koji	Geo-Research Institute	Japan
YAMAMURA	Kazuhiro	Toa Corp.	Japan
YAMANE	Nobuyuki	Toa Corp.	Japan
YAMAUCHI	Hiromoto	Penta-Ocean Construction Co., Ltd.	Japan
YAMAZAKI	Hiroyuki	Port and Harbour Research Institute	Japan
YAMAZAKI	Nobuo	Rinkai Construction Co., Ltd.	Japan
YANO	Kouichirou	Ocean Consultant, Japan Co., Ltd.	Japan
YASHIMA	Atsushi	Gifu Univ.	Japan
YASUHARA	Kazuya	Ibaraki Univ.	Japan
YASUTAKE	Keiki	The City of Yokohama	Japan
YOON	Gil Lim	Korea Ocean Research & Development Institute	Korea
YOSHIDA	Takaaki	Toyo Const. Co., Ltd.	Japan
YOSHIIZUMI	Naoki	Kawasaki Geological Engineering Co., Ltd.	Japan
YOSHIMINE	Mitsutoshi	Tokyo Metropolitan Univ.	Japan
YOSHIMOTO	Norimasa	Yamaguchi Univ.	Japan
YOSHINO	Hiroyuki	Yachiyo Enginnering Co., Ltd.	Japan
ZEN	Kouki	Kyushu Univ.	Japan

ZHANG	Wohua	Zhejiang University	China
ZHAO	Weibing	Hohai Univ.	China
Zhussupbekon	Askar		Kazakhstan

Coastal Geotechnical Engineering in Practice, Nakase & Tsuchida (eds)
© 2002 Swets & Zeitlinger, Lisse, ISBN 90 5809 151 1

Author index Volume 2

Coastal Geotechnical Engineering in Practice, Nakase & Tsuchida (eds)
© 2002 Swets & Zeitlinger, Lisse, ISBN 90 5809 151 1

Author index Volume 1

Kamata, R. 163, 363
Kamon, M. 629
Kanazawa, H. 771
Kang, M.S. 51
Kanno, Y. 325, 557
Kanorski, S. 249
Kantharaj, M. 3
Karunaratne, G.P. 465
Katagiri, M. 95, 167, 299, 307, 507
Kato, Y. 571
Katsumi, T. 629
Kawaguchi, M. 763
Kawano, K. 709
Kazama, M. 63
Kikuchi, Y. 375, 393, 635, 681
Kim, J.C. 577
Kim, K.H. 449, 583
Kim, S.K. 21
Kishida, T. 287, 703, 757
Kitahara, S. 641
Kitamura, R. 277
Kitayama, N. 663
Kitazawa, S. 375, 507, 697
Kitazume, M. 647
Kobayashi, H. 641
Kobayashi, S. 315
Koda, M. 489, 653
Kodama, S. 599
Kogure, K. 715
Kondo, T. 193, 687
Kondoh, M. 321
König, D. 781
Koreishi, T. 663
Kozawa, D. 325, 557
Ku, C.-S. 57
Kubo, M. 495
Kuboshima, S. 27
Kudoh, Y. 77
Kusakabe, S. 271
Kuwabara, M. 521
Kuwazuru, Y. 687
Kwon, H.S. 583

Le Ba Vinh, 199, 543
Lee, D.-H. 57

Lee, J. 623
Lee, K.H. 63
Lee, N.K. 21
Lee, S.L. 465
Lee, Y.S. 449
Leong, P.-L. 363
Leroueil, S. 733
Leung, C.F. 333
Liang, X. 381
Liao, J.-M. 437
Locat, J. 733
Loh, Y.-H. 363
Lohani, T.N. 67
Lovell, C.W. 623

Manivannan, R. 333
Masuda, T. 657
Matsuda, H. 663
Matsumoto, H. 325, 557
Matsumoto, K. 471, 495
Mimura, M. 339
Mishima, O. 73, 147
Mitachi, T. 77, 125, 173
Mitarai, Y. 9, 669
Mitsukuri, K. 709
Miura, N. 119, 515, 605
Miyaji, Y. 675
Miyata, Y. 715
Miyazaki, Y. 725
Mizuno, K. 113
Mizuno, T. 489
Mizutani, H. 571
Mori, N. 477
Moriwaki, T. 83
Mukunoki, T. 681
Murakami, S. 771
Muranaka, T. 657
Murata, O. 653

Nagaraj, T.S. 605
Nagatome, T. 681
Nakajima, H. 687
Nakakuma, K. 471, 495
Nakano, M. 113
Nakano, Y. 663
Nakanodo, H. 325
Nash, D.F.T. 483

Nguyen Van Tho, 199
Nishie, S. 89
Nishimura, M. 95
Nishimura, S. 101, 587
Nishimura, T. 107
Nishino, N. 587
Nishio, Y. 529
Nitao, H. 529
Noda, S. 697
Noda, T. 113
Nomura, T. 501
Noriyasu, N. 119
Nozu, M. 563

Ogino, T. 125, 131
Ohmaki, S. 345
Ohta, H. 351
Oikawa, H. 125, 131
Okazaki, M. 641
Okumura, T. 675, 693, 697
Orr, T.L.L. 261
Otani, J. 681
Oyadomari, M. 135

Patawaran, M.A.B. 429
Pipatpongsa, T. 351
Porbaha, A. 703, 757

Rahman, Md.Z. 489
Recker, K. 611
Redana, I.W. 357
Ryde, S.J. 483

Saeki, K. 345
Saeki, S. 599
Saitoh, K. 95
Sakai, S. 217
Sakakibara, N. 153, 157
Sakamoto, A. 719
Salgado, R. 623
Salim, W. 357
Sami, R.G. 3
Sandanbata, I. 495
Sano, T. 477
Santhaswaruban, V. 3
Sasagawa, T. 421
Sasahara, T. 617